STRATHCLYDE UNIVERSITY LIBRARY
30125 00336526 8

KV-119-738

ANDERSONIAN
LIBRARY
- 9 NOV

Plastics Forming

Production
Engineering Series

Macmillan
Engineering Evaluations

Plastics Forming

Edited by John D. Beadle

Macmillan Engineering Evaluations

Published by
The Macmillan Press Limited
Technical and Industrial Publishing Unit

Managing Editor William F Waller
AMITPP AssIRefEng

General Manager Barry Gibbs

The Macmillan Press Limited
Brunel Road Basingstoke Hampshire UK

Price £3.50

SBN 333 125 452

Printed Photolitho by Page Bros (Norwich) Ltd.

Foreword

The rapidly increasing use of plastics has resulted in a diverse range of sophisticated materials, forming methods and products, thus prompting this evaluation which reflects these changes and outlines the current practices and trends. Due to these developments in plastics materials and the equipment used to form them, the barriers between techniques intended to form thermosets or thermoplastics are becoming less clearly defined and the ability to produce good plastics products has improved at all levels

The first chapter covers the basic polymers, their availability and the forms in which they are supplied, and gives information on their processing and typical applications. Most plastics require some additives, and chapter 2 discusses the function and use of additives, including colourants. The materials are discussed further in the next two chapters, which detail the properties and applications of thermosets and thermoplastics individually.

Frequently, the selection of a forming method is determined by economics and chapter 5 outlines the factors involved before some of the major forming techniques are dealt with (chapters 6–12). Methods of applying plastics as coatings are then presented in chapter 13, whilst chapter 14 considers the more specialised topic of cellulose film. Fabricating techniques are also discussed (chapters 15–17) with practical information on machining, welding and assembly of plastics products. The development and growing use of reinforced and expanded forms of plastics result in their separate coverage in chapters 18 and 19.

Chapter 20 deals with material wastage and the need to recover plastics by reprocessing. The subject of pollution is now receiving much attention and chapter 21 puts the problem of plastics disposal into perspective and indicates the current practices and lines of research. The amount of data available to the designer is constantly growing and the final chapter takes a fresh look at the subject of design and the approaches that have evolved.

To complete the book a glossary of terms used in the Plastics Industry is included, together with a product guide to some of the material suppliers and plastics processors in the UK.

Throughout the book the forthcoming metrication of industry has been a consideration and the majority of data are presented in SI units. However, flexibility has also been a factor and a number of non-preferred SI units are used and, occasionally, where it is felt the subject warrants it, British units are used and an SI conversion factor given.

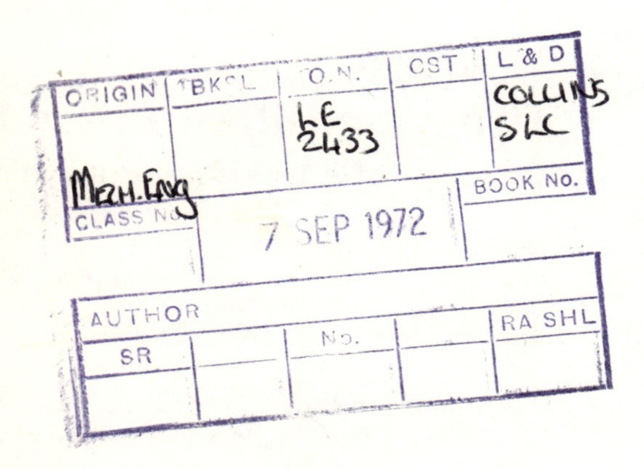

Contents

Chapter — Page

1 **Available Plastics** 7
P Clifton *MA* and **W S Pinney**
Market Research Manager and *Market Research Officer*
Bakelite Xylonite Limited

2 **Additives** 15
E W Thomas *BSc* and **J E Todd** *AMCT*
Technical Services Officer—Plastics Chemicals
Industrial Chemicals Division and
Head of Plastics Laboratory Pigments Division
CIBA–GEIGY (UK) Limited

3 **Thermosetting Plastics** 29
G W Parry *ARIC*
Senior Sales Development Representative
BP Chemicals International Limited

4 **Thermoplastics** 39
C C Gosselin *BSc*
Plastics Division
Shell International Chemical Company Limited

5 **Economics of Forming** 45
J H Briston *BSc ARIC API MInstPackaging*
Technical Development Division
Shell International Chemical Company Limited

6 **Compression Moulding** 51
J E Rogers *AMIED*
Chief Designer/Estimator
Minerva Mouldings Limited

7 **Transfer Moulding** 63
E D Quiddington
Managing Director
Ariel Pressings Limited

8 **Injection Moulding** 67
V E Moore
Technical Director
Invicta Plastics Limited

9 **Extrusion** 71
K J Braun *BSc CEng MIMechE*
Engineering Manager
Bone Cravens Limited

10 **Sheet and Vacuum Forming** 81
J C Evans
Product Manager Sheeting Department
Telcon Plastics Limited

11 **Calendering** 91
K J Hardman *CEng MIMechE MIPlantE AMCT*
Paints Division
Imperial Chemical Industries Limited

Chapter — Page

12 **Rotational Moulding** 97
T Russell
Deputy Managing Director
Thermo Plastics Limited

13 **Coating Techniques** 103
G E Barrett *BSc PhD*
Technical Director
Plastic Coatings Limited

14 **Cellulose Film** 111
D O Richards *BSc* and **W Scott** *BSc ARIC*
Chief Chemist and *Chief Analyst*
Rayophane Division
British Sidac Limited

15 **Machining Plastics** 115
T Lawrence and **R Godwin**
Chief Engineer and *Sales Engineer*
Tufnol Limited

16 **Welding Plastics** 121
H R Stilton
Director
General Industrial Plastics
(Research & Development) Ltd

17 **Assembly Techniques** 129
F A Dixon *FIWM*
Works Director
Creators Limited
Production Director
Creators Continental SA, Belgium

18 **Reinforced Plastics** 135
D Pickthall
Marketing Manager — R & I
Fibreglass Limited

19 **Expanded Plastics** 143
H Gibson *ARIC*
Industrial Chemicals Division
Shell International Chemical Company Limited

20 **Reprocessing Plastics** 155
G Cheater
Productivity and Planning Group
Rubber and Plastics Research Association of Great Britain

21 **Disposal of Plastics** 161
Professor G Scott *BA BSc*
The Department of Chemistry
The University of Aston in Birmingham

22 **Designing for Plastics** 167
V E Yarsley *MSc DSc(Tech) CEng FRIC MIChemE FPI*
Chairman
Yarsley Research Laboratories Limited

Product Guide 172

Glossary of Terms 203

Chapter 1

Available Plastics

P Clifton ***MA*** **and W S Pinney**
Bakelite Xylonite Limited

A wide range of plastics materials is now available to industry, each having its own combination of special properties. They aid production through the rapid and repetitive manufacture of complicated shapes which would otherwise require assembly from several parts. In many cases, important weight advantages can be achieved by the use of plastics. The engineer needs to know some basic facts about plastics and the forms in which they are available to him, and something of their performance characteristics.

All plastics can be made to flow by the application of heat and pressure to assume the shape of a mould or form into which they are pressed. Plastics materials are commonly classified, according to their characteristics after the processing stage, into two main groups—thermosets and thermoplastics.

Thermosets are usually processed by compression or transfer moulding or, more recently, direct injection moulding. During the moulding process the material experiences an irreversible chemical change called curing. After curing the thermoset moulding is rigid and hard and insoluble and relatively unaffected by heat up to decomposition temperature.

Thermoplastics, by contrast, can be remoulded or reshaped several times subject to the limits of thermal fatigue and degradation. This not only allows the re-use of scrap and trim but also lends versatility to the processes by which they can be formed. Thus, injection moulding and extrusion techniques are supplemented by blow-moulding, rotational moulding, dip-moulding, calendering and several methods of forming from flat sheet.

Generally, the service temperatures of thermoplastics products are lower than for thermosets, but this feature allows post-forming and welding techniques to be used which are quite impossible for thermosets.

Although the range of properties is considerable throughout the plastics field it can be said that, generally, thermosets are notable for rigidity, dimensional stability, and resistance to abrasion and solvents, while thermoplastics exhibit flexibility in varying degrees and an extensive colour range, including transparency. Almost all are good electrical insulators and resistant to chemicals.

THE AVAILABILITY OF PLASTICS

Moving from the general to the particular, a brief description of the more important plastics follows and Tables 1 and 2 summarise the principal characteristics affecting the designer's choice (including relative costs). In each case, unless it is otherwise stated, the plastics is readily available in the United Kingdom. Newly developed plastics have sometimes posed supply problems for the engineer as the materials make their transition from pilot production with extensive publicity to manufacture on a full commercial scale. It is always worthwhile investigating the supply situation for plastics that have not yet been on the market for five years.

THERMOSETTING PLASTICS

Long established, dimensionally stable, and familiar materials in many user industries, thermosets usually incorporate fillers which provide a variety of characteristics in processing.

The **phenol formaldehydes** were among the first commercially available thermosetting plastics, dating back to 1907, and have remained the mainstay of this group. They can be formulated with many and varied fillers which impart distinct characteristics to the finished product. Knowledge of the fillers used is a guide to the properties and performance to be expected. Thus, fibrous fillers are used to increase impact strength, mineral powders will raise working temperatures and increase electrical resistance, while graphite will confer low friction characteristics. These are just three examples.

Phenolic moulding materials are available as granular powders and as flakes in a limited colour range, mainly dark, for processing by compression, transfer, and injection moulding. Applications vary according to filler type. Wood filled phenolics are employed for meter cases and domestic electrical sockets, plugs, lamp holders, saucepan handles, iron handles, and distributor caps; cotton filled phenolics are used for washing machine agitators; and asbestos filled moulding materials for switch bases and terminal blocks.

The limited colour range of phenolic moulding materials, usually dark colours only, is a disadvantage overcome by the **urea formaldehydes** and, later, the **melamine formaldehydes,** collectively known as aminoplastics. As these are also odourless they are particularly suitable for domestic products and bottle closures.

Aminoplastics are available in an unlimited colour range, both translucent and opaque. Processing is mainly by compression moulding, but some grades are suitable for injection moulding. Applications include domestic electrical, toilet seats, bottle closures, and tableware.

The **alkyd** materials are notable for rapid curing, roughly three to four times faster than phenolics, and exceptional stability of electrical low loss factor. Alkyds are available as granules, 'putty' and 'rope' in a fair colour range, including some light shades. They are moulded by compression and transfer techniques, including encapsulation of electrical capacitors and resistors. Other applications include cooker controls, pan handles, and knobs, domestic electrical fittings, and sewing machine parts.

Polyesters

These are available as transparent, straw coloured to amber liquids for laminating, encapsulation and casting, and as dough moulding compounds for compression moulding. Polyesters are mainly used as liquid binders for glass fibre and other fibrous materials in reinforced plastics. Glass reinforced plastics (GRP) are based on polyesters. Applications include:

1. *Laminating:* boat hulls, vehicle bodies, chemical plant and piping, building components.
2. *Casting:* buttons, flooring mixes, embedded specimens, encapsulated electrical parts.
3. *Moulding:* trays, luggage, instrument cases, electrical switchgear housings.

Body fillers and repair kits for cars use polyesters.

Epoxides

These are available as liquid and solid resins and moulding compounds, into which large amounts of filler can be incorporated. Processing includes laminating, casting and compression moulding. Common applications are electrical castings and encapsulated components, castings for tools, high performance laminates for printed circuits, aircraft parts, and sports equipment. The particular properties of the moulding compound makes it suitable for exacting electro/mechanical applications.

Polyurethanes

Solid urethane plastics have become available very recently. These are tough, high temperature performance materials with good electrical properties and resistance to solvents, dilute acids, alkalis and grease. They have exceptional adhesive strength which ensures good wetting of reinforcements and firm grip for insets and moulded-in attachments.

Polyurethanes are available as resins for casting or low pressure moulding. The supply situation should be reviewed by the engineer for proposed applications. Applications currently being investigated include automobile door panels, boot lids, bonnets and seats, and domestic furniture. Suggested applications include boat hulls, industrial wheels, engine covers, fan blades and toys.

	Cost comparison index	SG	Impact strength (ft lb/in notch)	Tensile strength (000 lbf/in²)	Water resistance	Max. service temperature (°C)	Moulding temperature (°C)	Moulding pressure (000 lbf)	Trade name examples
Alkyds (Granules)	72	2.10–2.20	0.10–0.15	3–5	Good	155	130–165	1.0 –1.5	AMC and BAKELITE
Aminoplastics (Granular)									
Woodfilled Urea	10	1.40–1.5	0.22–0.28	5.5– 8.0	Fair	75	135–165	1.5 –2.5	BEETLE and
Woodfilled Melamine	15	1.50–1.55	0.25–0.35	5.5– 8.5	Good	120	140–170	3.0 –8.0	NESTORITE
Phenolics Granular GP	12	1.35–1.40	0.13–0.18	7.0– 9.0	Good	130	140–180	1.0 –1.5	BAKELITE,
Medium shock	40	1.37–1.42	0.27–0.32	6.5– 8.5	Good	130	140–180	1.0 –1.5	STERNITE, NESTORITE
Polyester DMC GP	33	2.0	2.00–8.0	6.0–10.0	Fair–Good	150	120–160	0.25–1.5	AMC and BEETLE
Silicone	500	1.86–1.88	2.0	5.0	Excellent	350	150–180	1.5 –2.5	—

NB. 1 lbf/in² = 68.95 mbar, 1 lbf = 0.454 kg.

Table 1. Comparison of costs and properties of some thermosetting plastics.

Silicones

Silicone compounds are non-flammable dielectrics with excellent retention of mechanical and electrical properties for long periods at high temperatures (300°C) though their original mechanical strength is not exceptional. They are resistant to acids, alkalis and solvents. Silicone compounds are available in granular form, mostly dark in colour, and usually filled with glass fibre and other mineral fillers. They are processed by transfer or compression moulding. Applications are mainly electronic devices for performance in exacting environments, including encapsulated components.

THERMOPLASTICS

Thermoplastics are distinguished by the fact that they do not change chemically when subjected to heat. However, as will be seen, there are significant differences in their particular plastics properties.

The earliest thermoplastics were the **cellulosics.** Celluloid or cellulose nitrate, followed by cellulose acetate, cellulose acetate butyrate and, more recently, cellulose propionate, have all found their special uses, but cellulose nitrate is used very little currently because of flammability risks.

Cellulosics

Cellulose nitrate (celluloid) is a tough, rigid material mostly available in sheet form. High, almost explosive, flammability is its main disadvantage. Applications include table tennis balls, slide rule covers, and covers for books, drums and other musical instruments.

Cellulose acetate is also tough and rigid, but unlike celluloid is not highly flammable (but will burn when ignited). It possesses good electrical properties including arc resistance but is susceptible to moisture absorption which lowers these properties. Cellulose acetate is used mainly as sheet material in packaging, but is also employed as toothbrush handles, knife handles, combs, bicycle pumps, fountain and ballpoint pens.

Cellulose acetate butyrate (CAB) is less affected by moisture than other cellulosics, and has better resistance to solvents and oils. Applications include strippable coatings, tool handles and lighter fuel containers.

Cellulose propionate is hard and dimensionally stable and has good surface finish. The product is comparatively unused in the UK, but in the USA it is employed in portable radios and domestic equipment housings.

Polyolefins

Perhaps the most important, certainly the most widely known and used thermoplastic must be low density polyethylene (often called polythene). This is the forerunner of the polyolefins group which now includes high density polyethylene and polypropylene. The actual difference in density between high and low density polyethylene is quite small but high density (HD) polyethylene is noticeably harder and stiffer and the two types exhibit different permeability characteristics and resistance to chemicals. Polypropylene has the lowest density of the three, the best rigidity and surface hardness and is able to withstand higher temperatures.

Low density polyethylene (LDPE) is a tough waxy solid of white translucent colour. Its softening point varies with density. LDPE has good electrical properties including low power factor and it is resistant to moisture and most chemicals other than oxidisers. Resistance to hydrocarbons and oils by LDPE is not good, and a satisfactory resistance to some environments is seriously impaired when mouldings are stressed.

LDPE is available as natural and coloured granules or powder, and in sheet and film form. Low density polyethylene can be processed by injection moulding, blow moulding, extrusion for film, sheet, pipe, profiles, coatings and wire covering, and by calendering and rotational moulding. Sheet material is formable by all techniques.

Applications include bottles and other containers, toys, kitchenware, water tanks, packaging film, coated materials for packaging, water and chemical pipe, wire insulation and sheathing, and film for agricultural and building applications.

High density polyethylene (HDPE) is harder and more brittle than LDPE and resistant to some of the solvents which attack the low density product. HDPE is available in the same forms as LDPE and processed by the same methods, although temperatures and pressures required are generally higher.

Applications are in the same range as for LDPE but have tended towards those requiring increased rigidity (or equal rigidity for less thickness) and where resistance to oils and solvents is important. HDPE is also used for monofilaments.

Polypropylene (PP) has the lowest density of the three well-known polyolefins (0.90 gm/cm^3) and the greatest rigidity and surface hardness. It has a low coefficient of friction and excellent resistance to water and most chemicals, detergents and oils and is not subject to environmental stress-cracking. Electrical properties are good but PP loses impact strength at low temperatures and is affected by outdoor exposure unless stabilised by carbon black.

Excellent resistance to mechanical fatigue gives PP its unique 'hinge' property. A service temperature of 120°C allows sterilisation. Polypropylene is available in granules or powder and as sheet in natural (off-white) translucent to opaque colours. It is processable by injection and blow moulding and the sheet can be formed by all techniques in use. PP is also produced in monofilaments and fibres.

Applications are multiple, and include housewares, laboratory and hospital equipment, 'under the bonnet' automobile components (such as cooling fans), electrical connectors, chemical plant linings and valves, packaging containers and closures, moulded chairs, toys, combs, shoe heels, instrument cases (with integral hinge), one-piece umbrella frames (integral hinge), brush fillings (monofilaments), woven sacks (fibrillated film), ropes (monofilament), and baler twine.

Polyvinyl chloride (PVC)

PVC is probably the most versatile of the thermoplastics, being available in many forms over a wide range of hardness and flexibility. It possesses good physical strength and resistance to water and chemicals. By careful selection of plasticisers many requirements can be built into the end product produced from this material. For instance, one compound may be formulated for resistance to impact at low temperature; another for continuous performance at 100°C. Mouldings and extrusions in contact with other materials must have non-migratory plasticisers, while those in contact with food must be odourless and non-toxic. Flooring and surfacing materials must resist staining. All these requirements and many more can be met.

In general, PVC materials are difficult to ignite, burn slowly and are self extinguishing when the flame is withdrawn. They are good electrical insulators.

PVC is available as dice or pellets, and as plastisols (resin dispersed in solvent) and as sheet. All colours can be provided, as well as clear transparent material. Polyvinyl chloride is processed by injection moulding, extrusion, calendering, blow and rotation moulding, spreading, dipping and foaming. Sheet material can be formed by all standard techniques.

Applications include flooring, footwear, pipes and ducting, chemical plant, bottles, gramophone records, domestic appliance components, gaskets, furniture, wall-covering, tapes, brush fillings (monofilaments), gloves, handbags, luggage, upholstery trim, hoses, cable insulation and sheathing, electrical connectors, clothing, and buoyancy aids.

Polystyrenes

Polystyrenes are available at different levels of impact strength from brittle to tough and ductile. Copolymers and terpolymers such as styrene-acrylonitrile (SAN) and acrylonitrile butadiene styrene (ABS) have been developed as extensions of the styrene range. SAN is at the low impact end but suitable for packaging and tableware while ABS is a tough, exceptionally durable material relatively unaffected by hostile environments.

General purpose polystyrene (GP PS) is rigid but brittle on impact and has low elongation on breaking in tensile stress. It is affected by solvents and natural oils but resistant to acids and alkalis. General purpose polystyrene has electrical properties of a very high order which are unchanged in humid environments. Dimensional stability is good but outdoor exposure causes embrittlement due to ultra violet attack. The maximum service temperature is 70°C.

The product is available in granules or pellets which are natural transparent (water-white) through a complete colour range of translucent and opaque colours. Expandable beads are also available. General purpose polystyrene is processed by injection moulding, blow moulding and extrusion and foaming.

Applications cover cosmetics, pharmaceutical and food packaging, disposable syringes and other hospital and laboratory ware, refrigerator boxes, lighting fittings, advertising, time switch covers, toys, fancy goods, ballpoint pens, monofilaments for brushes filling, clothes pegs, slide holders, recording tape reels, ceiling tiles, expanded sheet and foams for packaging.

Medium-impact polystyrene has improved toughness while retaining the characteristic rigidity of general purpose polystyrene. Impact strength is at least doubled and elongation at break is considerably improved. There is some sacrifice of electrical properties, but otherwise medium-impact polystyrene is similar to GP. Available in the same forms as GP, medium-impact polystyrene grades are processed by the same methods. Applications are also similar but tend towards those where improved mechanical strength is required and some sacrifice of clarity and sparkle can be tolerated.

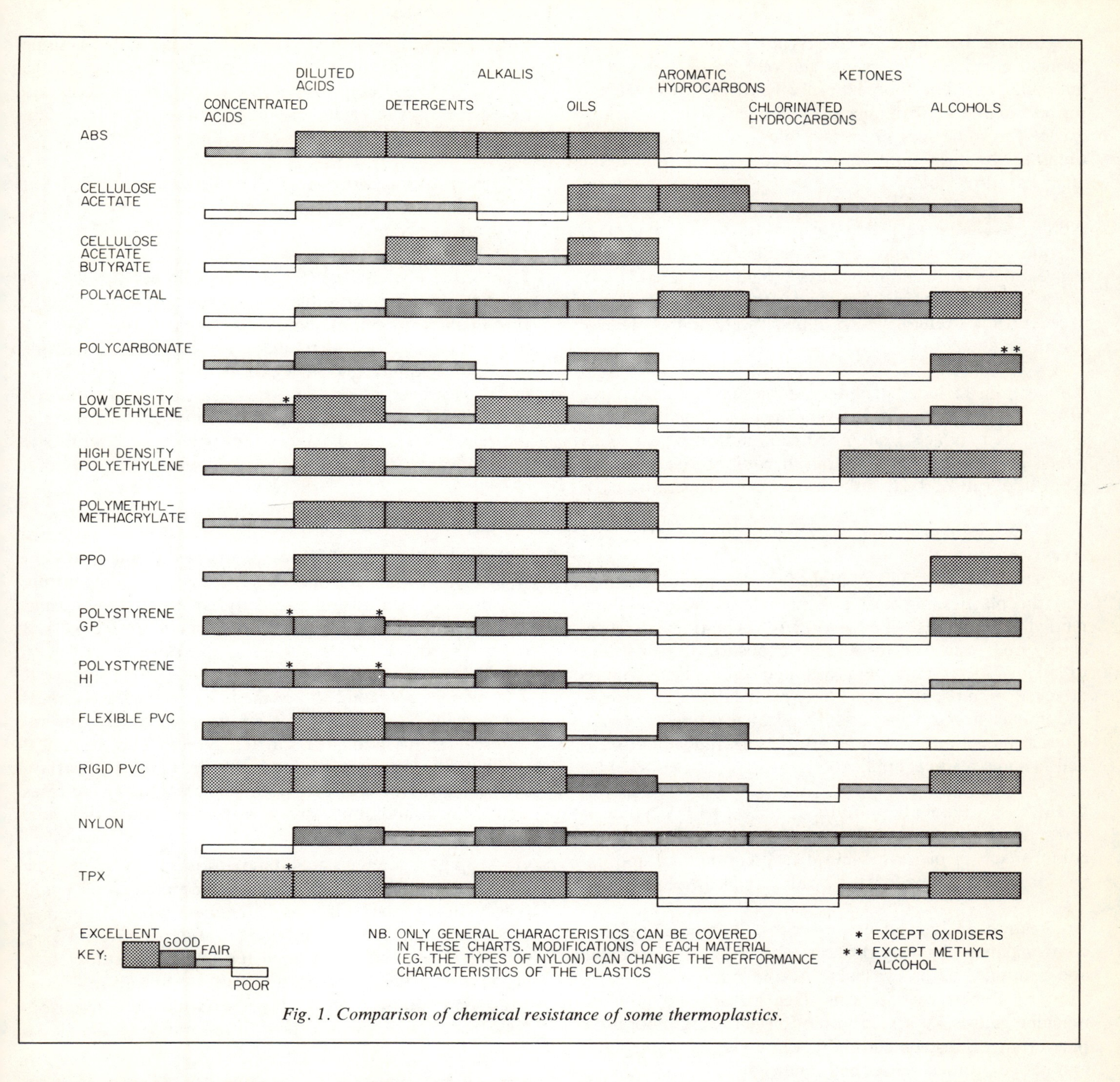

Fig. 1. Comparison of chemical resistance of some thermoplastics.

High-impact polystyrene is a tough ductile material deriving its strength from the incorporation of finely divided rubber particles or rubber latex chemically combined with the polystyrene molecules. Impact strength five times that of GP is obtained and elongation at break is further improved along with resistance to environmental cracking. Thermal properties are little different from GPs but there is further loss of electrical properties. Dimensional stability is retained. It is available in pellets in translucent to opaque colours, and processed as for GP polystyrene with precautions against overheating.

Applications include business machine covers, toilet cisterns and seats, drawers in furniture, containers and disposable cups for fruit juices and dairy products, portable radio cases, deep freeze and microwave oven food trays, tape and wire reels and bobbins, lighting fittings and toys.

Styrene-acrylonitrile (SAN) copolymer is a brittle material but has improved surface hardness and chemical resistance when compared with polystyrene and has excellent resistance to weathering. It is available in granules from transparent pale straw colour through a full range to opaque colours. SAN is processable by injection or blow moulding and extrusion. Applications include table, picnic and kitchen ware where resistance to fruit juices and natural oils is important, cosmetic and pharmaceutical packaging for similar reasons, refrigerator parts, and radio controls and scales.

Acrylonitrile butadiene styrene (ABS) has good resistance to most chemicals, detergents and oils, but is attacked by hydrocarbons. Outdoor exposure has little effect on the properties and performance of ABS over extended periods except for some loss of surface gloss. Protection can be obtained by pigmentation or lacquering to reduce the effects of ultra violet.

ABS is available as pellets or powder in translucent to opaque colours. Grades are available for metal plating. Processing is by injection or blow moulding and extrusion. Sheet is formable by the usual methods, and exhibits excellent hot strength which allows the handling of large formings.

Applications include telephones, domestic appliance parts, office machine parts and covers, luggage, helmets, automobile dash panels and instrument housings, car body panels and radiator grilles, aircraft components, chemical pipes and valves, boat hulls, toys, and refrigerator parts and liners.

Acrylics

The general term acrylics is applied to both surface coatings and the plastics material polymethylmethacrylate. Polymethylmethacrylate (PMMA) has exceptional optical clarity and resistance to outdoor exposure. It is resistant to alkalis, detergents, and oils and dilute acids but is attacked by most solvents. Electrical insulating properties and impact strength are good. Its peculiar property of total internal reflection is useful in advertising and similar signs and in some medical applications.

PMMA is available in granules and sheet from glass clear transparency through a full range of translucent and opaque colours, and is processed by injection moulding and extrusion. Sheet can be formed by all standard techniques.

Applications include advertising signs, illuminated notices, control panels, lighting diffusers and fittings, domelights and canopies, baths, wash basins and bathroom screens, shower cabinets, windscreens, face guards and goggles, machine guards, TV screen implosion guards, gramophone pick-up arms, instrument dials, watch glasses, nameplates and badges, optical lenses and fancy goods.

Fluorocarbons

These products, known by the initials PTFE, PCTFE and FEP, combine low friction characteristics with exceptional resistance to chemicals, good electrical insulation properties and resistance to high temperatures. These materials are relatively costly and difficult to process but they have an important place in high duty applications and, nearer home, as non-stick coatings. All are extremely tough and durable.

Polytetrafluoroethylene (PTFE) is available as a white powder or dispersion. Processing is not possible by the usual methods owing to the exceptionally high melt viscosity of the material. Mouldings are produced by sintering. Extrusion employs high temperatures and pressures and the material has to be treated with a volatile lubricant. Applications include high duty bearings, seals, valves, gaskets, linings, pipes in chemical plant, constructional bearings in bridge building, piston rings, insulation on heater cables, and non-stick cooking surfaces.

Polychlorotrifluoroethylene (PCTFE) is similar to PTFE but can be injection moulded (although not easily). Applications are also similar.

Fluoroethylenepropylene (FEP) is also similar to PTFE in properties but is processable by more conventional methods. It is available in granules and dispersions. Applications are more in the electrical than chemical fields, including cable insulation, connectors, and base material for printed circuits. However FEP's anti-sticking and anti-corrosion properties are used in control valves and machine parts in the paper, textile and bakery industries and in domestic kitchen ware.

Nylons

Nylon exhibits low friction properties giving excellent resistance to abrasion and wear with good temperature resistance and dimensional stability especially when formulated with mineral or glass fibre fillers. Bearings and moving parts are among nylon's applications.

The nylons (polyamides) are made by way of a chemical reaction between acids and amines. There are several different amine and acid starting points. The numerical suffix in a nylon description denotes the number of carbon atoms in the reacting chemicals; a single number indicates that the acid and amine are present in the same molecule.

Good impact strength, resistance to abrasion and a low coefficient of friction are properties common to all nylon types although the former varies between types. Resistance to most chemicals is fair to good but differs between types. In general nylons are not good for extended outdoor use, becoming brittle in sunlight and liable to absorb moisture, although endurance up to three years has been reported. Nylons are, of course, excellent fibre forming materials.

Nylon 6 and nylon 6.6 are very similar in properties and performance, and available in granules which are translucent cream white as well as a range of opaque colours. Processing takes place by injection moulding, blow moulding and extrusion. All nylons have a relatively high and sharply defined melting point, a feature which has to be taken into account. Applications include bearings and moving parts in industrial and domestic machines and instruments, in cameras and projectors, in door and in window latches and locks, furniture castors, curtain runners, tubing, kitchen spatulas and pot scourers.

Nylon 6.10 has lower impact strength than 6 and 6.6 but improved dimensional stability and slightly lower moisture absorption. It is available in pellets of translucent creamy white or pigmented. Applications are similar to those of nylon 6 and 6.6 with the accent on dimensional stability.

Nylon 11 and 12 are similar in properties although from different base materials. Both have high impact strength. These nylons are available as pellets or powder, and in filled versions in granular form in natural translucent to opaque colours. They are also available as film and sheet. Processing methods include injection moulding, blow moulding, rotational moulding and extrusion. Applications include bearings and moving parts, electrical components, and flexible tubing.

Acetal polymers and homopolymers

The polyacetals rival diecast metals in their performance as industrial components and fixtures. They are hard, rigid and tough materials retaining their excellent mechanical properties for long periods at high temperatures. They also have a low coefficient of friction. Resistance to most solvents, and oils and grease is good but resistance to acids and alkalis is only fair. Polyacetals are available in granular form with a range from translucent 'off-white' to opaque colours. (The homopolymer is not translucent.)

The polyacetals are processable by injection and blow moulding and extrusion. Applications include automobile components and fixings such as choke controls, gear levers, instrument housings, door and window handles, safety harness fasteners, bushes, gears and bearings, industrial and domestic machine parts, aerosol containers and valves, tool handles, heating and ventilating controls.

Polycarbonates

Polycarbonates are transparent, tough and resilient over a wide range of temperatures up to 145°C. They have good resistance to water, chemicals, greases and oils, but are attacked by strong alkalis such as ammonia and hydrocarbon solvents. Good electrical properties and fairly low coefficient of friction are shown.

Polycarbonates are available in granules from glass clear transparent through a range of translucent and opaque colours. Glass fibre reinforced material is also available. Processing takes place by injection moulding, and blow moulding and extrusion.

Applications include cable connectors, camera and projector components, pump parts and control valves, instrument covers, safety helmets and guards, hair curlers, filter bowls, lighting diffusers and covers, lamp holders and fittings, feeding bottles, kitchen ware such as measuring and blending fitments on mixers, golf club heads, and chocolate moulds.

Polyphenylene oxide (PPO)

PPO has good electrical and mechanical properties and excellent dimensional stability over a wide range of temperatures. Impact strength, hardness and rigidity are particularly good. Resistance to acids, alkalis and detergents is good and PPO can be sterilised repeatedly, but it is attacked by solvents. A modified type contains some polystyrene.

PPO is available in granular form, transparent amber coloured and opaque beige and black. Some other colours are offered. Processing methods comprise injection moulding, blow moulding and extrusion. Injection temperatures are relatively high except for the modified version. Applications include hospital ware and medico/surgical parts and instruments, electronic components, hot water controls, washing and washing-up machine components, cable insulation, and industrial machine parts. PPO is produced in the USA, and is available in the UK.

Polysulphones

Polysulphones (polysulfones) have high tensile strength, good electrical properties and excellent resistance to acids, alkalis and oils, and these properties are retained for long periods at high temperatures. To ensure this long term

Table 2. Comparison of costs and properties of some thermoplastics.

	Cost comparison index	SG	Impact strength (ft lb/in notch)	Tensile strength (000 lbf/in²)	Water resistance	Max. service temperature (°C)	Injection temperature (°C)	Extrusion temperature (°C)	Trade name examples
ABS	40	1.02–1.07	5–20	4.7–7.25	Excellent	94	210–240	190–210	CYCOLAC
Cellulose Acetate	36	1.23–1.33	2.25	4.25	Poor	65	160–230	140–240	ERINOID
Cellulose Acetate Butyrate	50	1.15–1.25	1.7	5.0	Fair	70	180–220	150–250	TENITE
Polystyrene GP	15	1.04–1.07	0.25–0.40	6.0–6.5	Excellent	60–70	180–280	150–200	STYRON, CARINEX
Polystyrene High Impact	20	1.04–1.07	1.0 –1.6	3.0–4.0	Excellent	60–70	230–290	180–210	ERINOID, STERNITE
Polymethylmethacrylate	45	1.18–1.20	1.0 –4.5	10.0	Excellent	80	210–240	160–220	DIAKON
Low Density Polyethylene	18	0.92–0.94	No break	1.4	Excellent	80	190–250	150–180	ALKATHENE
High Density Polyethylene	20	0.94–0.96	5.0	3.7	Excellent	80–90	220–260	170–190	RIGIDEX
Flexible PVC	21	1.10–1.70	n.a.	1.5–3.5	Excellent	105	150–200	140–170	BREON, VYBAK
Rigid PVC	23	1.35–1.50	0.7 –5.0	5.0–7.5	Excellent	80	140–180	140–170	WELVIC
Nylon	70–150	1.10–1.15	1.45	10.0	Fair	100	225–280	260–310	MARANYL
Polycarbonate	80	1.20–1.25	14.0	10.0	Excellent	140	275–320	—	MAKROLON
Polyacetal	75	1.41–1.43	1.85	10.0	Excellent	125	200–220	200–215	DELRIN
TPX	85	0·83	1–2	4.0	Excellent	160	260–300	250–280	TPX
PPO	80	1.06	1.5–1.8	9.90	Excellent	120–160	320–345	280–320	—

NB. 1 lbf/in² = 68·95 mbar, 1 lbf = 0.454 kg.

thermal stability an annealing period after moulding is advised. Polysulphones are available in granular form in transparent light amber to opaque colours. Fillers can be incorporated.

These plastics are processable by injection moulding, blow moulding, extrusion and sheet forming and can be plated. Applications include electronic connectors, TV components, printed circuitry, projector lamp housings, and filter bowls. Produced in the USA, it is worthwhile monitoring the availability of polysulphones as applications are planned.

Polyphenylene sulphides

Polyphenylene sulphides (PPS) are available in both thermosetting and thermoplastic versions depending on whether the polymer chain is branched or linear. This material is a white powder in its natural form but it can be formulated with fibrous fillers such as glass and asbestos and with mineral powders.

Processing is by injection moulding, compression moulding or blow moulding followed by annealing to effect cross-linking. It is also used for powder coating. PPS shows excellent resistance to acids, alkalis, and solvents and has good high temperature resistance. Applications are not well documented for this relatively new product, and the availability of polyphenylene sulphides should be confirmed as applications are planned.

Ionomers

These are tough, transparent, low friction materials, resistant to alkalis, and solvents and oils, but not to acids. They have good low temperature flexibility. Ionomers are processed by injection moulding, blow moulding and extrusion from transparent granules for applications such as medical phials and packages, tool handles and heads, and road marking 'cats eyes'.

Ethylene vinyl acetate (EVA)

EVA is a toughened copolymer similar to polyethylene with increased flexibility in proportion to the amount of vinyl acetate in the formulation. It has a lower service temperature than polyethylene, but good long term resistance to stress. EVA is available as a powder or in granular form for processing by injection moulding, blow moulding and extrusion. Applications include tubing and pipes, cable sheathing, closures and gaskets, packaging film and toys.

Chlorinated plastics

The properties and performance of **chlorinated polyethylene** depend on the degree of chlorination of the product up to an optimum 27% for maximum flexibility. It generally possesses excellent resistance to acids, alkalis, oils and aliphatic solvents, but poor resistance to aromatics. It resists thermal deformation better than PVC up to 100°C but then falls off rapidly.

Chlorinated polyethylene is available in powder form and processed in similar ways to polyethylene and PVC. Applications include footwear mouldings, gaskets, respirator parts, toys, weather protective sheeting, and textile interlinings.

Chlorinated polyether possesses outstanding resistance to acids, alkalis, oils, detergents and most solvents and has excellent retention of mechanical strength under prolonged outdoor exposure. It is available as translucent off-white or coloured granules and is processed by injection moulding and extrusion, and also used for powder coating. Applications are mainly where corrosion resistance is important: control valves, tank and pipe linings and fittings, and wire coatings.

Poly (4-methyl pentene—1)

This product, known by its trade name of TPX, is another transparent material with a high softening point and high service temperature. In some respects it is similar to polyethylene but it has lower impact strength. Electrical properties are good as are resistance to acids (except oxidisers), alkalis, detergents and oils. TPX is attacked by hydrocarbon solvents but is unaffected by repeated sterilisation. For outdoor use TPX requires protection from ultra violet attack. The density is the lowest known for a plastics material (0.83).

TPX is available as a white powder or transparent granules, and is processable by injection moulding, blow moulding and extrusion. The low density and high thermal capacity of the material calls for special attention in processing. Applications include laboratory and hospital ware, meal trays, lighting fittings, textile machinery parts, electronic components, and sight glasses in chemical plant and pipelines.

LENGTHS OF RUN

Any plastics which can be injection moulded, blow moulded or extruded is suitable for long runs. While low price provides some indication of the mass-production plastics, the unique properties of more expensive materials provide a low-cost solution to the problem of fabricating for specific applications, and the cost of production of the specific product is of most importance.

THE PROBLEM OF DEVELOPMENTS

Almost every year sees the introduction of another new plastics material, each one finding a special place by reason of its own combination of properties. Currently there is a certain amount of overlapping as several materials can be equally suitable for one application. In some cases availability or final cost is the decisive factor in making a choice, while in others ease of processing may be the criterion. The designer or intending user or the engineer with a particular problem to solve cannot do better than to seek the advice of the material manufacturers. Meanwhile, a regular study of the leading plastics journals is recommended to keep the engineer abreast of new developments.

Chapter 2

Additives

E W Thomas *BSc* and **J E Todd** *AMCT*
CIBA-GEIGY (UK) Limited

The basic polymers to which the plastics industry owes its existence are in most cases unsuitable for processing into useful articles, in their pure unmodified form. This is especially true of the thermoplastics which are mainly responsible for the explosive growth of the plastics industry since the Second World War. Polyolefins, for example, are extremely susceptible to oxidative degradation, particularly under the conditions of heat and pressure necessary for their extrusion or injection moulding, and must be protected with a suitable antioxidant before becoming commercially viable materials. Similarly, polyvinyl chloride (PVC), another high tonnage thermoplastic, is in its pure form a hard horny material, difficult to process and extremely susceptible to rapid degradation at the temperatures necessary for its processing. However, the addition of suitable plasticisers, stabilisers, lubricants and other additives can transform it either into a soft flexible material with pronounced rubbery characteristics, or alternatively into a tough rigid material suitable for water pipes, roof sheeting and a host of other applications.

These two examples are sufficient to illustrate the fact that the plastics industry depends for its existence, in its present highly sophisticated form, on the availability of suitable additives to modify or eliminate undesirable properties of the raw polymers while developing and enhancing their useful characteristics.

Plastics additives may be divided into two main categories —those which modify polymer properties by physical means, such as plasticisers, lubricants, impact modifiers, fillers, pigments, *etc*, and additives which achieve their effect by chemical reactions such as PVC heat stabilisers, antioxidants, ultraviolet absorbers and flame retardants. In selecting them, however, care must be taken to match the additives to the polymer they are intended to protect, taking into account such factors as compatibility, efficiency and possible synergism of two or more additives in the particular polymer under consideration.

PLASTICISERS

Many of the newer engineering plastics, such as nylon, polyacetates, polycarbonates, *etc*, are valuable for such properties as tensile strength, rigidity and high flexural modulus. Nevertheless, there remain large areas of raw material requirement where greater flexibility, softness, easy formability and toughness are essential. This is particularly so in the case of PVC, which the correct choice of plasticiser can transform into a soft supple material with a wide choice of possible physical properties. Although other polymers such as cellulose acetate are often plasticised, it is estimated that about 90% of world plasticiser production is used by the PVC industry.

A plasticiser may be defined as a susbstance which when incorporated into a polymeric material improves flexibility and processability, alters temperature-dependent properties such as softening point and low temperature flexibility, and has a significant effect on tensile strength, elongation at break and impact properties. The plasticiser achieves these effects by reducing the physical intra-molecular forces between the polymer chains, thereby increasing chain mobility which results in softening. Chemically, plasticisers are generally high boiling, non-volatile solvents, usually ester type organic compounds, polar, and with a relatively high molecular weight. The degree of polarity requisite in a plasticiser depends to some extent on the attractive forces between the polymer chains, so that PVC possessing strong intra-molecular forces requires plasticisers of reasonably high polarity.

A primary plasticiser for PVC is regarded as one which is fully compatible with the resin up to at least 100 parts per hundred parts resin (phr) and can be satisfactorily used for many applications as the sole plasticiser. Phthalate and phosphate esters and certain polyesters come into this category. A secondary plasticiser has limited compatibility with PVC, but is normally incorporated in a PVC compound as a partial replacement for a primary plasticiser to impart a specific desirable physical property. Low temperature plasticisers such as sebacates and adipates are typical of this category. Plasticiser extenders are plasticisers of limited efficiency and compatibility such as chlorinated paraffins and certain hydrocarbon oils, which are included in the composition to reduce costs.

The choice of a plasticiser for a given application is usually a matter of compromise between performance and economy, and the desired combination of properties can often only be obtained by using a mixed plasticiser system. A typical example would be a PVC leathercloth formulation for foul weather clothing which might contain a phthalate as general purpose plasticiser, with a polyester for greater permanence and a low temperature plasticiser, such as an adipate, for improved performance under cold conditions.

A number of factors must be assessed in choosing a plasticiser for a given application. These include solvating power, efficiency, compatibility, volatility, extraction and migration resistance, heat and light stability, flame retardant properties, toxicity, low temperature performance and, of course, cost. Plasticisers with good solvating powers, such as phosphates, give quick gelation and easy processing characteristics. Efficient plasticisers plasticise a PVC compound to the requisite softness at a low level of incorporation. In general, high efficiency is desirable in a plasticiser, although when PVC polymer is expensive, and plasticisers are cheap, it is often financially advantageous to incorporate an inefficient plasticiser, so that at a given softness a greater proportion of the PVC compound is made up of the cheap plasticiser than the more expensive polymer. Similarly, an inefficient plasticiser may be of advantage in a rigisol formulation giving reduced paste viscosity while still maintaining the desired hardness of the final product.

The ideal plasticiser should be compatible with the resin in all proportions. In practice, the level of plasticiser incorporation is very seldom greater than 150 parts per hundred parts resin (phr) and many common plasticisers, such as phthalates and phosphates, can be used at this level satisfactorily. A plasticiser is incompatible with a PVC compound if it exudes to the surface of the manufactured article, giving a greasy or sticky feel. In cases of gross incompatibility actual droplets of plasticiser appear on the surface of the article. Compatibility of a plasticiser with a resin is a function of polarity, molecular weight and structure. Most secondary plasticisers are marginally incompatible so that they can only be used in a PVC compound in conjunction with a highly compatible primary plasticiser, and even then only in limited proportions below the compatibility limit. Highly compatible PVC plasticisers, such as phosphates, can tolerate higher proportions of secondary plasticisers than less compatible primary plasticisers such as the higher phthalates, DiDP (di-isodecyl phthalate) for example.

The volatility of a plasticiser determines the degree of plasticiser loss during processing, as well as the subsequent loss during use at elevated temperatures. Highly volatile plasticisers such as DBP (dibutyl phthalate) are of limited use in PVC compounds as the loss of plasticiser during service leads to hardening and embrittlement of the compound. Plasticisers with low volatility are polyesters, trimellitates and to a certain extent the higher phthalates such as DTDP (di-tridecyl phthalate). This property is extremely important in the case of electrical insulation designed for high temperature operation, and in automobile upholstery where plasticiser volatilisation in hot climates can lead to a fogging effect on windscreens and windows.

For some applications, resistance to extraction of plasticisers by oils, greases, detergents or other agents is most important. In general, monomeric plasticisers such as phthalates show relatively poor extraction resistance, and best results are obtained with polyesters or other high molecular weight plasticisers.

When a PVC compound containing a plasticiser is placed in intimate contact with another plastics or rubber, it is important that the plasticiser should not migrate from the

PVC compound into the other polymeric material. As in other situations involving permanence, polyester plasticisers are usually the preferred materials.

Plasticisers should ideally be mobile free flowing liquids, for ease of handling under plant conditions. Most plasticisers satisfy this requirements, although some of the very high molecular weight polymerics used for maximum extraction resistance are more difficult to handle at room temperature. A plasticiser should also be colourless to give maximum versatility in pigmentation, and of low density for economy.

PVC plasticisers must be able to withstand normal processing temperatures up to 200°C without breakdown. Most commonly used plasticisers are reasonably heat stable, although the stabiliser content of a compound usually has to be increased when chlorine containing plasticisers such as chlorinated paraffins are used. Light stability and weathering resistance are also important.

PVC polymer is itself inherently flame-resistant but when plasticised with a plasticiser, such as a phthalate which supports combustion, the resulting PVC compound may not be flame-retardant. Flame-retardance is therefore a very desirable property of a plasticiser and phosphates are the preferred materials in such applications as flame-retardant wall coverings, conveyor belts and PVC coated metal or wood.

For some applications, in particular food packaging or toy manufacture, non-toxic plasticisers are required. Highly toxic plasticisers would also create problems on the factory floor during processing. Most phthalates and polyesters, however, have very low toxicity, although the compounder should satisfy himself of the particular requirements before formulating for applications where non-toxic properties are important.

Ideally a plasticiser should maintain the flexibility of a polymer over as wide a temperature range as possible. In practice, a certain proportion of low temperature plasticisers such as sebacates or adipates is often added to the phthalate or other primary plasticiser to give optimum low temperature properties, although certain phthalates derived from straight chain alcohols are appreciably better than DOP (di-octyl phthalate) or DiOP (di-iso-octyl phthalate) in this respect.

Phthalates are by far the most widely used plasticisers, possessing a reasonable combination of the required properties. They are used for large volume outlets such as flooring, cables, sheeting, leathercloth, footwear and general extrusions. The most popular are the octyl phthalates, with DiOP and D79P (di-alphanol 7-9 phthalate) being the most popular in the UK and DOP (di-2-ethylhexyl phthalate) being the most widely used in the United States. For most practical purposes these three can be regarded as interchangeable, apart from minor differences in volume resistivity, and the choice between the three is dictated by local cost and availability. They are efficient and cheap with low toxicity, high compatibility, reasonably low volatility and ease of handling. However, their extraction properties are poor, they are inflammable and unsuited for continuous use at elevated temperatures.

As the alcohol chain length of a phthalate plasticiser is increased volatility is improved, but low temperature properties, compatibility and efficiency are reduced. Consequently DiDP (di-iso-decyl phthalate) and DTDP (di-tridecyl phthalate) are used for some high temperature cable applications, but for the more stringent high temperature requirements it is necessary to use a polyester, a dipentaerythritol ester or a trimellitate ester.

DBP (di-butyl phthalate) was one of the earliest plasticisers for PVC, but is little used today on account of its high volatility. New contenders for large volume use are the phthalates based on straight chain alcohols, of which those based on the Linevol 7-9 and Linevol 9-11 alcohols are typical. They have better low temperature properties and volatility characteristics than their branched chain analogues D79P and DiDP which they otherwise resemble, at the expense of slightly inferior volume resistivity.

Self extinguishing PVC compounds may be formulated using non-inflammable plasticisers such as phosphates and chlorinated paraffins. However, chlorinated paraffins are secondary plasticisers of limited compatibility and cannot be used as the sole plasticiser, except in low percentages to give hard compounds. Aryl phosphates are therefore the standard plasticisers for these applications. In addition, their resistance to extraction by oils and petrol is superior to that of other monomeric plasticisers.

The original phosphate plasticisers were TTP (tri-tolyl-phosphate), also known as TCP (tri-cresyl-phosphate), and TXP (tri-xylyl-phosphate). TXP is not quite as efficient as TCP and is slightly inferior in low temperature flexibility, but in most applications the two products are interchangeable.

Aryl phosphates such as TTP and TXP are manufactured from tar acids (cresols and xylenols), arising as byproducts from coal gas production, coke production, and petroleum refining operations. Due to the increasing scarcity and high cost of such tar acids, in 1968 Geigy (UK) Limited introduced completely new triaryl phosphate plasticisers (*Reofos*) made from a wholly synthetic oil based feedstock. Their properties are such that *Reofos*$^{(R)}$*95* can be used where TXP types of plasticiser are used and *Reofos*$^{(R)}$*65* where TTP (TCP) types are used. The vastly improved light fastness of the *Reofos* products has led to their use in applications where the traditional phosphates have proved unsatisfactory, such as PVC wall coverings, clear wear layers on laminated PVC floorings, non-flammable thin PVC films and agricultural sheeting. Compared with the traditional tar acid phosphates which they replace, they have similar flame retardant action, lower odour and toxicity, better low temperature properties and give lower plastisol viscosities.

Low temperature plasticisers impart greater flexibility and resistance to cracking at low temperature, when used in relatively small proportions with a phthalate or other primary plasticiser. In order of performance (and also unfortunately of price) the best performance is given by sebacates such as DOS (di-octyl sebacate) followed by azelates, adipates, nylonates (esters of mixed adipic glutaric and succinic acids) and succinates.

Polyester plasticisers or polymerics are used in PVC compositions to confer the property of permanence, *ie* resistance to extraction by petrol or oil (important in tubing, oil seals, oil resistant industrial footwear, *etc*), low migration into other materials in intimate contact with the PVC (*eg* polystyrene in the case of refrigerator gaskets or the adhesives used in PVC insulation tape or decorative sheeting), and low volatility (important in high temperature electric cables and 'anti-fogging' automobile upholstery and trim).

The original polyester plasticisers were high molecular weight unmodified polyesters such as polypropylene sebacate. The higher the molecular weight of the polyester, the better the permanence, but attendant disadvantages such as high viscosity, leading to difficulty in handling and processing, the need for higher gelation temperatures, and poorer efficiency and low temperature properties, led to the development of modified polyesters of intermediate viscosity (5–50 stokes), chain end stopped by mono-carboxylic acids such as lauric acid or alternatively mono-hydric alcohols such as iso-octanol. Such polyesters are easier to handle and process and have better low temperature and gelation properties while not possessing quite such high resistance to extraction as the unmodified variety. They are also usually cheaper, and have a good all round balance of properties.

Trimellitate plasticisers are very low volatility plasticisers with reasonable low temperature properties and efficiency. As esters of C_7 to C_{10} alcohols and trimellitic anhydride, they find their main application in high temperature electrical cables, although their exceptionally low volatility also makes them useful for 'anti-fogging' automobile leathercloth. They have largely taken over from di-pentaerythritol esters due to their better processing characteristics and better hydrolytic stability.

Extender plasticisers (hydrocarbon oil extenders and chlorinated paraffins) reduce PVC volume cost without degrading physical properties to an unacceptable level. Normally, they cannot be used as sole plasticisers due to their limited compatibility. Chlorinated paraffins have better physical properties then hydrocarbon extenders, and are sufficiently compatible to be used in fairly high proportions with phthalates or phosphates in certain applications such as general purpose cable compounds or flame-retardant conveyor belting. Usually, they are 40–55% chlorinates of C_{13}–C_{17} normal paraffins or C_{20}–C_{23} paraffin wax. Their use usually demands increased heat stabiliser levels due to their high chlorine content and their light stability is still not good enough for applications where light fastness is of prime importance.

Although PVC is undoubtedly the most important polymer commercially, many other polymers often require plasticisation. Among these are polyvinyl acetate (PVA), nitrocellulose, cellulose diacetate and cellulose triacetate.

The most common plasticiser for PVA is DBP (di-butyl phthalate) as higher molecular weight phthalates are not compatible. Superior non-migratory properties are conferred by certain polyester plasticisers, such as *Reoplex*[(R)]-*400*, which is particularly useful for PVA-based glass fibre size used to maintain coherence of the glass fibre mat before reinforcing with polyester resin, as well as in adhesive applications.

Most PVC plasticisers are compatible with nitrocellulose, but phosphates give superior gloss and reduce the inherently high flammability of the material. In nitrocellulose lacquers the superior light stability of the *Reofos*[(R)] phosphates makes them particularly suitable.

Dimethyl phthalate and diethyl phthalate are traditional plasticisers for cellulose acetate, but due to their high volatility, triphenyl phosphate (TPP) is preferred for applications where permanence is important. TPP also improves the flame retardance of cellulose acetate and is generally used in conjunction with dimethoxyethyl phthalate to improve the melt flow properties of the compound. Dimethoxyethyl phthalate is also used as a bonding agent for cellulose acetate fibre waddings. About 5–15% of the plasticiser is sprayed on to the matted fibre which is then bonded at approximately 175°C to produce a non-woven fabric.

PVC STABILISERS

Unstabilised PVC polymer is susceptible to very rapid decomposition at the temperatures encountered in normal processing. This poor heat stability has been highlighted by modern conversion machinery in which faster throughput and shorter processing cycles have led to even greater thermal stresses on the material, particularly in the case of unplasticised or rigid PVC.

Fortunately, the PVC compounder today has a very wide choice of heat stabilisers at his disposal. For technical reasons certain classes of stabilisers have become associated with particular areas of the industry, *eg* lead stabilisers for cables and tin-based materials for blow moulded rigid PVC bottles and transparent rigid sheet extrusions.

It is common practice to combine two or more categories of stabilisers to take advantage of synergistic effects, *eg* barium/cadmium stabilisers with phosphite chelators, and calcium/zincs with pentaerythritol. The activity of both Ba/Cd and Ca/Zn stabilisers is also enhanced by combination with epoxidised oils, particularly epoxidised soya bean oil.

Lead stabilisers are one of the oldest categories of PVC stabilisers, and are still very widely used, particularly in the cable industry and in certain rigid applications such as the extrusion of soil pipes, rainwater goods, *etc*. Basic lead carbonate (white lead), tribasic lead sulphate (TBLS), dibasic lead phosphite, dibasic lead phthalate and lead stearate are the most common materials, but 'liquid leads'—organic lead compounds—have recently been developed which are claimed to offer advantages in combating porosity during high speed cable extrusion. The toxic dust hazard associated with the use of powdered lead stabilisers can be overcome by using paste dispersions of the stabilisers, generally in phthalate plasticisers, which are readily available from most suppliers.

Lead stabilisers are cheap, efficient and have good electrical properties, but they are toxic, opaque and of high specific gravity. Level of addition is usually from 4–8 parts per hundred resin (phr). Due to their toxicity, they are not suitable for food packaging, and in the rigid PVC field they are losing ground to the more efficient and less toxic tin based varieties, particularly where processing conditions are severe, *eg* in the injection moulding of thick walled pipe fittings.

Barium/cadmium (Ba/Cd) and barium/cadmium/zinc (Ba/Cd/Zn) based stabilisers are general purpose stabilisers for plasticised PVC. Large quantities are consumed in leathercloth, calendered sheeting, extrusions, injection moulded footwear and flooring. They are not suitable for cable extrusion due to their inferior electrical properties. They exhibit synergism with phosphite chelators such as tris nonyl phenyl phosphite, and are often used with 3–5 phr of epoxidised vegetable oils as co-stabilisers.

Their toxicity precludes them from use in food packaging applications, although as a class they are efficient, reasonably cheap and have good light stability. The ratio of barium to cadmium in a particular stabiliser is chosen to obtain the required balance between good initial colour and long term heat stability. The inclusion of a small amount of zinc improves initial colour at the expense of long term heat stability.

Ba/Cd stabilisers are fatty acid soaps and the choice of organic acid (lauric, stearic, ricinoleic, *etc*) is critical in calendering application to minimise the risk of plate-out. Ba/Cd and Ba/Cd/Zn stabilisers are often sold as a tailor-made complex package for a particular application, and may contain antioxidants such as diphenylol propane, phosphite chelators and other ingredients in addition to the metal soaps. Usual levels of addition are 1–3 phr, the lower levels being used in certain plastisols subject to fairly mild curing conditions.

Calcium/zinc stabilisers (Ca/Zn) are chiefly used in applications such as food packaging where their non-toxic properties are of paramount importance. Although cheap, their heat stabilising efficiency is relatively poor, and they are often used as co-stabilisers with aminocrotonates and epoxidised vegetable oils, particularly in food grade extruded or calendered film. Their use in blow-moulded rigid PVC bottles is declining due to the greater efficiency and clarity of the non-toxic octyl tins. Ca/Zn stabilisers are also used in certain plastisol applications, such as rotationally cast toys, where processing conditions are relatively mild. Most members of the class, particularly Ca/Zn stearates, have a pronounced lubricating action, which must be allowed for in compound design. Addition level is usually 1–3 phr.

The development of non-toxic octyl tin compounds conforming to the regulations of the US Food and Drug Administration regarding food packaging has led to a pronounced increase in the use of these highly efficient stabilisers in such fields as blow moulded PVC bottles and packaging films. Their high cost is partly compensated for by their superior efficiency which allows low addition levels of 0.5 to 1.5 phr and by other technical advantages such as the sparkling clarity which they confer on rigid PVC products. Technical grade thio-tin stabilisers are also used in the injection moulding of thick walled pipe fittings and the extrusion of flat and corrugated transparent roofing sheet. Similarly, their high stabilising efficiency can give higher outputs and permit the extrusion of thicker walled rigid pipe on modern multi-screw extrusion equipment. Tin stabilisers are rarely used in plasticised applications where the less onerous stabilising demands can usually be met by less efficient but cheaper materials.

Phosphite chelators such as tris-nonyl phenyl phosphite are used as co-stabilisers with Ba/Cd and Ca/Zn to improve heat and light stability at an addition level of about 0.5 phr. Similar co-stabilising effects are shown by the epoxidised vegetable oils such as epoxidised soya bean oil, used at 3–5 phr. Epoxies may also be used at higher levels of 10–15 phr as secondary plasticisers to improve low temperature performance, when their presence also improves compound heat stability. Epoxidised soya bean oils are regarded as non-toxic and are therefore widely used as co-stabilisers in rigid and flexible food packaging applications.

Other co-stabilisers are esters of β-aminocrotonic acid, usually used with Ca/Zn materials in gramophone records and food packaging applications. The use of α-phenylindole, with Ca/Zn stabilisers for oil-bottles, once popular in France, has declined due to its very poor light stability and poor initial colour.

LUBRICANTS

Lubricants are widely used in the plastics industry to facilitate the processing of a variety of polymers. They may be classified as internal or external lubricants depending on their mode of action: **internal lubricants** act within the polymeric material itself, by reducing the cohesive forces normally acting at molecular interfaces within the resin (*eg* monoglycerin esters, stearyl alcohol, stearic acid metal stearates); whilst **external lubricants** act by reducing the adhesion of the polymeric material to the hot metallic surface of the processing machinery (typical examples are waxes, and paraffin oils). In general, lubricants act either internally or externally depending on their compatibility and solubility in the particular polymer. In the main, external lubricants are low solubility, long chain materials with carbon chain lengths in the C_{22}–C_{32} range, and of low polarity. Internal lubricants tend to have lower carbon chain lengths in the C_{11}–C_{20} range, and to be of high polarity. Some lubricants may be of the internal-external type. Long carbon chain length materials of higher polarity than strictly external lubricants are an example—typical dual function types being fatty esters and partially saponified waxes.

It should be emphasised that a lubricant can only be categorised as internal or external in relation to the particular polymer in which it is to be employed. In many cases, a very careful balance of internal and external lubrication is required, particularly in the case of rigid PVC processing. This polymer is perhaps more critical of incorrect lubrication balance than any other, and in many cases it is necessary to establish a stabiliser-lubricant mixture carefully tuned to the characteristics of a particularly demanding application or machine. Certainly a lubricant system which is ideal for a single screw extruder will often be ineffective in a twin screw machine and *vice versa*. Care must also be taken not to exceed the haze limit, which can impair clarity, and attention must be paid to the heat stability and tendency to stress-whiten of a particular lubricant combination.

In view of the above, it is apparent that specific recommendations for lubricating particular polymers are difficult to make and may be misleading. Nevertheless, it is common practice to use the lubricants shown in Table 1 in the polymer systems indicated. Fatty amides, waxes and silica compounds are often used as slip additives in polyolefins. Typical examples are oleamides, bis stearamides and stearic acid esters.

LUBRICANT	POLYMER
Glycerine mono esters, stearyl alcohols, stearic acid, metallic stearates, waxes, fatty esters	PVC
Zinc stearate	Polystyrene
Waxes, paraffins	Polyethylene
Zinc stearate	Melamine
Calcium stearate	Polyethylene, polypropylene, phenolic resins, Melamine, PVC
Magnesium stearate	ABS resins Cellulosic materials
Sodium stearate	Polyamides

Table 1. Lubricants commonly used for various plastics.

ANTIOXIDANTS

Most polymeric materials are subject to oxidative degradation at the elevated temperatures necessary for processing. Antioxidants, usually phenolic compounds, are added to combat the effects of oxidation both in the processing cycle and in the subsequent service life of the manufactured article. Polyolefins, especially polypropylene, are particularly susceptible to oxidative attack, and best results are obtained with high molecular weight antioxidants, which are normally incorporated by the resin manufacturer. Such antioxidants should not be so volatile as to be lost during the processing cycle, leaving no protection for the article during subsequent use. Other polymers which normally contain antioxidants added by the manufacturer are ABS, polystyrene, polyphenylene oxide and acetals. PVC resin, on the other hand, does not normally contain an antioxidant, but phenolic antioxidants are often included to advantage in complex heat stabilisers based on metallic soaps added at the compounding stage by the processor.

Polypropylene, in common with many other plastics, is most effectively stabilised by high molecular weight phenolic antioxidants often in combination with thiodipropionate esters (*eg* dilauryl thiodipropionate) and a high molecular weight tertiary phosphite. Best results are obtained with specially developed hindered phenolic antioxidants such as Tetrakis [methylene 3-3′,5′-di-tert-butyl(-4′-hydroxyphenyl) propionate] methane, which are non-toxic, fully compatible with polypropylene and of sufficiently high molecular weight not to volatilise out of the polymer during the processing cycle. Such antioxidants also exhibit synergism with substituted benzotriazole UV absorbers and thus help to improve the weathering resistance of a polypropylene article in outdoor service. Such high molecular weight hindered phenolic antioxidants are replacing the traditional materials such as BHT (butylated hydroxy toluene) due to their much lower volatility, and are also suitable for other polyolefins, polystyrene and ABS.

Antioxidants are also vital in polyurethanes. In polyether foams, phenolic antioxidants are added to the polyol to guard against excessive temperatures developing during the exothermic cure reaction which takes place 4–8 hours after foaming, and which has been known to cause serious fires. In polyester based polyurethane coatings, phenolic antioxidants improve service life and light stability, and are especially effective in protecting materials subject to repeated flex cracking, *eg* 'poromeric' polyurethanes for shoe uppers.

FILLERS

Fillers are normally added to polymeric materials for two reasons — to improve physical properties and/or to lower the cost of the composition.

Flexible PVC compounds often contain 10–50 phr of fillers, normally calcium carbonate, which can be simply ground chalk or stearate coated precipitated calcium carbonate of finely controlled particle size. Examples are cable sheathing, footwear compounds and general extrusions. Other PVC compounds of a more specialised nature such as flooring compositions may contain from 100–500 phr of filler, often mixtures of calcium carbonate and asbestos flour. In this case the PVC compound is merely acting as a binder to hold the filler particles together. Electrical insulation compounds on the other hand often contain 10–40 phr of electrical clay (silicate filler) due to the excellent electrical properties of this filler.

8–10 parts of stearate coated calcium carbonate of a definite particle size is often added to rigid PVC extrusions, such as water pipe, as much to improve the impact strength of the finished product as to cheapen the compound. Similarly, chopped glass fibre is added to some engineering plastics, such as nylon or acetals, to improve flexural and tensile strength and raise heat distortion point. Exotic new materials such as carbon fibres have even greater reinforcing effects in both thermoplastic and thermosetting plastics, and talc or asbestos filled polypropylene is finding increasing 'underbonnet' applications in the motor car industry due to its improved heat distortion temperature and stiffness.

Thermosetting plactics are often reinforced by as much as 100 phr of various fillers, for example, wood flour, cotton flock asbestos and synthetic fibre used with phenolic thermosets, and silica flour, mica or talc with epoxies. Glass fibre reinforcement of polyester resins for structural, automotive, aerospace and marine applications is too well known to require comment.

ULTRA-VIOLET ABSORBERS

Most polymers are adversely affected by exposure to UV radiation, *ie* radiation of wavelengths between 4 and 400 nanometres (nm). The main source of UV radiation is sunlight, but significant amounts of UV radiation are also present in the spectra from fluorescent tubes and arc lamps.

The absorption of UV light by plastics results in discolouration, embrittlement and a marked fall-off in such physical properties as tensile strength, elongation at break and impact resistance. It is therefore obviously necessary to protect polymers from the degradative effects of UV radiation, particularly in the case of those plastics designed for outdoor service or certain severe indoor applications, such as wall coverings or guards for fluorescent tubes. As it is not usually practical to shield the plastics material from the source of radiation, protection is usually afforded by the inclusion of suitable additives in the polymer.

The most simple means of increasing the light stability of a plastics composition is to incorporate suitable pigments. High loadings of titanium dioxide in PVC compounds or carbon black in polyolefins are well known examples of this technique. However, such high loadings of pigments affect physical properties and restrict colour choice, and in any case cannot be used in transparent compounds. A more efficient way of tackling the problem is to include sophisticated UV absorbers which have been specially designed to protect polymeric materials against the effects of UV radiation.

Five main groups of chemical compounds show sufficient protective activity to be commercially important: substituted 2-hydroxy phenyl benzotriazoles, substituted 2-hydroxybenzophenones, aromatic substituted acrylates, salicylic esters and the so-called nickel 'quenchers'. Except for the nickel quenchers, UV absorbers act by absorbing a high percentage of the impinging UV radiation and converting it into harmless radiation of different wavelength or thermal energy. Nickel stabilisers, which are particularly effective in polypropylene thin films or fibres, act by a different mechanism, by passivating the effects of the radiation of the polymer and thus inhibiting the degradation of the substrate. The newer products such as Ni-3,5,-ditertiary butyl-4-hydroxy benzyl phosphonic acid monoethylate do not impart an undesirable green colour to the substrate.

The most powerful UV absorbers for a wide variety of substrates are substituted benzotriazoles and substituted benzophenones. Benzotriazoles are particularly effective in PVC, polystyrene, polyester resins of various types and acrylics, while benzophenones may be used in PVC, polystyrene and polyethylene, although, in general, somewhat higher concentrations of benzophenones are required to offer equivalent protection. Within each general class of compounds, such as benzotriazoles, specific compounds exist which are tailormade to ensure maximum compatibility with the particular substrates to be protected, since the full protective action of a particular UV absorber can only be achieved in a polymer substrate with which it is compatible at the optimum required concentration.

Addition levels vary between 0.05 and 1 phr, the higher concentrations being required in thin films, and lower dosages being adequate in certain polymers such as acrylics or polycarbonates which are inherently less sensitive to the effects of UV radiation. Non-toxic benzotriazole grades exist, which are suitable for use in applications involving food contact.

OPTICAL BRIGHTENING AGENTS

While the plastics processor is normally concerned with inhibiting the degradative effects of UV radiation on the physical properties of his product, there are certain instances where incident UV light may be turned to advantage. This occurs when optical brightening agents are incorporated in a plastics composition to enhance the apparent brightness of white or pastel shades.

Many transparent polymeric materials have a very faint yellowish cast in the natural state. This yellowish cast can also affect the brightness or apparent whiteness of pigmented formulations. Although the incorporation of trace amounts of violet or blue dyestuffs can improve matters to some extent, by reducing reflection of yellow light, the most effective solution is to incorporate small amounts of organic chemicals called optical brightening agents (oba's). These act by absorbing radiation in the ultra violet, and fluorescing or re-emitting radiation at the bluish end of the spectrum, *ie* between 400 and 500 nm. The eye thus perceives increased brightness due to the increased amount of visible light reflected from the plastics surface.

Some oba's produce intense whites with a bluish hue, others brilliant whites with bluish green or bluish red overtones, depending on chemical composition.

Many chemical compounds exhibit strong fluorescence under UV light, among them coumarins, naphthylimides, certain azolyl compounds and stilbene derivatives. It is of course not sufficient that a product should fluoresce strongly in the desired region of the spectrum—it should also be compatible with the polymer substrate, able to withstand high processing temperatures, have good solubility and chemical stability, and good light fastness. Materials are available commercially which satisfy these requirements for a wide range of polymers. In many cases they are available as the pure product or extended with inert materials to facilitate incorporation of trace amounts in the plastics composition.

Colourless oba's may also be used as 'invisible' markers in plastics substrates, which only show up when irradiated by ultraviolet light.

FLAME RETARDANTS

Most polymeric materials are flammable to a greater or lesser extent, and even such materials as PVC which are themselves non-burning may well support combustion when compounded with flammable materials such as phthalate plasticisers.

With increasing emphasis being placed by various governmental bodies on the use of flame retardant materials in the building, textile and automotive industries, the subject of flame retardance is assuming increasing importance in the plastics industry. Much legislation already exists; more may be expected to follow.

Flame retardance in polymeric materials may be achieved by two different mechanisms. The first is by using an inherently flame retardant material as a substantial component of the polymeric material or compound. This approach is exemplified by the use of chlorendic acid as a reactive component of flame retardant polyester resins, or the use of substantial quantities of phosphate plasticisers in non-flame PVC compounds, *eg* for automotive leathercloth.

The second approach is to incorporate small amounts of highly efficient flame retardant additives as minor constituents of the plastics composition. Such additives are generally based on antimony, boron, chlorine, bromine or phosphorus, or a combination of these.

A traditional additive which has been used for many years to confer flame retardant properties on a variety of materials including plasticised PVC, polyolefins and glass-reinforced polyester resins, is antimony trioxide. This is added at levels of 2–15 phr depending on the degree of flame retardance required. Its activity is much enhanced by the presence of chlorine, so that it is common practice to add approximately equal quantities of 70% chlorinated paraffin wax together with antimony trioxide to those polymers, such as polyolefins, which do not contain chlorine as part of their chemical make up. The high price, opacity and uncertain supply position of antimony trioxide in recent years has added impetus to the industry's search for alternative materials, so that increasing amounts of phosphate plasticisers with and without chlorinated paraffin extenders are being used by plasticised PVC compounders.

Complex compounds of chlorine, phosphorus and bromine have been developed for flameproofing polyester resins, polyolefins and polyurethanes. Unfortunately such compounds are usually toxic and expensive, and many of them, especially those containing high proportions of bromine, suffer from poor heat and light stability. Considerable research effort is being expended to develop a broad spectrum, highly efficient heat and light stable flame retardant additive for plastics materials which could be sold at an economic price, but none of the products currently on the market fully satisfy all the requirements of an ideal flame retardant.

IMPACT MODIFIERS

Impact modifiers are mainly used in rigid PVC at levels ranging from 5 to as much as 40 phr to improve the impact resistance of such articles as rigid PVC bottles, thus permitting the utilisation of thinner walled bottles with subsequent savings in raw material costs.

The most widely used are acrylics and acrylates, which are close to PVC in refractive index and readily dispersed, so that phase separation and stress whitening in transparent articles are minimised. ABS modifiers are used in a wide variety of rigid PVC articles, including pipe and other extrusions, mouldings and calendered sheet. Chlorinated polyethylene is also used in PVC film, and exhibits good weathering and chemical resistance.

Rubber modifiers, which may be SBR nitrile or natural rubber, have been used for some years, particularly as modifiers for polystyrene and other styrene based polymers. They are not favoured for PVC due to their generally poor weatherability.

In general, impact modifiers improve impact strength at the expense of rigidity and a reduced heat distortion temperature. The degree of compatibility between the resin and modifier must also be correct, *eg* in the case of modifiers for rigid PVC an additive which is too compatible is ineffective in improving impact strength, while a highly incompatible product will cause stress whitening on bending and opacity.

BIOLOGICALLY ACTIVE ADDITIVES

Plasticised PVC compounds vary widely in their susceptibility to degradation resulting from microbial or fungal attack. Phosphate plasticised PVC is highly resistant, but most phthalate containing compounds are susceptible to attack by certain micro-organisms, while those containing adipate plasticisers are readily degraded. A very large number of additives have been claimed as protective agents for plasticised PVC, based on a variety of active ingredients including arsenic, copper, tin, lead phosphorus mercury and various organic compounds. Unfortunately, many are highly toxic to human life as well as micro-organisms, others are incompatible or difficult to disperse in the PVC compound, and many have poor heat or light stability.

One of the most effective fungicides is probably copper 8-quinolinolate, which is effective in low concentrations, and resists weathering and leaching well. However, it imparts a green colour to PVC compounds and is thus only suitable for use in dark coloured compounds. A PVC compounder wishing to minimise microbial or fungal attack should therefore formulate using phosphate plasticisers, lead or tin stabilisers and avoid the use of low temperature plasticisers such as sebacates or adipates.

For some applications, such as baby pants, hospital mattress covers, drawer liners, curtains and flooring, it is desirable for plastics products to have a bacteriostatic finish. This can be achieved in PVC, polyolefins and other plastics by incorporation of a bacteriostat, which inhibits the growth of bacteria on the plastics surface.

A bacteriostat should be of low human toxicity, compatible with the relevant polymer, should withstand normal temperatures, and when incorporated in the polymer should withstand repeated washings with water and detergent solutions. Among products currently in use are certain mercury compounds (which may present a toxic hazard), hexachlorophene and 2,4,4′-trichloro-2′-hydroxy-diphenyl ether.

Hexachlorophene is satisfactory in many respects, but is only effective against gram positive bacteria. 2,4,4′-trichloro-2′hydroxy-diphenyl ether on the other hand, at concentrations of 0.1 to 0.5% on the plasticised PVC compound, is also effective against gram negative micro-organisms, such as those which can cause typhoid, paratyphoid, gastro-enteritis, dysentry and other diseases. It also shows bacteriostatic effect in low density polyethylene at concentrations as low as 0.01–0.1 phr.

ANTISTATIC AGENTS

Most plastics are prone to develop an electrostatic charge on their surface in use. This can be actively dangerous as in the case of underground conveyor belting where discharge of the built-up charge can cause a spark, or merely unsightly as in the case of dust build-up on consumer durables or packaging film.

A very large number of antistatic additives are on the market, among them fatty amides, quaternary ammonium compounds, glycol esters, sulphonates and other surface active agents. An antistatic agent is often slightly imcompatible with the polymer substrate so that it migrates at a slow controlled rate to the surface. Polyethylene glycol 200 monolaurate is widely used in PVC conveyor belting. This material acts as a secondary plasticiser and viscosity depressant for PVC in addition to its antistatic function.

Although new products are announced almost daily, no truly effective antistatic agent exists for the majority of plastics, polyolefins being particularly hard to treat. Many fabricators opt for the temporary solution of spraying the surface of the fabricated article with an aerosol surfactant to protect the surface against dust build-up while the product is on display in a store.

COLOURANTS IN PLASTICS

Additives are normally incorporated into plastics systems to modify the properties of the systems or to reduce the cost. The incorporation of colourants, on the other hand, increases the aesthetic appeal, so making the final product —especially in the consumer market—more saleable. In specific applications colourants may be used as a means of identification (cable insulation), to make the article more readily visible (high visibility clothing), or to simulate the colour of natural or traditional articles.

Plastics may be coloured using a mass colouration technique, or with specific polymers it may be possible to surface dye. Mass colouration is the method normally employed, and colourants are incorporated into the polymer along with other additives such as stabilisers or lubricants.

The colourants available to the technologist fall into two main groups: dyes and pigments. Dyes are defined as soluble and pigments as insoluble colourants. Dyes, as soluble colourants, promote colour in the system by dissolving in the polymer or some other component of the mix. Pigments, in theory, are insoluble, therefore to obtain colour value they must be dispersed in a fine particle size throughout the mass.

High transparency and strength are the major advantages to be gained from the use of dyes, but in many polymer applications, especially in the vinyl and polyolefin fields, many other factors prohibit their use. Pigments, therefore, find a much wider application. The range of pigments available can be divided into the two broad chemical classes of inorganic and organic pigments. There is also a miscellaneous class of metallic, phosphorescent and fluorescent types. The inorganic and organic classes are each typified by a wide range of chemical types. When selecting a pigment the criteria used for judging its usefulness in a plastics system will obviously vary according to the visual requirements, the processing conditions to which it will be submitted and the envisaged application.

Visual requirements

Colour may be characterised by three properties: hue, value and chroma, terms which may be broadly translated as shade, brightness and strength.

Shade selection is purely subjective, but maximum brightness is a desirable characteristic as it provides the facility to match a wider range of shades and it is always possible to reduce brightness. The strength of a pigment or the percentage addition required to achieve a specific colour should not be considered in isolation but should be related to its price. The opacity or transparency of a pigment is often an important consideration.

Processing requirements

Dispersibility. The problem of obtaining and reproducing a good dispersion of pigment in plastics media, and the many other problems which arise from it, is probably one of the major pitfalls the colourist meets. Before determining what is meant by a good dispersion, it is worthwhile discussing the nature of a pigment particle.

During the production of pigments, crystals of between 0.005–1 μm are formed initially but these particles may build up to form aggregates (compact collection of crystals) and agglomerates (aggregates cemented together). These agglomerates may reach sizes up to 3 000 μm. Although products are ground by the pigment manufacturer, the ultimate reduction in particle size is never achieved in practice.

During the processing of plastics, the compounder aims to break down any agglomerates present to as fine a particle size as possible and distribute them throughout the mix as homogeneously as possible. To achieve this aim, the two requirements are shear forces sufficient to overcome the attractive forces binding the crystals together and a wetting stage where the pigment/air interface is replaced by a pigment/system interface.

Inadequate dispersion may cause variations in hue and will produce variation in the colour strength. It may result in large discrete particles and streaks appearing which will not only detract from the appearance of the finished article but may cause processing problems and affect the physical properties of the product. Examples of these problems could be blockage of filter packs, tearing of polyethylene film during the blown film extrusion stage or 'sparking' and resultant reduction in the insulation properties of PVC cable.

The speed and degree with which a pigment will disperse in a system is governed by three main factors:

a. Method of processing.
b. Type of compound.
c. Texture of the pigment.

Methods of processing vary considerably, depending upon the polymer system, the economics of the operation and upon the quality of dispersion required. As the percentage of pigment present is normally of a low order, it is essential in all methods to have an efficient premixing stage prior to processing in order to achieve homogeneity. Concentrations of pigment are normally in the region of 0.01% to 2.0% depending on the depth of colour required and the pigment in use. To improve the evenness of distribution a low percentage of oil may be added to the polymer prior to pigment addition.

The most simple method, commonly known as dry colouring, involves tumbling polymer, pigment and any other additives in a drum or mixing in a simple paddle mixer. This operation is followed by the forming operation which may be extrusion, injection moulding or rotational moulding to form the finished article. Although rather poor dispersion is achieved, this technique has become very popular for a number of reasons, including:

i. Smaller capital outlay resulting in lower cost products.
ii. Reduced heat history on pigment and polymer system.
iii. Flexibility of process.

A method of achieving improved dispersion is to high-speed mix all the components including the required percentage of pigment and extrude a rod which is then chipped. This chip, known as coloured compound, can then be extruded, injection moulded or calendered to form the finished product.

The most sophisticated method involves the production of a masterbatch which may then be used to produce a coloured compound. The masterbatch is produced by mixing the polymer and a high concentration of colour in a heavy duty internal mixer, the resultant mix being extruded and chipped. The high concentration of pigment, varying normally between 10 and 50% depending on the pigment and polymer involved, produces a high viscosity mix at the processing stage. The shear forces imposed on the pigment are sufficient to break down most of the agglomerates present.

Compound. Two of the conditions required to obtain a good dispersion are shear on and 'wetting out' of the pigment. Both conditions are determined by the type and viscosity of the system, other factors are the method of processing and the surface of the pigment.

In liquid systems such as paints or inks, the presence of solvents facilitates the 'wetting out' action but plastics systems rarely contain such additives, although certain compositions contain plasticisers which may act in a similar manner. Polymers which are well known for their inability to 'wet out' pigments are rigid PVC, ABS and polyolefins. As a result of this it is often necessary to modify the process or use specially prepared pigments.

Texture. The measure of the texture of a pigment is not simply the size of the agglomerates in the powder but the ease with which they can be broken down in the polymer system to the required particle size.

The texture of pigments is to a large degree dependent upon the final conditioning treatments applied to them. Great advances have been made in recent years in pigment technology, enabling pigment manufacturers to produce pigments with much superior physical form. Nevertheless, pure pigments may still be unsatisfactory for specific applications, and therefore they are often sold in a modified form.

Modifications of pigments. The most simple of these modifications is the *reduced pigment*, which is normally produced by grinding the pigment with inert filler such as calcium carbonate, barytes or china clay. These reductions often contain between 10–25% pigment and find their major outlet in flooring compounds.

Plasticiser pigment dispersions are in the main prepared by triple roll milling pigment into plasticiser or special oils. The pigment is 'wet out' by the plasticiser and aggregates are broken down by the action of the mill. The percentage of pigment may vary between 10–70% depending on the absorption properties of the pigment and the viscosity requirements of the application. The choice of plasticiser or other liquid media is also determined by the application. These dispersions are widely used in flexible PVC, polyurethane foams and polyester resins.

Master batches are normally produced by companies outside the pigment manufacturing industry. The polymer used in their production will vary according to the envisaged application. The most commonly used products are based on PVC, polyolefins and rubber.

These modifications when produced from conventional pigments may still fail to satisfy the more critical dispersion requirements such as arise in polyolefin blown film, fibre and vacuum formed PVC. Consequently, pigment manufacturers have taken a completely different approach to the problem. Instead of relying upon shear forces to break down agglomerates present in the pigment powder, they are attempting to produce *agglomerate- and aggregate-free pigments*.

The principle is to encapsulate in a resin the primary particles as they are formed, thus preventing them from coming together and forming aggregates. When these preparations are incorporated into plastics systems, the mixing action distributes them throughout the mass whilst the resin either melts or dissolves away, leaving the primary particles to be homogeneously distributed throughout. A variety of resins are used for the encapsulation process, each being selected specifically for a particular application.

Resistance to processing conditions. Processing temperatures for the range of commercially available plastics may vary between 130°C and 310°C and therefore temperature resistance of the pigment must be considered. However, the problem is not quite so simple and there are a number of factors which may affect the heat stability of the total system. Reduction in concentration of pigment may reduce the temperature at which processing can take place without colour change. Chemical interaction may take place between pigment and polymer or the polymer itself could yellow, both effects resulting in colour change.

Service life requirements

Compatibility. Problems of incompatibility may occur if wrong pigment selection is made. These problems fall into one of three categories: chalking, migration or rub fastness.

Chalking is the name given to the phenomenon of pigment particles forming on the surface of the plastics into which the pigment had previously been satisfactorily incorporated.

The time lapse between production of a satisfactory product and this problem occurring varies depending on pigment type, temperature of processing and compound formulation. When this phenomenon does occur it is normally in plasticised PVC, polyolefins or rubber.

Migration. The problem of colourant transferring from one article to another with which it is placed in contact is defined as migration.

Rub fastness. Even though no pigment is visible on the surface, a colour stain may be formed on a piece of wet or dry fabric with which the article is rubbed.

These three defects are normally associated with solubility of the colourant in a mobile component of the mix such as lubricant or plasticiser.

Light fastness and weather-stability. Plastics are finding increasing usage in applications where weathering and light fastness are of prime importance and perhaps most prominent in this field is the building industry, where guarantees of 10 and 15 years weathering resistance are being sought. Often it is difficult to give an opinion on the light fastness of a pigment without taking into account a number of factors. It is more appropriate therefore to quote the light fastness of a pigmented system rather than that of a pigment. Amongst the factors which may affect the light fastness of a system are:

a. Chemical structure of the pigment.
b. Concentration of pigment in the system. Reduction in concentration of pigment may result in reduced light fastness.
c. Presence of white pigment. Incorporation of white pigment may reduce light fastness.
d. Light fastness of polymer system. Yellowing of the base polymer on exposure to light will alter the total colour of system, an effect which is particularly noticeable in blue shades. The yellowing of the base may be corrected by inclusion of ultra violet absorbers as mentioned previously, although, in general, UV absorbers only tend to improve the light stability of fluorescent type pigments.
e. Conditions of exposure. Hydrogen sulphide or sulphur dioxide present in industrial atmospheres could cause problems with certain pigments.
f. Heat history.

Table 2. A general guide to suitable colourants for the major plastics.

Polymer	Pigment properties required	Suggested colourants		
Flexible PVC, calendered, extruded or injection moulded	Heat stability up to 200°C Non migratory, non chalking light stability depending on application	Cadmium pigments, red and yellow toners Azo-condensation	Polycyclics Lead pigments Phthalocyanines Bis-arylamides	Isoindolinones Iron oxide Dioxazin Special monoazos
Rigid PVC	Heat stability up to 200°C Good light stability Easy dispersion important	Cadmium Yellow toners Azo condensation Polycyclics	Lead pigments Phthalocyanine Isoindolinones	Selected Bis-arylamides Dioxazin Lead pigments Special monoazos
PVC leathercloth	Heat stability up to 200°C Ease of dispersion. Resistance to dry cleaning solvents, detergents and wet and dry rub. Light fastness depending on outlet.	Cadmium pigments Yellow toners Azo condensation Lead pigments Polycyclic	Isoindolinine Selected Bis-arylamide Special monoazos Dioxazin Iron oxides	Special monoazos
LDPE	Heat stability normally up to 240°C Moderate light stability for many applications Easy dispersion important	Red and yellow toners Selected Bis-arylamides Isoindolinone Azo condensation	Phthalocyanine Green Phthalocyanine Blue Cadmium pigments Iron oxides	Lead pigments Ultramarine Blue Vat pigments Special monoazos
HDPE	Heat stability normally up to 280°C Moderate light stability for many applications Easy dispersion important	Yellow toners Selected red toners Polycyclic Isoindolinone	Azo condensation Cadmiums Phthalocyanine Blue Phthalocyanine Green Iron oxides	Ultramarine Blue Special monoazos
Crystal PS	Good light stability Processing temperatures up to 240°C	Red and yellow toners Polycyclics Azo condensation Isoindolinone	Metal complex Phthalocyanine Blue Phthalocyanine Green Selected dyes	Ultramarine Blue Cadmium pigments Lead pigments Special monoazos
Impact PS	Good light stability. Must withstand temperatures up to 280°C	Yellow and red toners Polycyclics Isoindolinone Selected Bis-arylamides Phthalocyanine Blue Phthalocyanine Green	Metal complex Selected dyes Cadmium pigments Ultramarine Blue Azo condensation	Special monoazos
ABS	Good light stability. Temperature resistance of up to 260°C is required but many pigments which will withstand this temperature in polyolefins fall down in this polymer	Azo condensation Selected polycyclics Phthalocyanine Blue Phthalocyanine Green	Cadmium pigments	
Polypropylene	Resistance to temperatures of up to 310°C. Good light fastness. Ease of dispersion important especially for fibres	Yellow toners Selected red toners Polycyclic Isoindolinone Ultramarine Blue	Azo condensation Cadmiums Phthalocyanine Blue Phthalocyanine Green	Special monoazos
Polyurethane leathercloth	Ease of dispersion Resistance to dry cleaning solvents and detergents and wet and dry rub. Moderate light fastness	Azo condensation Polycyclic Special monoazos		
Phenolic moulding powders	Temperature requirements moderate; up to 170°C but reducing atmosphere present. Only moderate light fastness requirements as compounds become brown.	Spirit and water soluble dyes Red toners Simple monoazo pigments		
Amino plastics	Good light stability but only moderate heat stability (170°C) is required.	Cadmiums Phthalocyanine Blue Ultramarine Blue	Selected monoazo pigments Selected Dis-azo pigments	

Toxicity. As coloured plastics are widely used in the production of toys and in applications where they come into contact with foodstuffs, it is necessary to consider the 'safety in use' of the incorporated pigment. Most European authorities have issued regulations covering the maximum permitted heavy metal content of colourants and restricting certain chemical compounds.

Effect of pigment on polymer

Pigment addition may affect the properties of a polymer adversely or advantageously. Carbon black is the classical case of a pigment which, when incorporated into polyolefins, has been found to improve their ageing properties. Phthalocyanine pigments and certain cadmium reds may also enhance polyolefin ageing properties although to a lesser extent than carbon black.

Electrical properties. The addition of pigments to PVC cable insulation compound may reduce the volume resistivity of the compound. Pigments for this application must, therefore, be specially selected.

Dimensional stability. One of the most intriguing problems that may occur is that addition of pigment to polyolefins may cause them to shrink more than would normally be anticipated. The degree of shrinkage could be sufficient to cause warping of articles, especially in containers which vary in thickness of section.

Pigment selection

Having considered the problems which may arise when colouring plastics, the problem now occurs in selecting a pigment for a specific application. It would be a time consuming task to evaluate all the pigments available and therefore the first approach when selecting a pigment should be to the pigment and masterbatch suppliers who have precise information readily available on the performance of their products in a variety of polymer systems. It is then only necessary to evaluate under the appropriate conditions the technical/commercial performance of their offers. Table 2, however, gives a general guide to the types of colourants used in some of the major plastics systems.

The number of pigments sold to the plastics industry is extremely large; the main chemical types and their general properties are listed below. This must only be considered as a guide as there could be considerable variation between individual pigments in the same class and the properties of a pigment may vary from polymer to polymer.

Inorganic pigments

Iron oxides. These pigments exist naturally and can also be produced synthetically in both anhydrous and hydrated form. The hydrated forms, which are yellows, are of less interest as when processed at high temperatures they tend to lose water and change shade to give browns. They are cheap, light fast, and have moderate heat stability. Their main disadvantages are low tinctorial strength, lack of brightness, and in PVC they may promote degradation of the polymer.

Cadmium sulphide and sulpho-selenides. This group of pigments ranges from bright yellow through orange to red and maroon. They are non migratory and in general the light and heat stability of these pigments is excellent. The weathering resistance of the yellow is inferior to the oranges and reds. Problems may occur when used in systems containing lead and in systems exposed to industrial atmospheres. A further disadvantage of these pigments is their high price related to their moderate tinting power.

Lead molybdate chrome and lead chromes. This range of pigments consists of yellows, oranges and scarlets. They are weak tinctorially but are low in price. In general their heat stability is moderate, suitable for use in low density and certain high density polyethylene applications. Some grades darken considerably on exposure to light and all are susceptible to industrial atmospheres and alkaline attack. One of their major disadvantages is their toxic nature.

Ultramarine pigments. There is available a limited range of red-shade blue to violets possessing good heat and light stability. They are non migratory and non chalking and although low in tinting power they are relatively cheap. Their main disadvantage is their poor resistance to acid.

Organics

Metal toners. This range of pigments includes yellows and reds. They are metal salts (normally barium, calcium, manganese or strontium) of sparingly water soluble acid dyes. The reds are tinctorially strong and bright and are moderately heat fast and non migratory. Their light fastness is variable depending upon the particular acid dye employed. The light and heat stability of the yellows is superior to that of the reds. One of the disadvantages of all grades is their poor resistance to alkalis.

Metal complexes. In this range the metal is chelated rather than being used to form a salt. The two best known examples are ferro-nitroso-beta-naphthol green and nickel azo yellow. Both pigments are dull and sensitive to acids. The nickel azo pigment has excellent light stability.

Phthalocyanine pigments. A limited colour range of blues and greens is available. The basic structure is an extremely chemically stable copper complex with outstanding heat, light and non migratory properties.

Mono-azo pigments. This range is represented by reds, oranges and yellows. The classical mono-azo pigments due to their low molecular weight have a tendency to migration and chalking and their heat stability is poor to moderate. There have appeared on the market, however, in recent years a range of mono-azo pigments with much higher molecular weights. These pigments have excellent heat, light, migration and rub fastnesses.

Bis-arylamide yellows and oranges. The increased molecular weight of these pigments make them more suitable for use in plastics than the classical mono-azos. In this range the molecular weight and therefore performance is varied. The heat and light stabilities vary from moderate to good and the lower molecular weight types may exhibit migration and chalking problems.

Azo-condensation — yellows, oranges and reds. The molecular weight of this range is higher than that of the bis-arylamide types. They exhibit excellent all round properties of heat, light and non migratory properties.

Polycyclics — violets, reds and blues. This range of pigments exhibits in general excellent light stability and their heat stabilities range from good to excellent depending on their chemical structure.

Isoindolinone. This range of yellows, orange and reds represents one of the recent advances in pigment technology. They exhibit excellent light stability and their heat stability in many polymers ranges from good to excellent.

CONCLUSION

It is hoped that this brief outline of the major categories of plastics additives available, their nature, mode of activity and disadvantages will serve as an initial guide to the materials available and their efficiency in overcoming certain drawbacks inherent in many of the more widely used plastics materials.

Continuous improvement in currently available additives is taking place, and many materials are in the final stages of development throughout the world. Exciting new developments are on the horizon, among which may be mentioned the work in progress at the University of Aston on additives to give controlled degradation of plastics 'throw away' items such as packaging materials, which may help to overcome the rapidly growing problem of environmental pollution by waste materials (see Chapter 21).

Chapter 3

Thermosetting Plastics

G W Parry *ARIC*
BP Chemicals International Ltd

Thermosetting plastics are materials which once set in their final shape cannot be re-shaped by means of heat or pressure. They are intractable materials in their processed state. This intractability is due to their chemical structure which consists of a chemically bound three dimensional network. This network is the result of a chemical reaction during processing which can be initiated either by thermal or chemical means. These curing reactions will be discussed in greater detail later in the chapter (see phenolics and polyesters).

Thermosets differ from thermoplastics in their greater resistance to temperature and creep. However, in many cases they cannot be prepared in light colours or as translucent materials and suffer from insufficient impact resistance.

Due to their intractability they cannot be re-shaped like thermoplastics nor recovered and reprocessed. It is therefore necessary to ensure that the component being produced is of the correct dimensions before commencing a production run.

Most thermosetting materials are available in two forms, *ie* resins and moulding compounds. The resins may be used alone as in encapsulation processes but they are normally used in conjunction with other materials. The resins may be used with reinforcement, *eg* polyester/glass laminates, or with fillers and reinforcement, *eg* moulding compounds. This differs from the thermoplastic field where, generally, the polymeric material is used without fillers and/or reinforcement.

Thermosetting materials are normally processed by compression, transfer or injection moulding. In some cases, such as polyester, 'hand lay-up' techniques are applicable. The choice of process will depend on the component design and choice of material.

The most common classes of thermosetting materials are alkyds, aminos, epoxides, phenolics, polyesters and silicones. The approximate consumption of these materials in the United Kingdom during 1970 was as follows[1]:

Alkyds	63 000 000 kg
Aminos	140 000 000 kg
Epoxides	13 500 000 kg
Phenolics	76 000 000 kg
Polyester	42 500 000 kg

These figures include resins used in the production of moulding powders. The two main moulding powders are those based on amino and phenolic resins, the consumption of which in 1970 was 43 500 000 kg and 31 000 000 kg respectively.

ALKYD RESINS

These materials are a special type of polyester. They are normally prepared by reacting together an acid, an alcohol and the fatty acids of certain oils giving a resin having the following main structural unit:

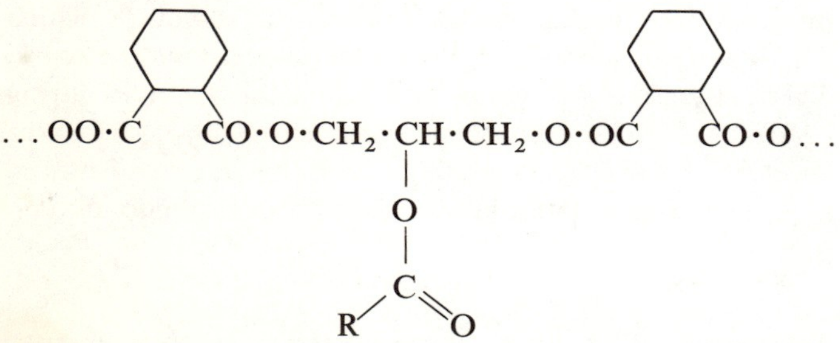

$$\ldots OO\cdot C_6H_4 \cdot CO\cdot O\cdot CH_2\cdot \underset{\underset{\underset{\large R-C=O}{|}}{O}}{CH}\cdot CH_2\cdot O\cdot OC\cdot C_6H_4\cdot CO\cdot O\ldots$$

where R is a long chain containing unsaturated groups. The resin is hardened by reacting together the unsaturated groups. This reaction being initiated by a drier, *eg* lead soap.

The main use of these resins is in surface coating applications. They provide relatively low cost, flexible coatings having good gloss retention. As this field is well documented and not within the scope of this book it will not be discussed further.

Certain alkyd resins are used for the production of moulding powders. In this case the resin is formulated on an unsaturated acid and alcohol producing an unsaturated polyester (see polyester resins). This polyester is reacted with a non-volatile monomer, usually diallyl phthalate, to give a hard cross-linked structure.

Alkyd moulding powders are supplied in four basic forms, granules, putty, fibrous and rope. Typical properties of these materials are given in Table 1.

All four forms can be processed using conventional compression moulding and transfer moulding techniques. These materials can also be handled satisfactorily on reciprocating screw injection moulding machines. However, it is necessary to modify the feed hopper in order to injection mould successfully the rope and putty types.

As the curing rate is rapid and proceeds without evolution of volatile matter it is possible to produce void free mouldings using a short cycle time. The rapid drop in viscosity of the material with temperature allows moulding to take place under relatively low pressure (15.5 MN/m^2 maximum, see Table 1).

The main applications for alkyd moulding powders are in the electrical field, where good insulation properties coupled with strength and heat resistance are required, and in the decorative field when mouldings in pastel shades with good heat resistance and surface hardness are required.

AMINO RESINS

There are two types of resin which fall into this class, urea formaldehyde and melamine formaldehyde resins.

Table 1. Typical properties of alkyd moulding powders[2].

Property		Granular	Putty	Fibrous	Rope
Shrinkage	m/m	0.005–0.007	0.008–0.010	0.002–0.004	0.003–0.005
Moulding pressure	MN/m^2	6.9–10.35	2.76–5.52	10.35–15.50	4.14–6.9
Moulding temperature	°C	130–165	130–165	130–165	122–160
Specific gravity		2.1–2.2	2.13–2.17	2.2–2.2	2.1–2.2
*Water absorption	mg	55–65	25	50	20
Impact strength	J/m	5.9–8.0	6.4–8.0	160–320	106–160
Flexural strength	MN/m^2	48.5–69	52–76	104–208	104
Tensile strength	MN/m^2	21–27.5	21–35	35–69	35–48.5
**Youngs modulus in tension	MN/m^2	10 350–11 750	8 250–9 650	7 600–8 950	—
Youngs modulus in flexure	MN/m^2	13 800–17 000	13 000–15 000	13 000–16 000	11 000–12 500
Electric strength	MV/m	10.25–11.8	9.45–11.0	7.8–9.85	9.05–11.8
Dielectric strength at 1 MHz	pF/m	31–37	31–35	28–38	35–45
Power factor at 1 MHz		0.018–0.022	0.018–0.023	0.015–0.020	0.012–0.014
Surface resistivity	ohm	$> 10^{13.0}$	$> 10^{13.5}$	$> 10^{12.5}$	$> 10^{14}$
Volume resistivity	ohm m	$> 10^{14}$	$> 10^{14}$	$> 10^{13.5}$	$> 10^{14.5}$
****Comparative tracking index		235	210	195	—
***Arc resistance	sec	190	182	185	180–200

* As per BS2782 except immersion in boiling water is 60 minutes.
** ASTM D790/58T (G.1.)
*** ASTM D495/58T
**** BS2781
All other tests to BS2782

Application	kg
Moulding Powders	43 500 000
Adhesives	45 000 000
Surface Coating	10 000 000
Textile Treatment	5 000 000
Paper Treatment	8 000 000
Laminating	4 500 000
Others, Foundry, Foam	8 000 000

Table 2. Amino consumption in the UK during 1970[1,4].

The first stage in the resin process is to produce long unconnected chains which then react together on further heat treatment giving the intractable product. This hardening process can also be carried out at room temperature using an acid hardener, *eg* phosphoric acid or an acid donor such as ammonium chloride. Details of the mechanism of resin formation and cure are given below.[3]

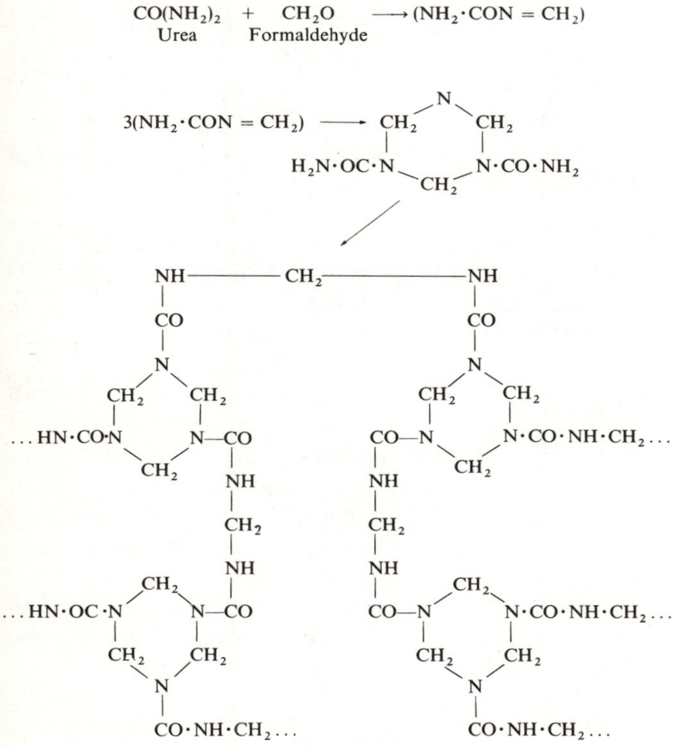

$$\underset{\text{Urea}}{CO(NH_2)_2} + \underset{\text{Formaldehyde}}{CH_2O} \longrightarrow (NH_2\cdot CON = CH_2)$$

```
                                N
                              /   \
3(NH2·CON = CH2)  ——→     CH2       CH2
                           |         |
                  H2N·OC·N             N·CO·NH2
                              \   /
                               CH2
                          ↙

      NH—————————CH2—————————————NH
      |                          |
      CO                         CO
      |                          |
      N                          N
    /   \                      /   \
 CH2     CH2                CH2     CH2
  |       |                  |       |
...HN·CO·N     N—CO        CO—N       N·CO·NH·CH2...
    \   /    |           |   \   /
     CH2     NH          NH   CH2
             |           |
             CH2         CH2
             |           |
             NH          NH
    CH2      |           |    CH2
  /     \    |           |  /     \
...HN·OC·N     N—CO        CO—N       N·CO·NH·CH2...
  |       |                  |       |
 CH2     CH2                CH2     CH2
    \   /                      \   /
      N                          N
      |                          |
  CO·NH·CH2...              CO·NH·CH2...
```

CROSS LINKED RESIN

As with other thermosets both these materials are supplied as resins or coupled with fillers, *etc*, as moulding materials. Table 2 gives an estimate of the main areas of use of amino resin in the UK for 1970.

From Table 2 it can be seen that in many instances these resins are used for applications which cannot be considered to be of an engineering nature, *eg* surface coating and textile treatment. However, composite materials, *eg* plywood, laminates, prepared using these materials may possibly be considered an engineering material. Moulding powders produced from these two resins are nearer true engineering materials in so much as they could be moulded into engineering shapes and components.

The properties and applications of these two classes of resin will be dealt with separately.

Urea formaldehyde resins

The main outlet for urea formaldehyde resins is in adhesives for plywood and chipboard. The ratio of urea to formaldehyde used will influence the properties of the adhesive. For example, a resin with a high formaldehyde to urea ratio will give an adhesive having the best clarity, water resistance and mechanical properties. The main disadvantage of this type of adhesive is its poor water resistance. Hence, it cannot be used for exterior applications where it is superseded by adhesives based on phenolic resins.

Urea resins are also used to impart crease resistance to fabrics, to improve the wet strength of paper and in paints. For the paper application it is necessary to use an 'ionic' resin. These 'ionic' resins are produced in such a way that the molecules carry an electrical charge. The resins can also be processed to produce a foam having a very low K value.

Urea resin based moulding powders are used in the following applications: domestic electrical accessories (plugs and sockets); toilet seats; bottle caps and closures. When cured these materials are very brittle and should not be used for load bearing applications. The major usage being in electrical and closure applications. Physical properties of these materials are given in Table 3.

Urea moulding powders have the advantage over phenolic materials in that they can be prepared in light colours and as an almost translucent material with the correct choice of fillers. The cured resin has a refractive index of 1.55 and so inclusion of cellulose as a filler, refractive index 1.55–1.56, will produce nearly translucent materials.

The moulding properties of the material can be altered by the choice of resin. For example, a high molecular weight resin will produce a moulding powder having a stiff flow but relatively long cure. The cure rate can be improved by leaving some lower molecular weight materials in the resin. Moulding powders based on high molecular weight resins also have the least shrinkage and this can be maintained and the flow increased by additions of up to 1% of a plasticiser.

In most cases there is an optimum temperature for cure as increasing the temperature will reduce the viscosity of the material but decrease the cure time. Therefore, if too high a moulding temperature is used, the material will cure before it has had time to flow and fill the mould cavity.

These materials are normally processed by compression moulding at temperatures of 135–170°C with pressures varying between 20 and 55 MN/m^2. Cure times can be as short as 20 seconds with a fast cure material.

Melamine formaldehyde resins

Two materials are available, *viz.* resin and moulding powders. The resins tend to be used only for special applications because of their cost.

Property		ASTM Test Method	Urea		Melamine		
			Wood-Filled	α-Cellulose-Filled	Wood-Filled	α-Cellulose-Filled	Glass-Filled
Mould shrinkage	m/m		0.006–0.010	0.005–0.010	0.007–0.008	0.004–0.009	0.002–0.005
Moulding temperature	°C		135–165	135–175	140–170	140–170	145–160
Specific gravity		D792	1.4–1.5	1.5–1.55	1.5–1.55	1.5–1.55	1.9–2.0
Water absorption	mg	24 hrs at 20°C	50–130	40–130	30–40	10–40	0–10
Impact strength	J/m	D256	12–15	13–16	14–20	13–17	19–27
Flexural strength	MN/m^2	D790	80–120	80–125	65–80	90–120	80–120
Tensile strength	MN/m^2	D638	40–55	50–75	40–60	50–85	42–70
Youngs modulus in tension	MN/m^2	D638	6 900–9 650	6 900–9 650	≃7 900	8 300–11 000	14 000–28 000
Electric strength*	MV/m	D149	12–17.5	12–16	14–16	12–16	6–10
Dielectric constant at 1 MHz	pF/m	D150	53.5–75.5	57.5–62	49–52.5	62–80	49–66
Power factor at 1 MHz		D150	0.025–0.040	0.020–0.030	0.025–0.040	0.025–0.045	0.020–0.030
Volume resistivity	ohm m	D257	10^{11}–10^{14}	10^{11}–10^{14}	10^{11}–10^{13}	10^{11}–10^{13}	10^{11}–10^{14}
Comparative tracking index	IEC 121		>600	—600	>600	>600	>600
Arc resistance	secs	D495	80–130	90–130	70–110	120–130	170–200

* Short time at 25°C

Table 3. Typical properties of amino moulding powders[2].

Melamine resins are used with various reinforcements to produce laminates for special electrical applications and for decorative purposes. The main use being in the decorative field where the laminates are used for household working surfaces and wall cladding. In this decorative application the resins are used in conjunction with a thin layer of a α-cellulose surface paper to produce a hard, scratch resistant, transparent coating. This coating acts as a preservative for the decorative pattern used in the preparation of the laminate.

As with resins, moulding powders based on melamine are used for specialised applications. The two main areas of application are tableware, where their improved water and scratch resistance compared to urea based materials makes them ideally suitable for this purpose, and for high performance electrical goods. Their electrical properties are superior to urea formaldehyde based materials under conditions of high temperature and humidity. They can be made in more attractive colours than phenolic moulding powders and have superior track resistance.

In the UK these materials are processed mainly in the same manner as urea materials (compression and transfer moulding) although some grades can be injection moulded on suitable machines. Care must be taken to establish the optimum moulding condition for each application as cure times can affect properties.[5]

Typical physical and electrical properties of melamine moulding materials can be found in Table 3.

EPOXIDE RESINS

The most important commercial resins of this type are those based on bisphenol A and epichlorhydrin. The general structure of the uncured material is as follows[6]:

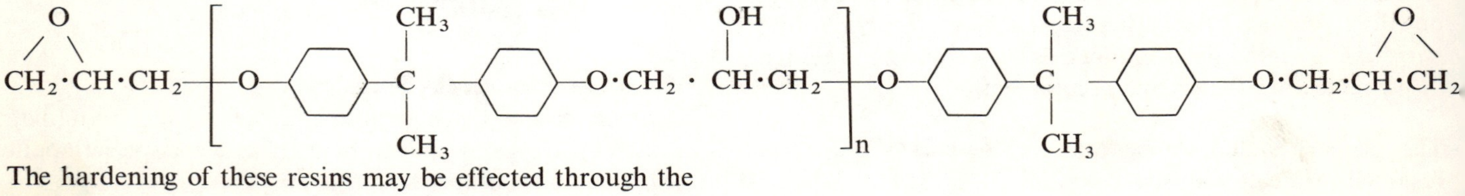

$$\overset{\overset{O}{/\;\backslash}}{CH_2\cdot CH}\cdot CH_2-\left[O-\langle\bigcirc\rangle-\underset{CH_3}{\overset{CH_3}{C}}-\langle\bigcirc\rangle-O\cdot CH_2\cdot\overset{OH}{CH}\cdot CH_2\right]_n-O-\langle\bigcirc\rangle-\underset{CH_3}{\overset{CH_3}{C}}-\langle\bigcirc\rangle-O\cdot CH_2\cdot\overset{\overset{O}{/\;\backslash}}{CH\cdot CH_2}$$

The hardening of these resins may be effected through the epoxide groups ($-\overset{\overset{O}{/\;\backslash}}{CH_2-CH}-$) or the hydroxyl groups (—CH—OH). Acids or amines may be used for the hardening process[7] and the properties of the cured product will depend, to a large extent, on the curing system used. Relatively flexible cured materials may be prepared using a curing system that produces a low cross-linked density.

Amine hardening systems are used at room temperature giving a fast cure rate and a product having good chemical resistance. Care must be taken with this type of curing agent as many of them are skin sensitisers.

Cure at elevated temperatures is normally carried out with acids or acid anhydrides. The anhydrides are preferred because they are more soluble in the resin and do not produce as much water as the acid during cure so that there is less tendency for foam formation. The advantages of this curing system, over an amine system, are a lower exotherm, less skin irritation and with the correct choice of acid a product with a higher heat distortion temperature. They do, however, produce cured products which are less resistant to alkalis.

The main applications of these resins are surface coating and encapsulation. Other applications are in adhesives, prototype tooling, flooring, laminating and, to a lesser extent, in moulding materials.

Apart from decorative coating, amine cured resins are used for protective coatings in chemical plant and oil refineries. Adequate chemical resistance is obtained by room temperature cure which obviates the necessity of heat treatment as is required in the case of phenolic resins.

Laminates prepared from these resins have improved mechanical properties, heat resistance and chemical resis-

		General Purpose	Heat Resistant	Flexibilised
Flexural strength	MN/m²	95	71	71
Flexural modulus	MN/m²	1 900–2 800	2 500–3 400	1 000–2 250
Tensile strength	MN/m²	62	42	62
Compressive strength	MN/m²	95	132	76
Elongation	%	1.2–4.0	0.6–2.5	6.0
Izod impact strength	J/m	10–26	10–24	42–80
The electrical properties of all three grades are very similar at room temperature				
Electric strength	MV/m		17.5	
Dielectric constant at 1 MHz	pF/m		26	
Volume resistivity	ohm m		10^{14}	
Surface resistivity	ohm		10^{15}	
Power factor at 1 MHz			0.015	
Arc resistance	secs		100	

Table 4. Typical properties of cast epoxide resins[2].

tance when compared with polyester materials, although they are more expensive.

When used for potting and prototype tooling, fillers can be added to reduce the exotherm and the thermal expansion. The low exotherm and the correspondingly low shrinkage makes these material very suitable for prototype tooling[7,8,9]. Aluminium filled epoxy resin moulds can be used for compression moulding at temperatures up to 200°C and pressure of 14 MN/m².

Moulding powders produced from epoxide resins can be supplied in granular or powder form. Typical properties are given in Table 6.

These materials can be processed in the same way as other thermoset materials. Moulding temperatures are normally higher, being in the range 250–350°C, in order to obtain economical cycle times. If moulded at normal temperatures, 150–165°C, a cure time of 3 minutes would be required for a 3 mm section. Their main processing advantage is that due to their fluidity they can be moulded at relatively low pressure 3.5–13.5 MN/m², a characteristic which enables them to be used to encapsulate delicate electronic components. They would, however, only be used where high temperature electrical properties are of paramount importance because of their high cost. The main application of epoxy moulding powders is for specialised electronic components.

Table 5. Typical properties of epoxy/glass laminate[2].

Flexural strength	MN/m²	531
Flexural modulus	MN/m²	24 000
Tensile strength	MN/m²	345
Compressive strength	MN/m²	352
Specific gravity		1.85
Dielectric constant at 1 MHz	pF/m	38
Dissipation factor	1 MHz	0.014
Resin content	%	35

Table 6. Typical properties of epoxy moulding materials[10].

		Glass-Fibre Filled	Mineral Filled
Tensile strength	MN/m²	97–207	34.5–48.5
Tensile modulus	MN/m²	20 700	
Flexural strength	MN/m²	138–180	70–105
Compressive strength	MN/m²	172.5–207	124–172.5
Impact strength	J/m	850–1600	26.5–48.0
Specific gravity		1.8–2.0	1.6–2.06
Volume resistivity	ohm m	10^{16}	10^{16}
Electric strength	MV/m	14.2	13.0–15.7
Dielectric constant at 1 MHz	pF/m	35.5–44.5	35.5–44.5
Arc resistance	secs	125–140	150–180
Water absorption	%	0.05–0.095	0.1
Mould shrinkage	%	—	0.4–1.0

PHENOLIC RESINS

These are resins based on phenol, cresol or modified phenols and formaldehyde. They can be supplied as solids, liquids, or solutions. The hardening of these materials is an example of a heat initiated process as opposed to a chemical initiated curing reaction.

A B (Hardener) $\xrightarrow{\text{heat}}$ A + B

Resin + A $\xrightarrow{\text{heat}}$ Hardened infusible product.

There are two forms of phenolic resin, novolacs and resols. A novolac is a fusible resin incapable of hardening without the addition of a hardening agent. A resol is a resin which is capable of hardening under the influence of heat without the addition of any other product.

A typical novolac structure[11] being

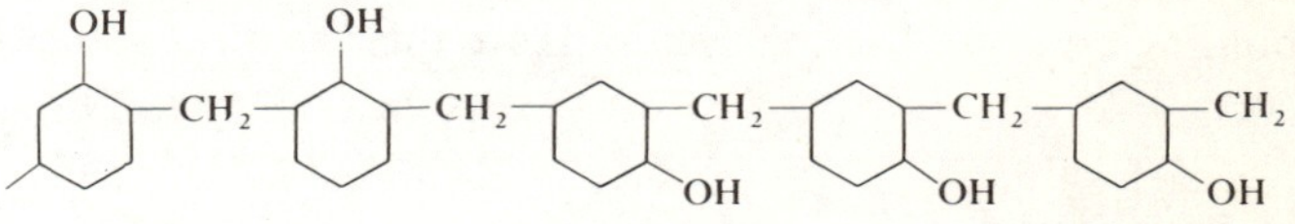

OH OH
—CH$_2$— —CH$_2$— —CH$_2$— —CH$_2$— —CH$_2$
OH OH OH

whereas a resol structure[6] would be

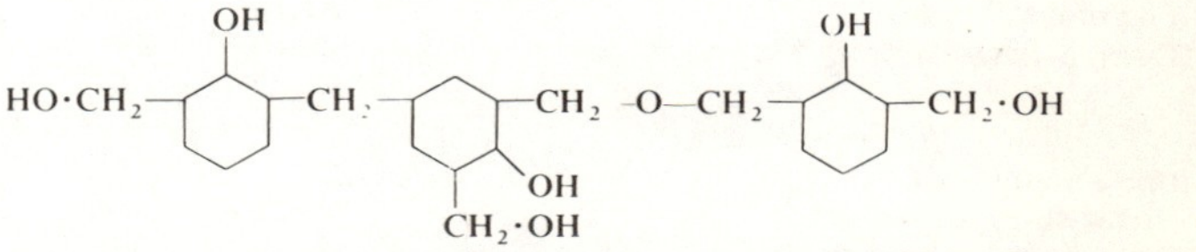

OH OH
HO·CH$_2$— —CH$_2$— —CH$_2$—O—CH$_2$— —CH$_2$·OH
OH
CH$_2$·OH

It is the presence of reactive methylol groups (—CH$_2$ OH) in the resol molecule which make it unnecessary to add a hardening agent in order to achieve a cured product.

Application	Kg
Abrasives	4 800 000
Foundry	15 000 000
Laminates	16 500 000
Friction materials	5 000 000
Adhesives	12 000 000
Moulding powders	15 000 000
Misc.	7 700 000
	76 000 000

Table 7. The main applications for phenolic resins and their UK consumption during 1970[2,4].

These resins can be used alone, for example, for food can lacquers, wire enamels, or blended with other materials to produce laminates and moulding powders. The main applications and approximate 1970 consumption are given in Table 7.

The main engineering applications of phenolic resins are laminates, foundry and friction materials. In the electrical laminating field, phenolic resins are used to impregnate kraft paper for the production of insulation boards, printed circuits and transformer bushings. The resin used in this case is usually based on cresol or a substituted phenol and used in a water/alcohol solution. The properties obtained would be a function of the reinforcement and the characteristics of resin used[12], *eg* high or low molecular weight.

For mechanical grade laminates the reinforcement can be paper, cotton fabric or asbestos fabric depending upon the strength requirements of the application. For this type of laminate a phenol based resin is normally used and the solvent chosen, for the resin, would depend upon the degree of impregnation required. Aqueous media is used to improve the degree of impregnation over alcohol solutions. Typical applications are gear wheels, wet and dry running bearings, and chemical plant. Typical properties of phenolic laminates are given in Table 8.

The resins are used in the foundry industry for the production of sand moulds and cores. Here the advantages of phenolic resin as the binder are a very good surface finish on the casting, little gassing during metal casting, good breakdown of binder after casting. Consequently, there is only a minimum of finishing to be carried out on the casting.

The important parameters for choosing these resins as the binding agent for friction materials are their heat resistance, and their resistance to abrasion over a wide temperature range. It is not unusual for slipping speeds to vary between 5 m/sec and 15 m/sec and for the temperature to reach a level which would normally be considered excessive for organic materials. Even at these excessive temperatures phenolic resins retain sufficient binding power to be the most suitable material available.

Phenolic resins are used in adhesives for marine and exterior grade plywoods where they are preferred over urea based materials because of their superior water resistance. They are used for binding paper-pulp for the production of automotive glove boxes and parcel shelves, television backs, and suitcases; also for a range of adhesive applications.

There are a wide range of moulding powders available based on phenolic or modified phenolic resins. Generally, these powders are based on novolac resins but resol resins are used for minimum odour grades. The flow of the moulding powder can be varied by choice of resin and/or fillers. For example, a long flow material can be prepared using a resin of low molecular weight or a short flow one from a high molecular weight resin. However, a short flow powder could have a long cure time due to its low reactivity but this can be increased by leaving some of the lower molecular weight materials in the resin during manufacture. It is possible by choosing the correct catalyst and reaction conditions to prepare a resin which, when converted into a moulding powder, would have a relatively fast cure and a medium-long flow. Minimum shrinkage grades are based on a highly condensed resin, and the flow of these grades can be improved without affecting the shrinkage by adding up to 1% of a plasticiser. Typical properties of four grades of phenolic moulding powders are given in Table 9.

The three main areas of applications of these moulding materials are electrical fittings, automotive, and domestic appliances. An estimated market breakdown is given below:

Automotive Applications	25%
Domestic Appliances	25%
Toilet Seats	11%
Electrical Fittings	30%
Miscellaneous	9%

In most of the applications advantage is taken of either the electrical or heat resistant properties of these materials. They are superior to amino resin based materials in water

Table 8. Typical properties of phenolic laminates[12,13].

Base Material		Paper	Cotton Fabric	Asbestos Fabric	Glass Fabric
Tensile strength	MN/m²	90–150	62–110	65	85–207
Flexural strength	MN/m²	124–200	110–186	127	170–276
Compressive strength	MN/m²				
(i) Flatwise		235–340	290–340	300	242
(ii) Edgewise		130–207	189–215	166	—
Impact strength	J/m	19.5–26.5	36–67.5	107	537
Shear strength flatwise	MN/m²	83–110	103–120	112	—
Electric strength at 90°C	MV/m	18.0–33.5	3.15–15.5	—	—
Power factor at 1 MHz		0.35–0.46	0.10	0.11	0.01–0.02
Dielectric constant at 1 MHz	pF/m	40–49	57.5	54	40–49
Water absorption	%	0.3–1.0	0.65–1.3	—	—

		General Purpose	Electrical	High Impact	High Temp.
Shrinkage	M/m	0.005–0.010	0.002–0.009	0.004–0.007	0.002–0.006
Specific gravity		1.3–1.42	1.3–1.75	1.35–1.44	1.5–1.75
Water absorption	mg	25–50	6–25	35–75	15–30
Tensile strength	MN/m^2	45–62	27.5–62	41.5–62	31–55
Tehsile modulus	MN/m^2	6 200–8 250	6 200–13 800	5 500–9 000	8 250–9 300
Flexural strength	MN/m^2	76–93	62–93	76–90	55–86.5
Impact Strength	J/m	9.0–12	6.4–12	10–32	7.5–12
Compression strength	MN/m^2	207–275	103–275	170–240	125–185
Electric strength	MV/m	4.7–8.0	4.3–11.8	2.5–5.1	3.75–7.9
Power factor at 1 MHz		0.03–0.05	0.015–0.040	0.04–0.06	0.04–0.07
Dielectric constant 1 MHz	pF/m	35.5–49	35.5–53	35.5–53	35.5–53
Volume resistivity	ohm m	$3–5 \times 10^{10}$	$10^{11}–10^{12}$	$10^{9}–10^{10}$	$10^{10}–10^{11}$
Surface resistivity	ohm	$10^{12}–10^{13}$	$10^{11}–10^{13}$	$10^{11}–10^{12}$	$10^{11}–10^{13}$

Table 9. Typical properties of phenolic moulding powders[2, 14].

absorption and heat resistance but have a disadvantage in that they cannot be produced in light colours.

Processes normally used for forming these compounds are compression moulding, transfer moulding, injection moulding and extrusion. Mould temperatures for compression and transfer moulding are normally in the range 140–170°C. However, higher temperatures can be used to reduce the cycle time on thin sectioned components, *eg* bottle caps. For injection moulding the barrel temperature should be in the range of 70–100°C with mould temperatures similar to those for the compression process. Although at present only 3% of phenolic moulding powders are processed in this way this figure is expected to increase rapidly as production techniques are up-dated on economic grounds.

Extrusion of the materials is relatively rare and is usually carried out on a reciprocating ram machine. The type of section which can be formed is limited and is normally restricted to rods and tubes.

POLYESTER RESINS

These thermosetting materials are produced by reacting together an alcohol and an acid and are more correctly classified as unsaturated polyester resin. Saturated polyester moulding materials are available but they are thermoplastic.

Unsaturated polyester resins are usually supplied as a solution in styrene, which acts as a solvent and also as the cross-linking agent joining together the chains of polyester molecules.

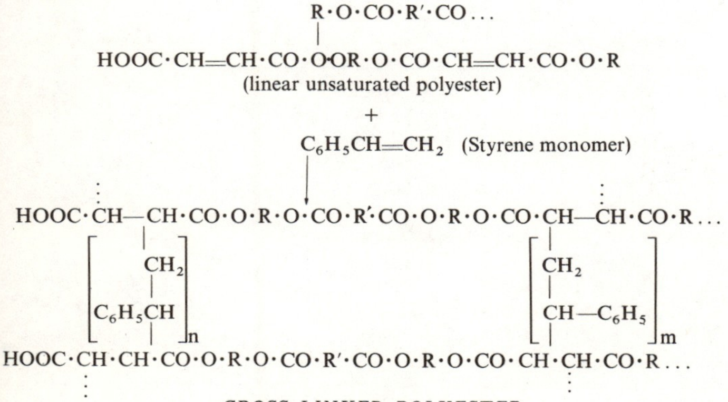

$$\text{R·O·CO·R'·CO...}$$

$$\text{HOOC·CH=CH·CO·O-OR·O·CO·CH=CH·CO·O·R}$$

(linear unsaturated polyester)

+

$C_6H_5CH{=}CH_2$ (Styrene monomer)

$$\text{HOOC·CH—CH·CO·O·R·O·CO·R'·CO·O·R·O·CO·CH—CH·CO·R...}$$

$$[CH_2—C_6H_5CH]_n \qquad [CH_2—CH—C_6H_5]_m$$

$$\text{HOOC·CH·CH·CO·O·R·O·CO·R'·CO·O·R·O·CO·CH·CH·CO·R...}$$

CROSS LINKED POLYESTER

The cross linking which can take place at room temperature or elevated temperature, is initiated by a catalyst. It is, however, necessary to activate the catalyst by heat or a chemical reaction. If the activation is carried out by a chemical reaction the subsequent curing process would be an example of a chemically initiated process as opposed to a heat initiated one.

CATALYST $\xrightarrow[\text{promoter at room temp.}]{\text{heat or}}$ activated catalyst

POLYESTER + STYRENE + ACTIVATED CATALYST ⟶ CROSS LINKED POLYESTER

Unsaturated polyester resins are used alone, in conjunction with glass reinforcement or compounded with fillers and glass strands to prepare moulding materials. Table 10 gives an approximate breakdown of polyester consumption in the UK.

These resins can be processed either by ‘hand lay up’ or by compression moulding. They can be used at room temperature and low pressures to give large void-free products which are relatively light and strong without the necessity of expenditure on capital equipment. This property of void-free materials is due to the absence of volatile components during the curing process, which is an advantage over amino resin and phenolic resin based materials.

The properties of the resin can be varied considerably by the choice of the chemical constituents. They can be made as hard, brittle materials, flexible materials or fire retardant materials by correct choice of acid and/or alcohol. Typical properties of cast resins are given in Table 11.

Table 10. Breakdown of approximate polyester consumption in the UK[2].

Application	% Usage
Transport	17.7
Body fillers	9.4
Building	21.0
Marine	17.7
Pipe seals	8.0
Chemical plant	4.7
Electrical	4.7
Insulated containers	5.8
Miscellaneous	11.0

Table 11. Typical properties of cast polyester resins[2,14].

		General Purpose	Chemical Resistant	Self Extinguishing	Flexible
Tensile strength	MN/m^2	41	46	31.5–40	2.07
Youngs modulus in tension	MN/m^2	3 820	4 000	3 1000–3 900	—
Flexural strength	MN/m^2	107	103	60–77	—
Youngs modulus in flexure	MN/m^2	3820	3660	3450–3740	—
Impact strength	J/m	13.5	15	6.5–12.25	29.4
Heat distortion temperature at 1.82 MN/m^2	°C	69–73	115	70–90	—
Electric strength	MV/m	13.8	13.4	14.8–15.8	—
Volume resistivity	ohm m	10^{13}	10^{13}	10^{13}	—
Surface resistivity	ohm	10^{15}	10^{14}	10^{14}	—
Power factor at 1 KHz		0.001	0.0025*	0.0004–0.001	—
Dielectric constant at 1 MHz	pF/m	28.5	26.5	27.5–29	—

* at 1 MHz

Table 12. Typical properties of cast polyester resins[15, 16]

		Woven Cloth	Chopped Strand Mat	Woven Roving
Tensile strength	MN/m^2	138–310	69–105	220–345
Flexural strength	MN/m^2	138–345	105–170	310–485
Youngs modulus in tension	MN/m^2	20 700		12 400–19 250
Youngs modulus in flexure	MN/m^2	20 700		—
Impact strength	J/m	800–1050	800–1050	—
Electric strength	MV/m	5.9–7.9	5.9–7.9	—
Power factor at 1 MHz		0.02–0.05	0.02–0.08	—
Dielectric constant at 1 MHz	pF/m	32–37	28.5–40	—
Water absorption	%	0.2–0.8	0.2–0.8	—

Table 13. Typical properties of DMC and SMC[2,14].

		DMC			SMC	
		General Purpose	Low Shrink	Non Shrink	Low Glass Content	High Glass Content
Shrinkage	m/m	0.0015–0.003	0–0.0015	Nil	0–0.001	0–0.001
Water absorption (1 day)	mg	10–30	10–30	20–40	25	25
Tensile strength	MN/m^2	41.5–69	41.5–69	41.5–62	69	114
Youngs modulus in tension	MN/m^2	8 300–11 700	8 300–11 000	6 900–9 650	10 350	11 030
Flexural strength	MN/m^2	83–124	83–138	83–110	160	207
Youngs modulus in flexure	MN/m^2	8 300–13 800	7 600–10 300	6 900–9 650	10 350	12 100
Impact strength	J/m	106–426	106–265	160–345	267	400
Compressive strength	MN/m^2	138–207	138–207	—		
Electric strength	MV/m	7.9–13.8	7.9–13.8	9.9–13.8	9.85	10.8
Power factor at 1 MHz		0.007–0.02	0.01–0.02	0.02–0.03	0.015	0.015
Dielectric constant at 1 MHz	pF/m	44.5–53.5	40–47.5	44–49	39.8	39.8
Volume resistivity	ohm m	10^{12}–10^{13}	10^{12}–10^{13}	10^{10}–10^{12}	$10^{12.5}$	$10^{12.5}$
Surface resistivity	ohm	10^{10}–10^{14}	10^{14}–10^{16}	10^{12}–10^{14}	10^{14}	10^{14}
IEC (112) tracking index		600–1000	300–1000	800–1000	1000	1000
Arc resistance	secs	80–160	70–170	100–130	125	120

Table 14. Typical properties of DAP and silicone compounds[2].

		Diallyl Phthalates			Silicone Transfer Moulded Material*
		General Purpose	Mineral Filled	Short Glass Fibre Filled	
Tensile strength	MN/m^2	31–41.5	39.5–48.5	34.5–69	36
Youngs modulus in tension	MN/m^2	6 900–9 650	6 900–9 650	12 400–13 800	—
Flexural strength	MN/m^2	48.5–62	55.0–76	62–117	50
Compression strength	MN/m^2	124–145	165–180	165–180	0.13
Impact strength	J/m	7.5–9.6	4.7–6.4	10.7–21.4	—
Flexural modulus	MN/m^2	—	—	—	10 650
Shrinkage	m/m	←	0.1–1.1%	→	
Electrical strength	MV/m	—	—	—	—
Dielectric constant at 1 MHz	pF/m	41.5	41.6	45.5	30
Power factor at 1 MHz		0.039	0.038	0.017	0.0013
Volume resistivity		10^{14}	10^{14}	$>10^{15}$	10^{12}
Arc resistance	secs	←	140–180	→	

* Post cured for 2 hrs at 200°C.

Resins can be used with woven glass cloth or chopped strand mat to produce laminates having properties as shown in Table 12. However, they are used without fibre reinforcement for automotive body filler compounds, imitation marble, encapsulation of various products, pipe joints and buttons.

There are two types of moulding material made using unsaturated polyester resins, dough moulding compounds (DMC) and sheet moulding compounds (SMC). Both compounds contain fillers and glass fibre reinforcement. In SMC the glass has a longer strand length and is normally used at a higher loading giving materials with higher strength properties.

Polyester DMC is of a putty-like consistency and has a very low viscosity at moulding temperatures. Consequently, it can be compression moulded using low pressure.

One of the outstanding properties of these materials is that they can be moulded with no shrinkage (see Table 13). This enables moulded components to be produced to tolerances which are smaller than those of diecast metals and other thermoset moulding materials. Hence, cost savings can be made by using moulded-to-size DMC components instead of machined castings. These newer sophisticated moulding compounds are now beginning to be accepted as true engineering materials and have not yet realised their full potential[17,18].

Another facet of the non-shrink DMC is that it can be moulded to give components having a very good surface finish and appearance. The surface finish is preserved even when the component contains ribs and bosses which, because of the absence of shrinkage, are not witnessed onto the fair face of the moulding. The general purpose and low shrink materials, like amino based materials, can be prepared in a range of colours including pale colours. The non-shrink material cannot be coloured but can be painted with air drying paints or stoving enamels, *eg* car finishes.

Sheet moulding compound as the name implies is a flat sheet of material supplied in rolls from which can be cut the required preform weight prior to moulding. This material is not as easy flowing as DMC, therefore, requires a higher pressure in order to mould complex shapes.

The main existing applications for these compounds are in the electrical industry, *eg* meter boxes, insulators, electric drill housings, and for industrial mouldings, *eg* machine guards, automotive components, valve handles, computer printer parts. It is anticipated that engineering applications of these materials will produce considerable growth in the consumption of polyester based compounds.

The compounds are usually processed by compression moulding at temperatures of 120–150°C and a pressure in the range of 1.75–69 MN/m^2. However, DMC can be transfer and injection moulded without difficulty but it should be noted that some degradation of the glass reinforcement will take place leading to a slight decrease in physical properties.

MISCELLANEOUS RESINS

Two other thermosetting materials can be used to prepare moulding compounds, *ie* silicone resins and polymers of diallyl phthalate. Chemically diallyl phthalate is an ester and it polymerises through the unsaturation of the allyl grouping. Compounds based on diallyl phthalate (DAP) are a special type of polyester material. They are used in specialised electrical and electronic equipment where high temperature and high humidity environments are encountered.

DAP compounds can be processed by compression and transfer moulding at temperatures in the range 135–165°C and pressures of 14–70 MN/m^2. Their hot strength is inferior to that of phenolic moulding powders and care must be taken when extracting these materials from a mould cavity. Typical physical and electrical properties are given in Table 14.

Compounds based on silicone resin are supplied in granular form and are either black or dark grey in colour. Silicones are inorganic polymers, whereas other materials which have been discussed in this chapter are organic in nature. The lengths of the polymeric chains are much shorter in silicones, hence, the molecular weight of the polymer is lower than that of many organic polymers, and the strength of resultant composite materials is lower. The inorganic nature of the compounds means that they are non-burning and have very good heat resistance.

Silicone moulding powders are used almost exclusively in electronic applications and especially for the encapsulation of electronic devices. They are processed by compression or transfer moulding at temperatures of 130–180°C with a minimum pressure of 2.74 MN/m^2. Typical properties of transfer moulded materials are given in Table 14.

Silicone resins can also be used with glass reinforcement to produce laminates which are electrically superior to phenolic resin and amino resin based laminates but are mechanically weaker than other thermosets. Silicone laminates are used where very good electrical properties are required under high humidity conditions.

REFERENCES

1. *British Plastics*, January 1971.
2. *British Plastics Materials Guide*, 1970–1971.
3. Encyclopedia of Polymer Science Vol. 2. *Interscience Publishers (Division of J. Wiley & Sons)*, New York, 1965.
4. *BP Chemicals International Limited*, unpublished information.

5. C P Vale and W G K Taylor. Aminoplastics, *Iliffe*, London, 1964.
6. J A Brydson. Plastics Materials, *Iliffe*, London, 1966.
7. *Ciba Limited*, Information Sheet T9a.
8. *Ciba Limited*, Information Manual T11d.
9. *Ciba Limited*, Information Sheet T3a.
10. H Lee and K Neville. Handbook of Epoxy Resins, *McGraw-Hill* 1967.
11. A A K Whitehouse, E G K Pritchett and G Barnett. Phenolic Resins. *Butterworth & Co Ltd*, 1967.
12. W J Brown. Laminated Plastics. *Plastics Institute Monograph.*
13. *Tufnol Limited*, published information.
14. *BP Chemicals International Limited*, published information.
15. Glass Fibre Reinforced Plastics, ed. P Morgan, *Iliffe*, London.
16. F.R.P. Design Data, *Fibreglass Limited* publication.
17. D R Hudson, D.M.C. in Automotive Applications, *British Plastics* Oct. 1969.
18. E A Davies. Non Shrink or low Profile Polyester Moulding Compounds for the Automotive Industry. *7th International Reinforced Plastics Conference*, Brighton, UK 1970.

BIBLIOGRAPHY

D Morgan and C P Vale. The Physical Properties of Polymers. *S.C.I. Monograph, No. 5*, London 1959.

C P Vale and W G K Taylor. Aminoplastics, *Iliffe*, London 1964.

J A Brydson. Plastics Materials, *Iliffe*, London 1966.

W G Potter. Epoxide Resins, *Iliffe*, London 1970.

S Oleesby and G Mohr. Handbook of Reinforced Plastics, *Reinhold* 1966.

H Lee and K Neville. Handbook of Epoxy Resins, *McGraw-Hill* 1967.

B Parkyn, F Lamb and B V Clifton. Polyesters Vol. 2—Unsaturated polyesters and polyester plasticisers, *Butterworth & Co Ltd*.

Chapter 4

Thermoplastics

C C Gosselin *BSc*
Shell International Chemical Company Limited

The techniques of forming thermoplastics are obviously influenced by the fact that the molecules of thermoplastics do not cross-link on heating and therefore can be maintained in a softened state while being made to flow under pressure into a new shape. The properties of each thermoplastic, however, sometimes limit the nature of the possible processes in which it will be involved. Some, for example, will be flexible enough to be ejected from a mould having undercuts in it; others will be too brittle to withstand such a treatment. It is therefore preferable to know about each plastics in terms of properties and applications in order to appreciate which forming techniques may be most suited to it.

THE POLYETHYLENES

Low density polyethylene (LDPE). This is the well-known plastics, often termed polythene, used for buckets and bowls destined for the less expensive end of the market, for aerosol caps and detergent bottles, and in films used for, say, dry cleaning bags, mulching film for agricultural usage and temporary protection for building sites open to weather. There are numerous further applications but, in these, other plastics are also used if circumstances dictate; nevertheless, cosmetic squeeze tubes, carboys, toys, cable and wire insulation, paper and cellulose coating, building pipe and cold water storage tanks all contribute to the enormous amount of low density polyethylene (LDPE) used every year.

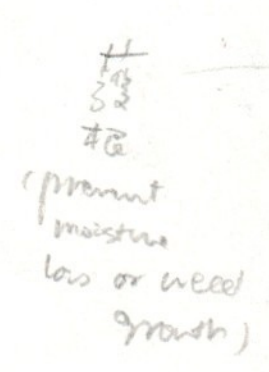

Polyethylene is flexible and very strong, it is not brittle except when the molecular weight is low (which in commercial usage is rare anyway). It burns only slowly, with a smell like burning candles. It softens, however, at about 85–87°C, so does not resist boiling water. It is translucent, almost transparent in thin sections, and is a very good electrical insulator — though it should be borne in mind that no thermoplastic can legally be used for electrical plugs and fittings because of its susceptibility to heat. It resists almost all chemicals except strong acids below 60°C, unless its product shape is such that it has internal strains within the moulding. In such cases, detergents and certain oils can produce cracks in the strained area. This liability, called environmental stress cracking, can nevertheless be eliminated by choosing a grade with a higher molecular weight.

High density polyethylene (HDPE). Sometimes commercially termed 'rigid polythene', this material is specially suitable for appliance parts, piping, sheet for fabrication, carboys, crates for bottles and other products, fish boxes, dustbins and bleach bottles. This last application gives a ready introduction to the difference in properties between low and high density polyethylene. Notice the comparative rigidity and springiness of the bleach bottle when squeezed, and the less waxy feel. High density polyethylene is also more resistant to heat; it is unaffected by boiling water, but softens at about 120°C. In its natural (unpigmented) form it is a little less translucent than an equivalent thickness of low density polyethylene. Whilst more rigid than LDPE, HDPE has a lower impact strength. Its resistance to chemicals is similar to LDPE but holds this resistance up to about 80°C.

Polypropylene. This material bears both a chemical and physical resemblance to the polyethylenes, and indeed is often thought of as a sort of 'polyethylene-plus'. It is, however, sometimes dangerous to think of polypropylene in this fashion, for there are properties possessed by the polyethylenes that cannot be matched by polypropylene. Nevertheless, it is in general a versatile and sturdy plastics, being used in domestic ware (especially consumer durables like vacuum cleaner casings, washing machine tops and agitators and so on), laboratory and hospital ware, automotive facia, textiles, toys, containers, and crates and film fibre applications. This last is a very fast developing field and sacks, ropes, twine and fine fibres for upholstery are appearing more and more before the public eye. The material has a high rigidity, a good impact strength (but not as good as low density polyethylene), does not soften until about 140°C and hence is especially suitable for steam sterilisable hospital ware. It produces mouldings of a high gloss (if the mould surface allows!) has electrical properties similar to the polyethylenes, and is not affected by stress cracking. Its outstanding property is its resistance to fatigue on flexing; it is possible to mould hinge sections into, say, a ventilation louvre without fear of flex failure. The design of such hinges is, however, important.

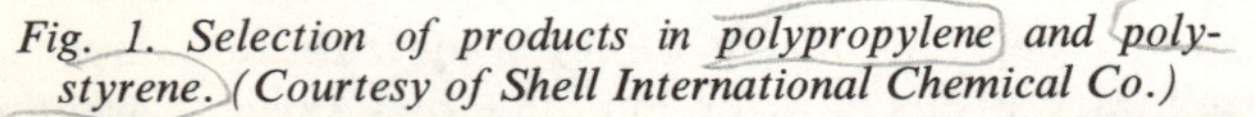
Fig. 1. Selection of products in polypropylene and polystyrene. (Courtesy of Shell International Chemical Co.)

Fig. 2. Soft (plasticised) PVC raincoat and boots illustrating the versatility of PVC. (Courtesy of Dick Swayne.)

Before proceeding to discuss other plastics there are two general points which can easily be spotlighted by reference to the three plastics mentioned above — which, incidentally, are often referred to, in a group, as the polyolefins. It will be seen first that the polyolefins present a range of plastics of increasing softening point, rigidity, gloss, chemical resistance, *etc*, and it might be asked, why, for example, polypropylene is not used for more applications usually associated with low density polyethylene. The answer is that, quite apart from the fact that some of the properties of the latter (for example, flexibility, which is the reverse of rigidity) may be preferable for certain applications, economics are also important. Polypropylene is more expensive than high density polyethylene, which is again more expensive than low density polyethylene. So perhaps price should be included as a 'property' when determining what plastics would be used for a given application.

The second point is that there are generally several grades of each thermoplastic, and the properties of the plastics vary, to a degree, according to the grade of material. For example, a detergent bottle that shows environmental stress cracking is simply being made in the wrong grade of LDPE; choosing a grade of higher molecular weight will obviate this; the second grade, however, has to be processed at slightly higher temperatures, and mouldings take slightly longer to make, so if high production speed is important and resistance to environmental stress cracking is not, then obviously the first grade should be used.

THE VINYL PLASTICS

The most important member of this group is polyvinylchloride (PVC). At first sight this material is puzzling, since it can exist as a rigid, high gloss plastics of the sort used for guttering or down pipes, or as the highly flexible material from which a raincoat or the covering of car upholstery is made. This is because PVC is one of the few plastics to which plasticisers can be added. Plasticisers convert the more rigid type into the soft, flexible type, by getting between the polymer molecules and forcing them apart so that they slip over one another. Unmodified, PVC is found in house pipes, in much profile and sheet production (perhaps lightly plasticised), and in bottles for concentrated fruit squashes, olive oil, *etc*. Plasticised, it is used in fashion and protective clothing, soft toys, handbags and shoes, electrical wiring and insulation, vinyl wallpapers, leathercloths, floor tiles and packaging film for such items as meat, where oxygen must be allowed to permeate in order to prevent darkening of the meat. A copolymer of PVC with polyvinylacetate is a basic material from which gramophone records are made. Polyvinylacetate can be polymerised by itself, but is not suitable as a moulding powder; it is nevertheless important both as an adhesive and as the basic film-forming agent in many emulsion paints.

Fig. 3. *Styrene-acrylonitrile (SAN) battery box. (Courtesy of British Plastics.)*

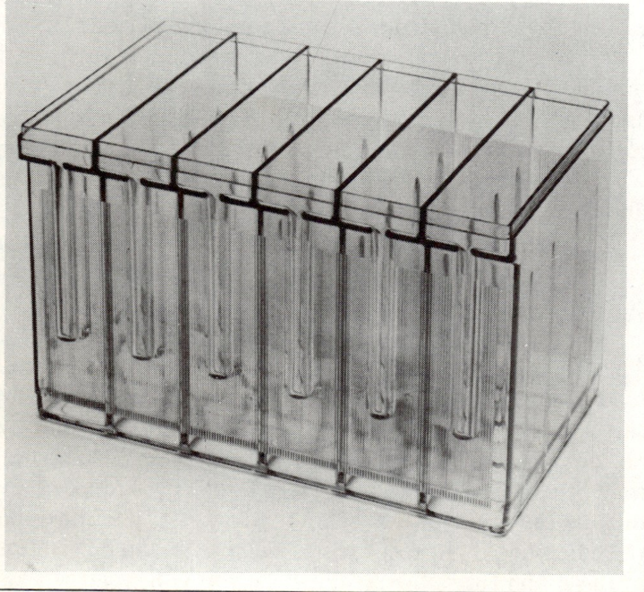

Unplasticised PVC is a hard, rather brittle plastics (but strong enough for domestic piping, *etc*), resistant to alkalis and many solvents, although soluble in ketones, esters and, to some extent, chlorinated hydrocarbons. It is fairly resistant to flame, and an excellent electrical insulator. It softens at about 80–100°C.

Polyvinyl alcohol is one of the few water-soluble plastics, and its main use is derived from its ability to be formed into a film for sachet production. This is useful for adding metered amounts of chemicals (dangerous or otherwise) to water in order to make up standard solutions.

POLYSTYRENE AND RELATED PLASTICS

Polystyrene itself can easily be recognised by the metallic clatter made when a polystyrene moulding is dropped on to a hard surface. This noise is even present when the material has been toughened by the addition of rubber. Basic polystyrene is found in refrigerator shelving, plastics tumblers, toys and model kits, lighting fittings, *rigid* talcum powder bottles and other decorative, often transparent, cosmetic packs, reels and spools for film or tape and, when toughened, in radio and TV cabinets, refrigerator liners, electric fan plates, battery cases and yoghurt pots. Basic polystyrene is colourless and transparent, hard and rather brittle. It softens at about 85–95°C, possesses a low power factor and high dielectric strength, resists acids, alkalis and aliphatic hydrocarbons, but is soluble in aromatics like benzene, in esters and higher alcohols. One of its best attributes is that of reproducing mould detail with extreme faithfulness, allowing clean edges and fine ornamentation to be reproduced without difficulty. Like polyethylene, it is an inexpensive plastics, and many toys are therefore made of it. When toughened, polystyrene loses its transparency and some surface hardness, but gains in impact strength and, to some degree, flexibility.

Polystyrene, PVC and all the polyolefins can also appear in a structural foam form, giving materials with a satin surface bearing random swirl surface patterns which are aesthetically appealing. The materials lose some impact strength and gain much rigidity when foamed, and are appearing in such fields of application as furniture, record players and TV cabinets. Items of much thicker section are generally employed to minimise the drop in impact strength, and the resulting mouldings are interesting in that they resemble neither plastics nor wood as these materials are traditionally imagined.

Structural foam polystyrene should not be confused, however, with expanded polystyrene, well-known in the form of ceiling tiles and packaging shapes which hold their often awkward shape and fragile contents safely throughout the roughest of journeys. This lightweight material (as little as $\frac{3}{4}$ lb/ft^3) is an excellent heat insulator below its softening temperature, which is the same as basic polystyrene, but it should be remembered that, since polystyrene dissolves in aromatic solvents, it should only be painted with emulsion paints and not spirit or oil-based ones.

Styrene-acrylonitrile (SAN) is a basic polystyrene-plus; possessing greater strength, a marginally improved heat resistance and greater resistance to weathering. It is also more chemically resistant to food oils such as orange peel oil, and is consequently used in food mixer liquidiser attachments, cups, tumblers and other kitchen equipment, as well as in hair curlers, some fountain pen barrels, electric razor parts, laboratory utensils and suchlike.

Acrylonitrile-butadiene-styrene (ABS) can be regarded as toughened polystyrene-plus. A strong, translucent material with good chemical resistance, it is unaffected by aromatics or chlorinated hydrocarbons (*cf.* polystyrene), and is widely used for tote boxes and trays, vacuum cleaner casings, luggage, safety helmets, telephone housings and various automotive trim components. It is also being used for piping, carrying corrosive chemicals and natural gas.

POLYMETHYL METHACRYLATE (PMMA)

Well known by the ICI trade mark of 'Perspex', PMMA is a highly transparent, hard and weather resistant material which softens at about 80–100°C. Its virtues lie in its suitability for outdoor signs and other glazing applications where scratch-resistance is not too important (minor imperfections can be polished out with silver polish) and its ease of forming into wind-shields for motor bikes and boats. During the Second World War it had great use in cockpit covers and gunners' blister windows. Currently, it is widely used in inspection windows, instrument panel shields and fascia, protective goggles and safety shields, not to mention its use in colourful baths and bathroom furniture. Whilst PMMA can be moulded, most of its applications stem from its prefabrication into sheet and rod; the latter section has the ability to 'pipe' light through a curved path to emerge with little loss of power at the other end of the tube — a use of considerable importance to the medical world.

NYLON

Apart from the multifarious uses of the plastics in the textile world, nylon is used as self-lubricating bearings (especially in food processing, where the presence of lubricating oils might lead to contamination), in door catches and in curtain runners. By filling it with glass fibre, nylon finds use in door handles, window fastenings and sliding door rollers. Its low gas permeability has given it a film usage for packaging commodities such as cheese slices. It has a high softening point because it melts more sharply than most plastics, at about 200–250°C. It has good chemical resistance to dilute acids and alkalis, strong alkalis and most organic solvents. Its water absorption is high, however, and this appears to affect its impact strength, leading to some difficulties in applications like dishwashers. This drop in strength, however, is obviously not enough to affect its use in textiles. It should also be noticed that nylon is embrittled by prolonged outdoor use.

Fig. 4. Large-scale cast nylon gear and pinion system. (Courtesy of British Plastics.)

FLUORINATED PLASTICS

These expensive plastics are mostly used where a low coefficient of friction or complete resistance to highly corrosive chemicals is required. The best known member of the group, polytetrafluoroethylene (PTFE), is used as the lining of non-stick frying pans and in food processing equipment handling sticky products like chocolate. It cannot be injection moulded; sintering and subsequent machining being generally employed where a 'solid' material is required rather than a lining of plastics. Injection moulding is possible, though not easy, however, with polychlorotrifluoroethylene (PCTFE) and fluorinated ethylene propylene (FEP), and extrusion can also be carried out. Sheet, profile and film, gaskets, electronic parts, liquid level indicators and similar products are all made from these materials. Of the two latter plastics, FEP has the higher impact strength.

POLYACETAL

It is not easy to give a 'recognition application' which will always typify this material, which is a pity, because it is a useful engineering plastics in the medium price range, possessing properties of rigidity and strength which it maintains at high temperatures for long periods of time under load. It has a smooth, hard surface, is resistant to organic solvents and only slightly attacked by strong acids and bases. Its softening point is high (175°C), as is its wear resistance, and it is consequently much used in small gear wheels and bearings in business machines, cash registers and computers, ball castors for furniture, pump casings and food mixer parts. An interesting future development (already underway in fact) lies in the manufacture of aerosol containers.

The foregoing mention of acetal's ability to work well under load brings up the subject of 'creep'. All thermoplastics can distort and set permanently when under load for long periods, even at ambient temperatures. Some do it more than others, and the study of creep tendency is involved and outside the scope of this chapter. However, it is important to remember that creep exists, because it comes into play in the design of say, beer bottle crates, which get stacked,

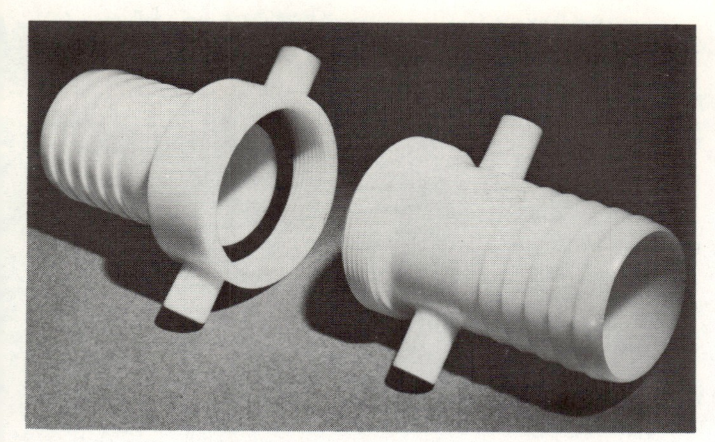

Fig. 5. Acetal coupling for industrial hosepipe. (Courtesy of Creators Ltd.)

full, one on top of the other and can produce crate distortion and even breakage. In this particular field it is interesting to note that crates for beer are made in polypropylene because this has superior creep properties to the less expensive HDPE, whereas HDPE is used for milk bottle crates because stacking is lower and HDPE has superior low temperature impact strength properties needed during cold early morning deliveries. The point is made to spotlight the necessity to examine all the working parameters applied to an application before choosing what material is to be used.

POLYCARBONATE

This is a medium priced material whose only disadvantage is a poor chemical resistance to alkalis and some organic solvents. It is hard, rigid, heat resistant (softening point 165 °C) and transparent, with good electrical properties and high impact strength. Uses are in babies' transparent 'unbreakable' feeding bottles, lighting covers (vandal-proof street lamps), safety helmets, sterilisable ware, golf club heads, business machine parts and draughtsman's instruments.

THE CELLULOSICS

Cellulose acetate is well-known in the textile world, but also has wide use in toothbrush handles, cutlery handles and combs, bicycle pumps and fountain pens. It is also very popular in packaging, as the 'blister' in blister packs and in fancy semi-rigid containers for chocolates and flowers. Cellulose acetate butyrate has improved impact strength over cellulose acetate, is more resistant to petrols and oils and has, therefore, been used for lighter fuel bottles. Other uses lie in metallisable parts (reflectors, *etc*), pipe and pipe inspection traps, tail light covers and tool handles. It has an unusual limitation in that it slowly gives off an odour of rancid butter, particularly on storage in a still atmosphere. Cellulose propionate is another variant, dimensionally stable in damp conditions, with a hard, glossy surface. Its main uses are in car parts, telephone housings, radio and TV parts and casings, toothbrushes and sunglass frames. All three members of the group are transparent.

OTHER PLASTICS

A number of other plastics are developing fast, but many of these are specialist materials whose virtues are often due to combining virtues found in other plastics. For example, TPX, a comparatively new material from ICI, has similar chemical properties to polyethylene and polypropylene, but is completely transparent. Its softening point is higher than polyethylene or polypropylene, but its impact strength is lower (though higher than that of polystyrene). Phenoxies combine good impact strength and transparency with good water vapour barrier properties and chemical resistance to all but aromatic hydrocarbons and esters. They also have very good adhesion to metal, wood, glass and other plastics, and can thus be used to join dissimilar materials with a waterproof joint.

Polyphenylene oxide will probably find wide use as an engineering material, having good strength, rigidity and hardness, and a high softening point, good electrical properties and resistance to acids, alkalis and alcohol. The important point, however, is that the mechanical properties are retained over a wide range of temperatures, and uses are found in washing machine parts, pipes, pumps and controls for hot water and aqueous chemical solutions, and may in places replace stainless steel in surgery and medicine because of its chemical resistance and easy steam sterilisability. It is, however. again not a cheap material. Polysulphone is another high strength, rigid material which retains its virtues at high temperatures and has extremely low creep resistance even when hot.

CONCLUSION

It will thus be seen in general that the spectrum of desirable attributes is being filled more and more fully, and that if one is prepared to pay the price for specialist materials and their forming costs, there is little that cannot be achieved in one plastics or another. It therefore becomes abundantly clear what a large part is played by the price factor, obviously there are closely defined limits to what the market will stand in a given application. The potential user of plastics must therefore take this point into consideration.

Fig. 6. Laboratory dispenser fittings in fluorinated ethylene propylene (FEP) resist all but molten alkali metals and elemental fluorine. (Courtesy of British Plastics.)

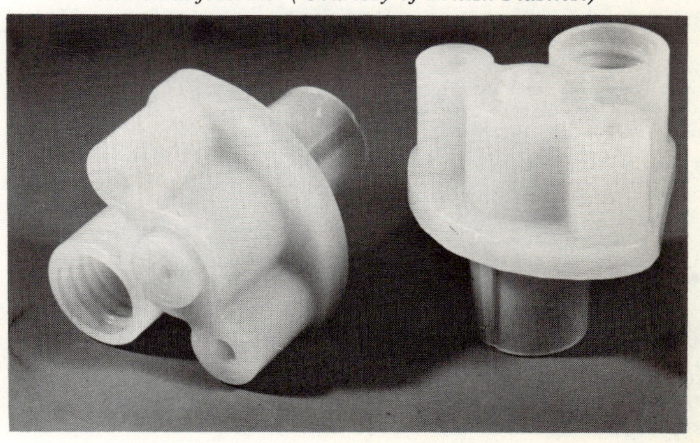

Finally, it is extremely important to reiterate that the designer, engineer or potential user of plastics should study all information available on the way in which the product will be used. Sometimes the information provided will miss out important points which it never occurred to anyone to include. For example, an under-the-bonnet automotive product might never reach high temperature in use, and a plastics might therefore be chosen with a comparatively low softening point as a result. However, the picture would certainly be changed if it was known that the completed car passes through a touch-up oven whose temperature is above the softening point of the plastics selected! Again, a bottle containing fungicide was discovered after some time to be *growing* fungus; this turned out to be because sulphur dioxide in the fungicide diffused out, allowing the fungicide to oxidise and become a virtual nutrient for fungus! It thus cannot be emphasised too much that a complete case history of the use to which a product would be put is essential before a choice of plastics can be made.

Chapter 5

Economics of Forming

J H Briston *BSc ARIC API MInst Packaging*
Shell International Chemical Co Ltd

Before discussing the economics of forming it should be emphasised that there may be other limiting factors apart from purely economic ones. For example, injection moulding cannot, in general, be used to mould objects having re-entry curves or reverse tapers. It is, however, possible to mould slight undercuts in certain materials, though not in others. Undercuts are often moulded in polypropylene screw caps, for instance, into which may be inserted imitation jewels in, say, polystyrene. Polypropylene is a sufficiently resilient material for this purpose but a similar attempt using polystyrene for the caps would lead to fracture when the cap was removed from the mould (see Fig. 1). Similar restrictions on reverse tapers and re-entry curves apply to articles thermoformed from sheet but here there is a little less restriction since the combination of a fairly thin sheet and a resilient material will allow some degree of freedom.

In general, hollow components with apertures smaller in diameter than the body are made by blow moulding (either extrusion or injection blow moulding) although it is possible, in the case of objects having a circular cross-section at some point, for them to be injection moulded in two halves and then joined by spin welding (see Fig. 2). Such a method is sometimes used if particularly good control of wall thickness is necessary.

Other limitations may be imposed by the material being moulded. It is difficult to thermoform nylon sheet, for instance, because nylon materials usually have fairly sharp melting points, thus rendering temperature control very critical — too critical for most commercial operations. Other materials, such as PTFE, are far too viscous for normal forming methods, however strongly heated, and articles must be made by powder sintering processes.

There are also, of course, forming methods designed for thermosets and others designed for thermoplastics, although the barriers between them are now becoming rather blurred. For example, compression moulding is sometimes

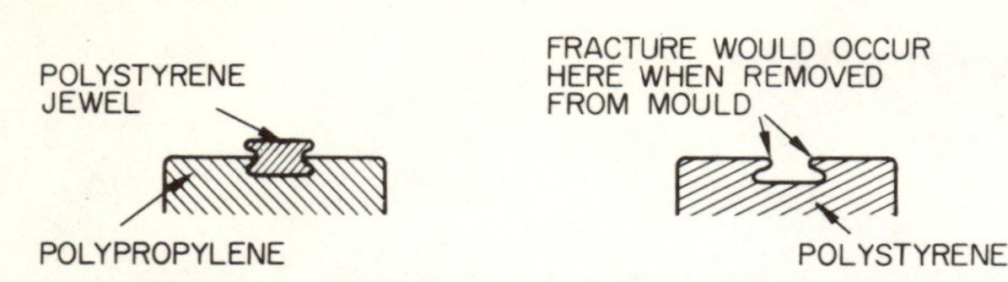

Fig. 1. Undercuts are typical limiting factors; they can be successfully moulded in polypropylene but not polystyrene.

Fig. 2. A component such as that shown could be injection moulded in two halves and then joined by spin welding, as opposed to producing the complete component by blow moulding.

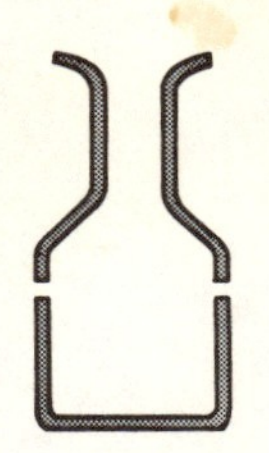

used for moulding laboratory test pieces in thermoplastics. However, the main commercial process utilising compression moulding of thermoplastics is the manufacture of vinyl records, where the fine detail in the micro-grooves militates against the use of injection moulding by virtue of the inability of the plastics to flow in such confined spaces. Conversely, both injection moulding and extrusion can be carried out using thermosets, the advantage being speed in the case of injection moulding, and the use of a continuous process in the case of extrusion.

MAJOR PROCESSES

With the foregoing in mind it is possible to look at the economics of the main forming processes, namely, compression moulding, transfer moulding, injection moulding, extrusion, blow moulding, thermoforming, and rotational moulding. Brief mention will also be made of the economic factors affecting solid-phase forming and the so-called structural foam processes.

Perhaps the most illuminating way of looking at forming economics is to first consider the various processes in isolation, in order to bring out the general principles, and then to look at a few items which can be made equally well by two or three methods, and consider the relative economics.

Compression moulding and transfer moulding

Compression moulding is carried out using high pressure hydraulic presses so that the capital cost of equipment is high. Because the moulds also have to withstand high pressures, and have to be fitted for heating, tooling costs are also fairly high. Cycle times tend to be rather long because the high temperature has to be maintained after the molten polymer has finished flowing, in order to complete the cross-linking reaction, or 'cure'. On the other hand the finished article can be removed from the mould while still quite hot because hardening takes place at the same time as 'curing'. The heating cycle can be reduced, and hence made more economic, by pre-heating the moulding powder. In many cases the powder is pre-compressed to form pellets or preforms. These are then placed at strategic positions in the mould to give better flow, especially in awkwardly shaped mouldings, and their use also makes pre-heating an easier process.

Transfer moulding involves heating the powder in a separate chamber and usually gives faster cycles since the material is already fluid when it enters the mould cavity (or cavities). In addition a certain amount of local heating is generated by frictional heat as the molten material is injected through a narrow gate into the mould.

One factor which affects the cost of both compression and transfer moulded parts is the fact that a certain amount of material may 'flash' between the two parts of the mould while the molten material is under pressure. The amount of 'flash' depends on the mould design but if it is appreciable it must be removed. In the case of small objects a simple tumbling process is usually sufficient but larger parts may have to be treated on a linisher or similar device.

Minimum production runs vary according to the size of article being moulded, but, as an example, half a million would probably be the minimum run for a screw cap. For a multi-impression tool for such a moulding (say one producing 40–50 caps per cycle) the mould cost would be of the order of £1 000–1 500.

Injection moulding

Injection moulding is still essentially a process for moulding thermoplastics, but is becoming increasingly popular for the moulding of thermosets. Equipment is generally more expensive than for a compression press of comparable size (about twice as much for the smaller machines and five to six times as much for the larger sizes). However, there are savings inasmuch as pelleting machines and pre-heaters are not required.

Cycle times for injection moulding of thermosets can often be reduced dramatically compared with compression moulding, sometimes to as much as a sixth. Times for transfer moulding would probably be somewhere in between.

Mould costs, too, are greater for injection moulding than for compression moulding, by a factor of about 4. For large and intricate items such as a dairy crate, the tooling cost is in the region of £10 000. With tooling costs of this order, it is important to have long production runs. The high cost of moulds is dictated by the fact that they have to withstand high pressures. These high pressures are necessary to cause the very viscous liquids to flow. In this connection it should be noted that new processes have been evolved for the injection moulding of foamed plastics, particularly the polyolefins, toughened polystyrene and ABS. The plastics are mixed with a chemical blowing agent before feeding to the injection moulding machine. In the machine the blowing agent decomposes with evolution of a gas (usually nitrogen) but expansion cannot take place in the constrained conditions of the injection moulding machine. A metered amount is injected into the mould, and filling of the mould occurs by the foaming action of the plastics/gas mixture. Because the pressures developed in the mould are comparatively low it is possible to use low cost aluminium moulds at anything from 25% to 60% of the cost of a fully machined steel mould. Mould costs can, therefore, be amortised over shorter runs than in conventional injection moulding. Production costs, on the other hand, are generally higher because the cooling cycles are longer. Another factor affecting economics is the fact that the use of a foamed structure gives greater rigidity for the same material weight, or equivalent rigidity for less material.

Extrusion

This varies in one important particular from the processes previously described — it is a continuous process and not a batch or intermittent one. It is all the more important, then, to have the equipment working for long, uninterrupted periods. The products of extrusion are diverse and range from film and sheet, pipe, rod, profiles and cable or wire covering. Extruders are, again, expensive pieces of equipment and will also require some sort of take-off equipment as well as an appropriately shaped die.

Another important point to be noted about extrusion is that scaling up will usually give substantial saving in investment costs per unit output. Thus, an 80 mm extruder costing around £40 000 could extrude nearly double the tonnage of sheet of a 60 mm extruder costing around £30 000. These figures are very much 'rounded-off' ones but will give some idea of the economics involved in extrusion processes, such as for pipe or sheet.

Blow moulding

Extruders also form part of the equipment in one type of blow moulding, namely, extrusion blow moulding. Equipment costs are high but mould costs are lower than those for injection moulding due to the lower pressures involved. Economic runs will vary according to size and shape of the article produced, but they are likely to be above about 40 000. Notwithstanding all this, if a hollow article, with a narrow orifice, is required then blow moulding of some sort is the only answer.

The other method of blow moulding is injection blow moulding where a rough shape (or parison) is moulded in an injection moulding machine, around a blowing stick. The parison and blowing stick are removed from the mould while the material is still hot and then placed in the final mould where air is blown down the blowing stick to inflate the molten plastics to its final shape.

The tooling costs are obviously greater than for extrusion blow moulding since two moulds are required for each article produced. Cycle times, too, tend to be higher because of the transfer time for moving the parison from one mould to another.

Thermoforming of sheet

The main factor affecting the cost of sheet thermoforming is whether the sheet is extruded and then thermoformed in-line, or whether the sheet is cooled and stored in between extrusion and forming. If the latter is the case then an appreciably higher manufacturing cost is involved because of the re-heating and extra handling.

Another important factor is the large amount of scrap involved, particularly during the forming of parts having a circular cross-section. This material is not completely lost but an allowance must be made for regrinding and re-blending.

Tooling costs can be extremely low because moulds can be made from plaster of Paris or wood. However, for large multi-cavity moulds, as used in high speed moulding, aluminium is the most likely material to be used because of its good thermal contact. Equipment costs can be quite high because both extruders and forming equipment are necessary.

Rotational moulding

This is a process where a split, hollow mould is supplied either with a PVC plastisol or some other plastics in powdered form. The mould is then rotated on two axes at right angles to each other and heated so that the thermoplastic flows and covers the inner surface of the mould. Tooling costs are, therefore, low because no great pressures are involved. Mild steel sheeting or cast aluminium moulds are usually used. Equipment capital costs are also reasonably low. Raw material costs are higher than for some other processes because the powders used have normally to be obtained by grinding pellets, while plastisols also have to be specially compounded.

Low pressure (contact) laminating

The moulding of articles such as boat hulls, car bodies, *etc*, from glass fibre reinforced polyesters is essentially a low cost one from the point of view of capital investment, both in equipment and moulds. It is, however, labour intensive and is, therefore, more suited to the manufacture of items required in fairly small quantities.

Fabrication techniques

In this context, fabrication is taken to mean production of complicated shapes from sheet or by joining two or more moulded items. Hot gas welding is expensive in labour costs but utilises low cost equipment whereas ultrasonic welding, heat sealing and spin welding can be automated to a greater or lesser extent but need more expensive equipment. Hot gas welding is more likely to be used, therefore, for items required in small numbers such as ducting and hoods for fume cupboards, *etc*.

CHOICE OF FORMING METHOD

As discussed earlier, the choice of a particular forming method may be dictated by factors other than the purely economic. However, there are many occasions when two or more forming methods are technically feasible and items such as capital expenditure, production costs, tool costs and the length of run then become of paramount importance.

	Injection moulding	Blow moulding	Rotational casting	Vacuum forming
Capital cost	High	High/medium	Low	Low/medium
Mould cost	High	Medium	Low (except for large multi-moulds)	Low
Cycle time	Fast/very fast	Fast	Slow	Slow
Finishing required	None	Yes (in extrusion blow-moulding, little in injection blow-moulding)	Little	Yes (trimming)
Residual strain	Some	Little	Very little/none	Some/much
Wall thickness control	Excellent	Poor/good	Good	Poor
Processing pitfalls	Weld lines	Damage at pinch-off	Fusion/crystallinity	Thinning

Table 1. Comparison of processing techniques.

The question of length of run can sometimes be the determining factor, one example being when an item can be made by either injection moulding or thermoforming. Refrigerator liners, for instance, can quite well be made by injection moulding or by thermoforming. Some years ago they were, in fact, injection moulded but when thermoforming techniques were developed that would handle such large and complex items, with good control of wall thickness, injection moulding was abandoned. This was because the numbers-off were comparatively small and the very large tool costs involved in injection moulding made the liners very expensive. Even though the number of refrigerators manufactured has increased over the years the numbers involved in any one model are still small. The much lower costs of thermoforming moulds are also coupled with lower equipment costs. Liners for the larger sized refrigerators, particularly, present a very large projected area and large injection moulding machines with high clamp pressures would be necessary.

The question of injection moulding versus thermoforming of small objects, such as the pots used for yoghurt and cream, is rather more complicated. This is because there is a lower limit to the wall thickness achievable by injection moulding, dictated by the flow properties of the particular thermoplastics being moulded. For the type of toughened polystyrene used in the manufacture of yoghurt tubs the lower limit (for this size tub) is about 450 μm. If, therefore, the performance requirements of a tub are such that a thinner wall is satisfactory then the use of injection moulding produces an unnecessarily thick container. Thermoforming is not subject to such limitations so that quite thin containers can be made by this method and savings in material cost are thus possible. For very small numbers per annum of, say, a $5\frac{1}{2}$ oz yoghurt pot, the injection moulding process would be uneconomic by virtue of the cost of the moulds but at moderate productions of around five million per annum, the mould costs in the injection moulding process are negligible while at these speeds it becomes necessary to use more expensive thermoforming equipment. The use of thermoforming machines with large forming areas permits high production rates per cycle but even at rates of fifty million per annum it is cheaper to produce these containers by injection moulding than by thermoforming, providing the containers are of the same wall thickness. Where thinner containers are permissible then thermoforming becomes a more economic process because of reductions in raw material costs.

There are other economic factors which also affect the argument of injection moulding versus thermoforming. One of these is utilisation of machine production times. Of the two methods of forming, the injection moulding process has greater flexibility of use so that, if the container market were to fluctuate in demand, the production capacity of injection moulding machines could be more easily filled than could that of a sheet extrusion/thermoforming line.

The above remarks apply, whether one is considering single or multi-cavity injection moulding machines. In fact, in the cases considered, the use of multi-cavity moulds does not show an economic advantage until rates of just under fifty million per annum are reached.

The question of single cavity versus multi-cavity injection moulding is of general interest and is worth considering in this chapter. Let it be said straight away that there is no clear cut answer to the question of what is the optimum number of mould cavities, since it depends on the size and type of machine available, the complexity of the moulding, the cycle times likely to be achieved from each possible number of cavities, and the number of mouldings required. There has been a tendency to use single cavity machines in recent years because of the development of very fast cycling machines coupled with super high flow moulding materials. The advantages of using single cavity moulds are even more marked when considering thin wall containers such as the yoghurt pot because of the difficulty of accurately centering multi-cavity cores when the clearances in each cavity are so small.

Several factors need to be considered when looking at the number of cavities required under more general conditions. If a fairly accurate estimate can be made of cycle times and mould costs, and assumptions made as to the running cost of the machine without material, then a break-even point can be calculated which can then be compared with the total production required. An example will serve to show the general idea although it is emphasised that the problem is even more complicated than has been assumed here.

Let us take the costs for one, two and three cavity moulds for a certain moulding as £500, £750 and £1 000 respectively, and cycle times of ten, fifteen and twenty-two and a half seconds. The number of mouldings per hour will then be 360 (single-cavity), 480 (two cavity) and 480 (three cavity). There is thus no economic advantage to be gained, in this case, from using a three cavity mould. It will be noted that cycle times do not increase *pro rata* with the number of cavities. This is because the limiting factor in a single-cavity mould is the cooling time of the moulding, whereas with multi-cavity moulds the controlling factor tends to be the plasticising capacity of the machine.

Having obtained the number of mouldings per hour, we can now calculate the piece price of the moulding, if the cost of running the machine is known. This is a complicated factor, depending on many variables, some of which will be individual to the particular moulding shot, but we will assume here a cost of £1.80/hour. The piece price for mouldings produced on the single-cavity mould is, therefore, 0.5p, while that for the two-cavity mould is 0.375p. The per piece saving is, therefore, 0.125p. The extra mould cost of £250 has, therefore, to be justified by the numbers to be moulded, multiplied by the savings of 0.125p/moulding. That is:

$$X \times 0.125\text{p} = 250 \times 100\text{p}$$

(where X = number of mouldings)

$$\text{therefore } X = \frac{250 \times 100}{0.125}$$

$$= 200\,000.$$

This should be taken as the minimum number since there are other factors which increase the cost of multi-cavity moulding.

Large mouldings

The choice of forming method for large thermoplastics mouldings is often a matter of deciding between three separate processes, namely, blow moulding, injection moulding and rotational moulding. Capital costs normally favour rotational moulding, a typical comparative costing being approximately, £15 000, £45 000 and £40 000 for rotational moulding, injection moulding and blow moulding, respectively, to produce a 90 litre capacity container.

Mould costs vary but in general injection moulding would be the most expensive, with rotational and blow moulding of similar magnitude where cast aluminium rotational moulds are suitable. Fabricated steel moulds may have to be used for the larger items and here they would be somewhat more expensive than blow moulds, although still much cheaper than injection moulds. Labour costs, however, are high for rotational moulding. Other factors favouring rotational moulding over the other two methods are the ease of colour changing (with no lost time and material), the absence of scrap and reclaim (*eg* sprues and runners in injection moulding and 'tails' in blow moulding) and low maintenance costs. All these factors complicate the issue and each case must be evaluated separately. Nevertheless, it would appear from a study by Arthur D Little Inc presented to a USI Symposium on rotational moulding at Chicago in 1965 that some general conclusions can be drawn in certain well-defined cases. The study in question was based on the manufacture of polyethylene containers in a range of sizes. Conclusions were that the costs of producing 5 gallon containers were virtually the same by all three methods, the 20 gallon containers were slightly cheaper by injection moulding and 55 or 250 gallon containers were more economical by rotational moulding. The above is true for relatively long runs ranging from 10 million × 5 gallon containers to 100 000 × 250 gallon containers. At lower levels (10 000 × 5 gallon to 1 000 × 50 gallon) rotational moulding was the cheapest in every case.

Film

Turning now to continuous processes, such as film production, the economics of one process versus another depends very much on the length of run. Taking the two usual processes for making polyethylene film, namely, blow extrusion and slit-die extrusion with chill roll casting, the capital costs for a blow extrusion line are much lower than for casting, but throughputs are usually less. Where a variety of film widths and thicknesses are required it is more economic to run several blow extrusion lines than a smaller number of casting lines because individual tonnages may still be relatively low even though the total throughput is high. A casting line usually needs a large run on one type of film to make it truly economic. Production of a range of widths to meet a varied pattern of orders would increase capital costs still further by the necessity to purchase more dies and increase production costs by time lost in die changes. With tubular film, the width of film depends on bubble diameter which can be altered to some extent by altering the blow-up ratio. Dies are also cheaper and easier to change. The issue is complicated by the fact that casting produces a rather different quality of film, being usually of superior clarity.

A better comparison can, perhaps, be made by looking at cast polypropylene film and a water quench blown film such as that produced by the TQ (tubular quench) process. This involves the downward extrusion of a tubular melt followed by rapid cooling on water-covered converging plates or boards. The rapid cooling prevents the formation of large crystallites and so yields a high clarity film similar to that produced by casting. It has been estimated that the TQ process is more economical than cast polypropylene film at productions between 100–500 tons/annum whereas cast film becomes more economical at tonnages of about 500 tons/annum.

Solid phase forming

One item which is worth separate mention here is solid phase forming. This is the forming of solid plastics billets at temperatures below their softening point. The economic advantage lies in the fact that less heat energy is expended while mould costs are also less. Production speeds are difficult to assess and depend very much on the particular method used and the design of the finished part. Another indeterminate factor at this stage is the production cost of the plastics billets.

INFLUENCE OF MATERIALS ON ECONOMICS

The intrinsic price per kilogramme of a polymer is not always the best guide to its economics in use. What has to be looked at is the cost per unit in terms of the properties which are most desirable in the finished moulding. In the case of a packaging film or bottle, for instance, it may be the barrier properties towards gases or water vapour. It may seem elementary to point out that a polymer costing 10% less but which has to be used at twice the thickness in order to provide equivalent protection is no bargain; yet similar mistakes are often made in other fields.

Processability is another factor which can reverse the apparent economics of two different materials. This is more likely to apply to different grades of the same material since the material choice itself will probably have been made on the grounds of the properties needed. In injection moulding a high flow grade may prove cheaper overall provided that other properties are still satisfactory, while for thermoforming the grade with the greatest rigidity at elevated temperature might be more economic, since the length of time for the heated forming to cool sufficiently for it to support its own weight is the governing factor. The decreased overall cycle in both cases can lead to substantial savings in a long run.

One point which should not be overlooked in plastics forming generally is the question of the specific gravity of the material. The polymer granules or powder are bought by weight but the mouldings are made by volume so that a 10% difference in density means a 10% difference in the number of mouldings per kg.

The question of the economics of forming, then, is not at all straightforward but some of the general principles have been given which is hoped will act as a guide. A summary of some the relevant factors governing the choice of a particular method is given in Table 1.

Chapter 6

Compression Moulding

J E Rogers *AMIED*
Minerva Mouldings Ltd

The first commercial exploitation of a thermosetting material of the phenol formaldehyde group was made about the year 1908, by Dr. Leo Hendrik Baekeland (from which the trade name of 'Bakelite' is derived). Other developments soon followed, such as the introduction of urea formaldehyde in the early 1920's, by Pollak and Ripper, until a whole range of materials suitable for compression moulding which has now become available.

The largest percentage of all moulding tools for thermosetting materials, despite the use of transfer and the much more recent 'direct-screw-transfer' system of moulding, are based on the well tried and proven compression method. This is the oldest means of moulding and modern developments retain much of the old art.

Basically, the compression moulding system, which will be enlarged upon later, requires a male and female die which are heated to between 127° and 160°C (260° and 320°F). Moulding material is placed between the two halves of the tool, which are then closed together by hydraulic means. A pressure of 2 to 4 tonf/in^2 (300 to 600 kgf/cm^2) is applied and, under the action of both heat and pressure, the moulding material becomes plasticised, fills the mould to the shape of the tool cavity and becomes polymerised, or cured. Temperatures and moulding pressures depend upon the type of material being moulded and on the complexity and depth of the component.

To give the reader a working knowledge of compression moulding, it is now proposed to discuss the various aspects of moulding and its general requirements.

MOULDING MACHINERY

Whilst the machinery, or presses, that are required for both compression and transfer moulding are basically the same, the latter type has the addition of an integral transfer ram mechanism. It follows therefore that the purchase of a press specifically for moulding by the compression process will show a small cost advantage.

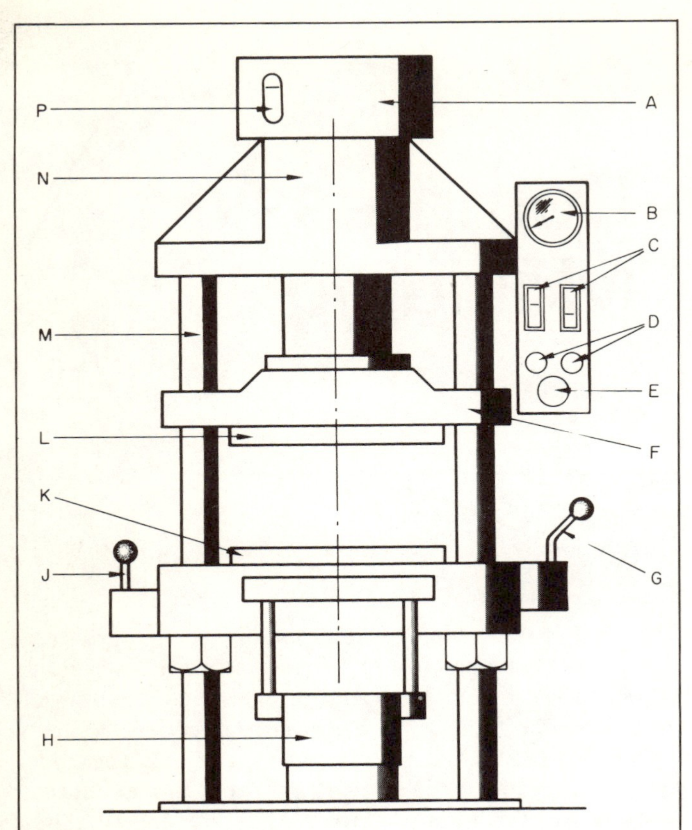

Fig. 1. A typical downstroking type of compression moulding press (guards not shown for clarity). A—hydraulic reservoir; B—main ram pressure indicator; C—top and bottom platen temperature indicators and controls; D—start and stop buttons; E—emergency stop button; F—moving head; G—main ram control lever; H—hydraulic ram for operating ejection mechanism; J—ejection control lever; K—heated bottom platen carried on a fixed head; L—heated top platen carried on a moving head; M—press pillars; N—main ram; P—oil level indicator.

Transfer moulding may actually be carried out on a compression press using a three-plate tool but this is not considered by the author as a wholly successful means of moulding, and should only be used for special applications.

The conventional press is hydraulically operated with an independent hydraulically controlled ejection system which lifts the mouldings clear of the tool, enabling the operator, or special take-off equipment, to remove the moulded components without difficulty.

Whilst the moulding press must be sturdily designed and constructed to perform the process with the mimimum of maintenance, it must also be accurate to align the tool. Four pillars are usually incorporated to give rigidity and to act as guides for the moving press head. It is not intended to represent any specific make but it may assist the reader to consult Fig. 1 for a simplified build up of a typical compression press. Three main types of press are available and these will now be generally described.

The most widely used type is probably the **downstroking** press (Fig. 1) where the top platen, making use of an overhead hydraulic cylinder, uses the downward stroke of the ram to move the platen and thus exert moulding pressure onto the stationary lower section, or bottom force, of the tool.

The **upstroking** press (Fig. 2) is similar in operation except that the hydraulic cylinder is mounted beneath the bottom platen and moves its platen towards the fixed top. Whilst little trouble is experienced with modern presses, the upstroking press is considered to be safer in operation as, in the event of hydraulic trouble, the bottom ram and its platen fall into the press—open stage.

For more specialised work the **side-ram** press (Fig. 3) is used. This normally has a downstroking top platen with additional hydraulically operated platens which move across the bottom platen in a sideways direction. This allows for coring moulding detail in vertical and horizontal directions.

Automatic presses (Fig. 4) are available but, whilst these produce mouldings at a rapid rate and save labour costs, one of the main disadvantages is that the loading of inserts is extremely difficult, if not impossible. Once set-up, these machines should run uninterrupted for several weeks, day and night, to be an economic proposition to the moulder. Production is mainly restricted to fairly simple shapes, such as covers and bases, which do not have complicated or fragile tooling. Whilst a wary eye is usually kept on automatic machines to ensure consistently sound components, one operator is capable of supervising several machines, filling their feed hoppers with moulding material and removing the resultant mouldings. The process of hopper filling and component removal may be made automatic if the quantities are sufficient, but here one must

Fig. 2. A bank of six 20-ton upstroking presses, giving economy of cost and manpower, using a common hydraulic supply. (Courtesy of Daniels Hamilton Ltd.)

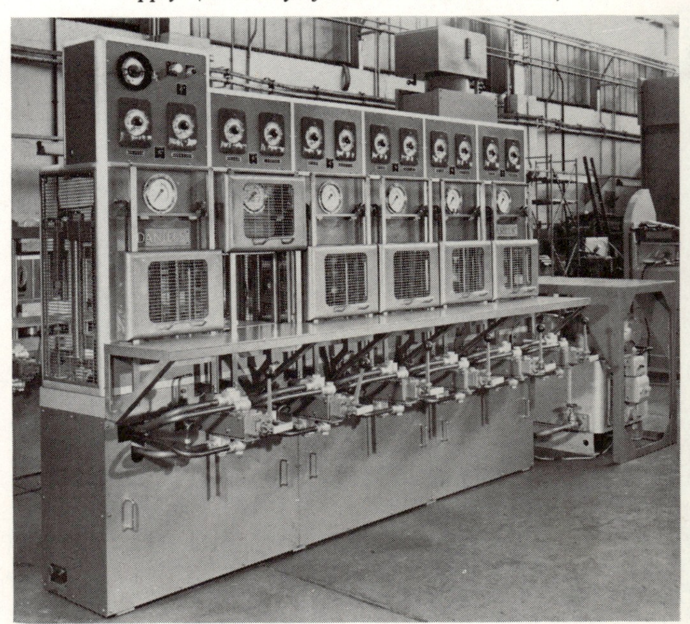

Fig. 3. 100-ton side-ram compression press, which though operator controlled, has automatic process control equipment. (Courtesy of Bradley & Turton Ltd.)

Fig. 4. 150-ton automatic compression press incorporating novel features including easy tool accessibility, quick change-over of top and bottom ejection. (Courtesy of British Industrial Plastics Ltd.)

Fig. 5. 75-ton press using a sliding table and a hinged top tool for easy access. A simple two impression flash-type tool is shown. (Courtesy of Daniels Hamilton Ltd.)

consider running this system endlessly, over perhaps several years, to recoup the high installation cost involved.

Versions and adaptations of the above press types may be seen, such as the sliding table or 'shuttle' press which uses two tools and allows one tool to be loaded, for instance, with inserts, whilst the other tool is moulding (Fig. 5). Rotary presses may be used for moulding high production quantities but require multiple tools and are not in great demand for thermosetting plastics materials.

COMPONENT DESIGN CONSIDERATIONS

Whilst the major component design points will be explained here, it is in the designer's interests that he should discuss moulded parts with the moulder, or his own production department, at the early stages of design. This may often eliminate design faults and produce a moulding tool and components far more economically.

a. The cheapest type or grade of moulding material should be used, consistent with its use.

b. Walls of all mouldings should be of uniform thickness wherever possible as the cure time of moulding, and hence the cycle time, is dependent upon this. Cure time may be cut by the reduction of heavy thickness sections as by the introduction of a hole or form. Section thicknesses will be

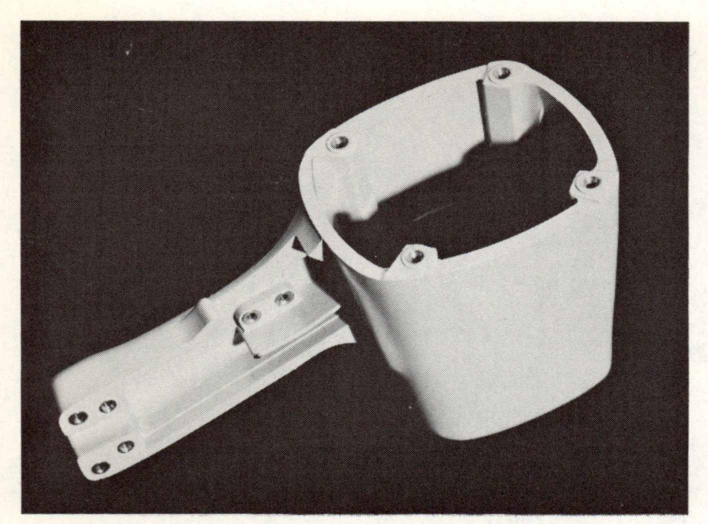

Fig. 6. A motor-frame for the Wolf 'Sapphire' drill incorporating both moulded-in and pushed-in inserts. This is moulded on a side ram press similar to the one shown in Fig. 3. (Courtesy of Wolf Electric Tools Ltd.)

governed by the size of the component and strength considerations, but walls of 3 mm ($\frac{1}{8}$ in) for small items such as control knobs up to 6 mm ($\frac{1}{4}$ in) for switch bases of up to 250 mm^2 (10 in^2) are to be preferred.

c. For cost considerations it is recommended that the natural colour of the material should be used. Normally this is black, but varying grades and types have natural moulded colours of beige, off-white and even green flecked with brown. Special colours and shades may be available but often they are only sold by the materials manufacturers in lots of one metric tonne mimimum. Annual moulding orders should, at least, cover this quantity.

d. For fixing purposes, it is often necessary to incorporate brass or steel inserts to meet electrical requirements for items, such as contact fixings, on switchgear parts (Fig. 6). Wherever possible these should be of the 'blind-insert' type, *ie* not drilled and tapped throughout their length. Through-inserts which, as the name implies, are drilled and often tapped completely through their length, are easier to produce but cause problems to the moulder due to flash creepage into the hole. Re-tapping moves are often necessary after moulding to remove this flash and may result in an excessively loose or damaged thread form. Typical dimensions of blind inserts are tabulated in Fig. 7 as a general guide. It is quite practicable to push-fit inserts into mouldings after the moulding process and a range of specialised types is marketed. A component cost reduction is usual but it is even cheaper not to use inserts of any type.

e. Due to variations, such as inconsistent temperatures and cure times which may well result during moulding, all tolerances should be as wide as possible. Close limits, hole diameters for instance, may be controlled by the aid of shrink rigs but overall lengths and widths may be affected by shrinkage. All materials shrink during the moulding process and the greater the shrinkage rate the more tolerance should theoretically be allowed. Table 1 gives the expected shrinkage of various compression moulding materials, but differences may be found from one material manufacturer to another.

f. Undercuts and side cored holes, being those which are not readily withdrawn from the tool, do complicate tool manufacture and may involve extra finishing moves to the mouldings. If at all possible they are to be avoided.

PROCESSING OF MATERIALS PRIOR TO MOULDING

Raw moulding materials, which are purchased in the form of powder or granules, for example, are loaded into the moulding tool by several methods. The best suited method is the one which reduces the press cycle time and ensures

Table 1. Expected shrinkage of various compression moulding materials.

Class	Type	Moulding shrinkage per Unit Length
Phenolics	General purpose	0.005 to 0.008
	Medium shock resistant	0.004 to 0.007
	High shock resistant	0.002 to 0.004
	Electrical low loss (Mica filled)	0.001 to 0.003
	" " " (Nylon filled)	0.006 to 0.010
	Heat resistant	0.001 to 0.002
	Shock & chemical resistant	0.006 to 0.009
Alkyds	Diallyl Phthalate	0.002 to 0.006
	Diallyl Isophthalate	0.002 to 0.004
	Polyester Dough Moulding Compound	NIL to 0.004
Aminos	Urea-Formaldehyde (Cellulose filled)	0.004 to 0.008
	" " (Wood flour filled)	0.004 to 0.008

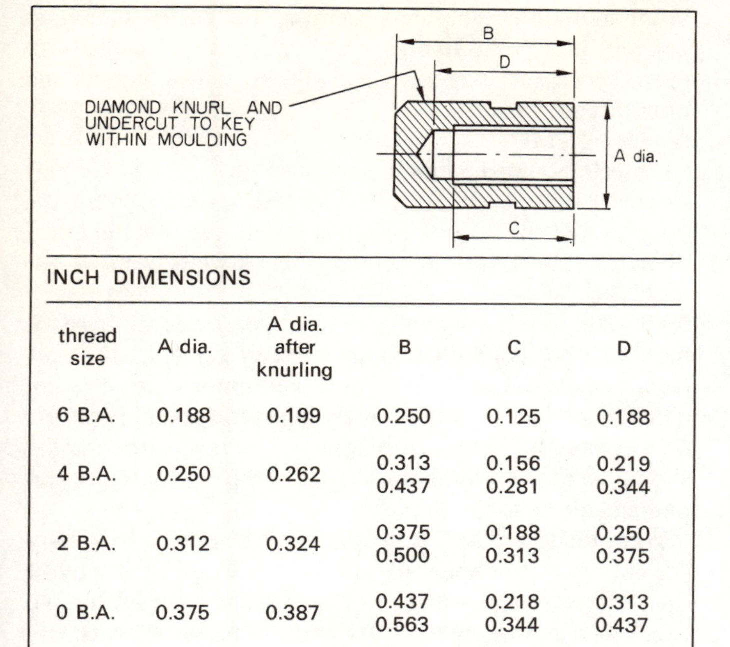

thread size	A dia.	A dia. after knurling	B	C	D
INCH DIMENSIONS					
6 B.A.	0.188	0.199	0.250	0.125	0.188
4 B.A.	0.250	0.262	0.313 0.437	0.156 0.281	0.219 0.344
2 B.A.	0.312	0.324	0.375 0.500	0.188 0.313	0.250 0.375
0 B.A.	0.375	0.387	0.437 0.563	0.218 0.344	0.313 0.437
1/4	0.375	0.387	0.500 0.625	0.250 0.375	0.375 0.500
5/16	0.437	0.449	0.563 0.688	0.313 0.437	0.437 0.563
METRIC DIMENSIONS					
M3	5	5.3	7	4	5
M4	7	7.3	9 12	5 8	6.5 9.5
M5	8	8.3	9.5 13	5 8.5	6.5 10
M6	10	10.3	11 14.5	5.5 9	8 11.5

Fig. 7. Suggested design and dimensions of a range of blind inserts.

consistently good moulding quality. Those most commonly used are as follows:

a. Loose powder which is usually weighed on scales to prevent excess material being loaded into the tool. Automatic compression presses use powder which is hopper fed to the tool by means of a sliding feed plate.

b. Preforms, pellets or 'pills' which are made to a determined weight and size on a special pelleting machine prior to moulding. These are easily handled and, if some particular items are found difficult to mould due to porosity, *etc*, a preform is easily broken and, by a process of trial and error, it can usually be placed in a position to correct this fault.

Production is greatly assisted by pelleting as any occluded air is removed from the moulding material, a consistent weight of charge is loaded, fear of contamination is reduced and preheating is more uniform. Wherever possible, pelleting machines are screened from other equipment and their feed hoppers covered to prevent the ingress of 'foreign bodies', such as materials of other colours and grades.

c. Loading cups or loading trays (see Fig. 8) enable the press operator to rapidly charge the moulding tool, especially when a loading tray is used to fill multi-cavity tools as might be used for pressbuttons, small covers and bases.

d. Whilst normal only to the dough moulding compounds (usually known at DMC) the material is weighed as a charge or cut-off to a specific length from a rope form.

e. A polyester mat is available for moulding but this necessitates a cutting operation with a knife on, or by, the press. Attempts have been made to guillotine to a rough shape but the shelf-life, that is the life from the opening of a roll in this particular application, is only a few hours, so that timing from cutting to use is critical.

MOULDING PROCEDURE

The press cycle, or the time taken to perform a series of operations until they are again repeated, varies according to the material being moulded and the requirements, such as the need to load loose cores and metal inserts into the tool. A brief description of the necessary operations on a semi-automatic press (Fig. 9) will now be given to acquaint the reader with the normal sequence of operations for a typical moulding on a compression press.

(*a*) Load inserts, or die cores for forming side holes and forms for instance, where these are required.
(*b*) Load moulding material into the die cavities in powder,

Fig. 8. A typical loading tray in this instance for charging a small six-impression tool with powder.

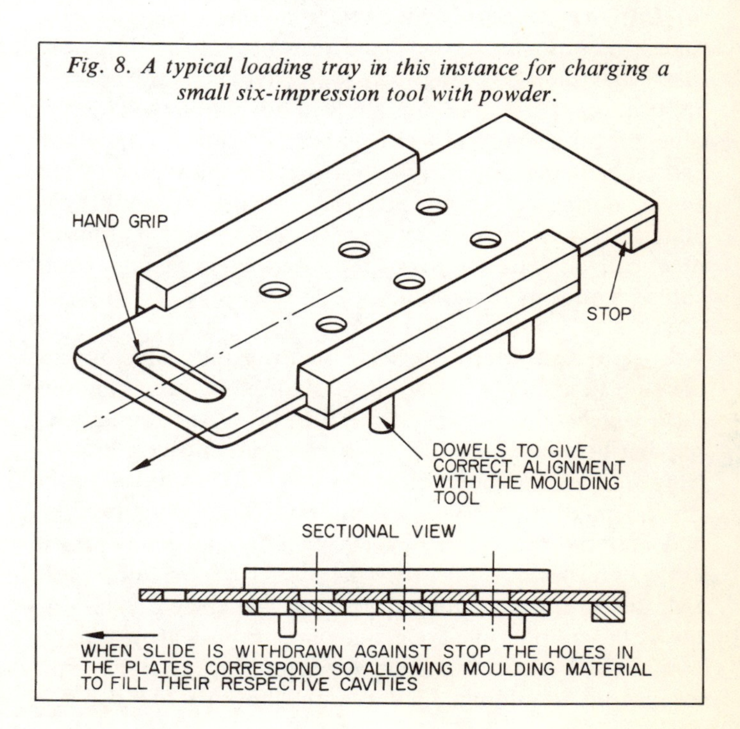

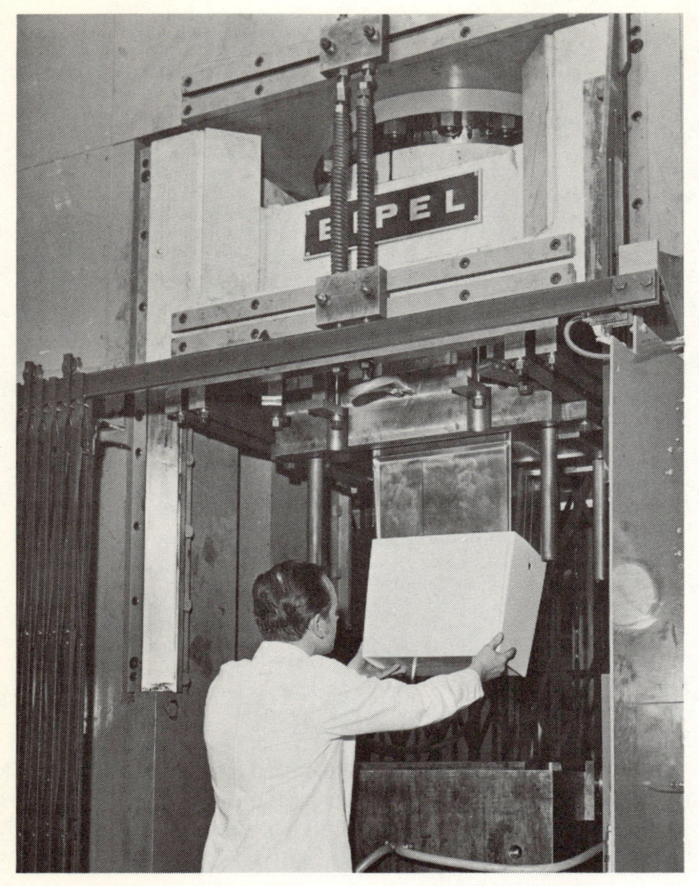

Fig. 9. 800-ton semi-automatic press showing a moulding for the Singer Carrying Case being removed by the press operator. (Courtesy of British Industrial Plastics Ltd.)

pellet or preform types. A combination of these forms may also be used to enable difficult shapes to be moulded with the minimum of trouble.

(*c*) Close press.

(*d*) Curing time, during which the moulding material flows under the influence of heat and pressure into an infusible mass, completely filling the tool cavities. Any further application of heat and pressure will not soften the component due to the resin, which is part of the raw material make-up, being polymerised during the cure. It may be worth noting at this point that, whilst thermoplastic materials may be theoretically softened and hardened practically indefinitely, all thermosetting materials, once cured, cannot be re-moulded.

Any spare time which the press operator may have during the cure time can be advantageously used by removing the flash which remains on the previous lift of mouldings, deburring holes and a large range of other finishing moves. Cure time varies between, roughly, $\frac{1}{2}$ and 5 minutes but is dependent upon the component wall thicknesses, weight and material type. The best curing time and temperature are generally determined by trial and error, but by experience the moulder is able to get near the correct values at the first attempt. Time cycles and temperatures are normally commenced on the high side as overcuring can only result in a scrap moulding. Undercuring, however, may result in the moulding adhering within the tool cavity. Once this happens it is often necessary to strip the tool on the tool-room bench and carefully clean. This can be a tedious and often needless task.

(*e*) Open press.

(*f*) Eject moulded items from their cavities. This is performed either automatically, semi-automatically by the operation of the press opening or, in the case of hand tools, often by taking the tool completely apart to remove the moulding.

(*g*) Clean tool by blowing out any particles of flash or moulding powder with a compressed air gun (Fig. 10). Any flash, or excess material, must be removed to prevent damage to the tool butting or sliding surfaces and to ensure a consistently sound moulding. Nozzles, incidentally, should be made of brass or a fairly soft material to prevent damage to the tool.

(*h*) Lubricate the tool if necessary. When walls of a component are virtually parallel to the line of draw, it may be that lubrication, possibly of a zinc stearate base, is required to act as a release agent. Care and discretion must be employed, however, to prevent contamination of the following lift of mouldings.

Whilst the above is provided as a general guide only, it is hoped that it will give the reader a wider understanding of the 'mechanics' of compression moulding. Transfer moulding is basically the same but the moulding material is loaded into a transfer sleeve or 'pot'.

PREHEATING

The practice of preheating moulding materials offers the thermoset moulder many advantages. Firstly, the cure time is reduced due to the part heating of the moulding material before it is loaded into the tool. An amount of moisture is also removed, which tends to give a uniform moulding without the undue flow marks, ripple and

Fig. 10. Close-up of a 26 × 26 in area moulding tool making twelve Lucas Headlamp Surrounds per cycle. The press operator is seen blowing excess flash from the tool with an air gun before loading with powder. (Courtesy of Daniels Hamilton Ltd.)

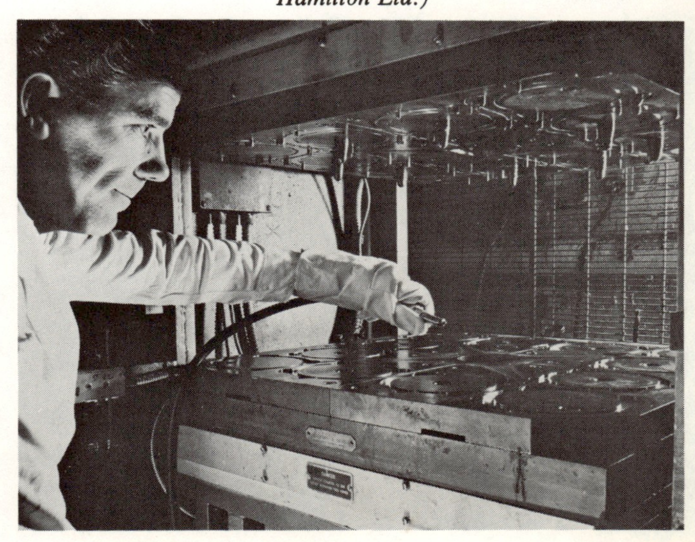

Fig. 11. A typical $2\frac{1}{2}$ kW preheater with the oven lid open showing pellets in position on the bottom electrode. (Courtesy of Radyne Ltd.)

'orange-peel' effects which are often caused by occluded air being trapped in the moulded item.

Modern preheating systems (Fig. 11) are of the electrical high-frequency types where the pellets, or loose powder contained in a cardboard box for instance, are heated between electrodes to approximately 110°C (230°F). Temperature varies according to the material being moulded, however.

MOULDING FAULTS

Thermosetting materials cover a large class of plastics and only very general faults which may be found in the moulding process, covering both compression and transfer systems, can be described here. It is to be hoped that the user does not find these faults as mouldings possessing them should be rejected on the press shop floor, or by inspection staff before packaging and despatch. Various materials in the thermoset range give differing results but the following summary may be regarded as typical.

(1) *Blistering,* shown as large swollen areas on a moulding surface may usually be traced to an insufficient cure or a low tool temperature. Blistering with the effect of small swollen areas, descriptively resembling small pimples or a rash, may very well be attributed to a tool running at an elevated temperature.

(2) *Short mouldings* having the appearance of minor indentations or poor surface gloss, or even more severely by parts of the moulding being not fully formed, especially in rib sections, is caused by a variety of reasons, such as the following:

a. Gas or air trapped within the tool cavity. The former is given-off from the moulding powder during the moulding process, whilst trapped air could very well be caused by rapid closure of the tool.
b. Insufficient charge of moulding material.
c. Incorrect material flow. 'Flow' in this context describes the stiffness of the moulding material grade and may be varied by the materials manufacturer to suit differing applications.
d. Tool not fully closing, which allows moulding material to freely escape causing a shortage in the tool cavity. Other reasons which prevent closing include the powder lands not being clean, or loose tool cores not being correctly located. It has even been known for guide bushes, especially when these are positioned in the female die or bottom force of the tool, to become partly blocked and thus prevent full closure.

(3) *Cracking of the moulding upon ejection* is usually to be blamed on incorrect tool design which allows uneven ejection, or for the moulding tending to hold onto the opposite side to the tool ejection. Inadequate cure and a rapid ejection rate may give the same results.

(4) *Poor gloss finish* on one or more faces of the moulded part indicates uneven temperatures of the tool, an insufficient cure time, or by attempting to mould with low moulding pressures.

(5) *Internal voids* are not visible to the eye but can be seen upon cutting through the moulding. Causes are usually insufficient cure or gas being trapped without adequate means of escape.

(6) *Scorched surfaces* identified by discolouration or a rippled moulding surface are caused by over preheating the moulding charge or running the tool at above normal temperatures.

(7) *Oversize mouldings* could be from tooling errors so that drawing tolerances are not met, but in dealing with moulding faults these are likely to be caused by thick flash at the parting lines due to the following:

a. Too great a moulding charge being loaded with insufficient tool clearance to allow the excess to escape.
b. Moulding material partly curing before the tool is fully closed, thus preventing complete closure. Further reasons which prevent the tool closing have been previously mentioned in the sub-heading regarding 'short mouldings' (see *2d*).

(8) *Soft mouldings* which are prone to distortion and often have a dull finish can normally be associated with an insufficient cure or low tool temperatures.

To the above list, which is not complete, it may be worth adding that swarf from metal inserts may contaminate moulded items and be a potential hazard on electrical equipment. All inserts should, therefore, be degreased, rotary tumbled to remove swarf and carefully handled at all stages.

TYPES OF MOULDING TOOLS

The various tooling types may be classified into three general designs, but before describing these it may assist the reader to describe the working operation of a typical tool in the moulding press.

The male and female dies, often termed the top and bottom forces, are bolted directly onto the heated press platens. Tool clamping is often used for small tools, but even these are considered by the author to be fraught with danger as damage can easily result to the tool, or the operator. if the clamps should be accidentally loosened. Unlikely though it may be, the unlikely does happen.

Movement of the platens is normally by hydraulic cylinders which may be operated from a self-contained or closed circuit system built into the press or, alternatively, by means of a hydraulic accumulator which may supply pressure to a complete line and range of presses. Hand operated presses are seldom used now and restricted to the small available pressures. They are, therefore, disregarded from this description.

The required pressure applied by the hydraulics system to the moulding area will vary considerably according to the general shape, depth and moulding material. For phenolic types of materials this may be considered to be within 2 to 4 tonf/in^2 (300 to 600 kgf/cm^2). Materials of the DMC classes will mould on much less tonnage however, and may very well mould within the 0·5 to 1 tonf/in^2 (75 to 150 kgf/cm^2) range.

The moulding operation requires tool temperatures of roughly 127° to 160°C (260° to 320°F) which is normally automatically controlled. Adjustment up or down the temperature scale is by a simple hand setting.

Moulding tools are classified according to their special construction and various types may use any, or a combination of more than one classification, to suit a specific design or production requirement.

Flash mould

Whilst this type is not in general use today it has the advantage of a low tooling cost. The tool cavity is easily accessible for cleaning and the loading of powder and inserts. The mould is loaded with an excess weight of moulding material which, when pressure is applied, will squeeze out of the cavity over the land area. This excess of material, as needed on a flash type mould, may well account for an extra 10% to 25% over other types. This extra material obviously increases the component price. As shown in

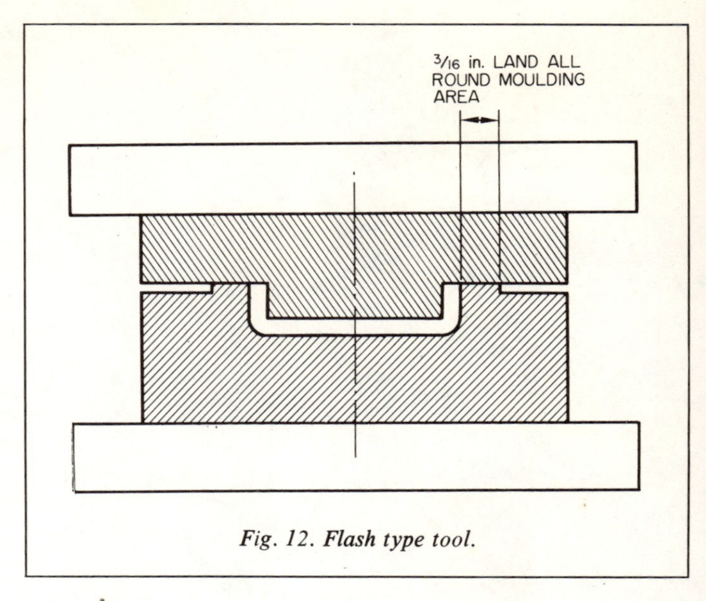

Fig. 12. Flash type tool.

Fig. 12, the land area is kept to about 5 mm (0·19 in) and it is only this area which opposes the flow of moulding material from the cavity. Due to this flow, which cannot be successfully limited, it is not possible to obtain a moulding of high density. Materials of more than a low bulk factor cannot be used due to the limited depth which is available to load the moulding material.

As no powder cavity is present, reliance for the alignment of the tool is on the guide pins. If accurate, uniform wall thicknesses are required, it must be stressed that this type of tool is not entirely suitable. If powder or pellets are not evenly distributed in the cavity, extra stress may cause a bending moment on the guide pins and bent pins result, together with an out-of-line or damaged tool.

Flash, being the name given to the excess material which escapes during the moulding process, is horizontal and in a properly designed tool using correct moulding pressures will amount to about 0.13 mm (0.005 in) when moulding general-purpose grades. Coarser grades will give a corresponding increase in flash thickness. Typical mouldings produced from this type of tool include ashtrays, plates, beakers and generally simple shapes (see Fig. 5).

Fig. 13. Semipositive type tool.

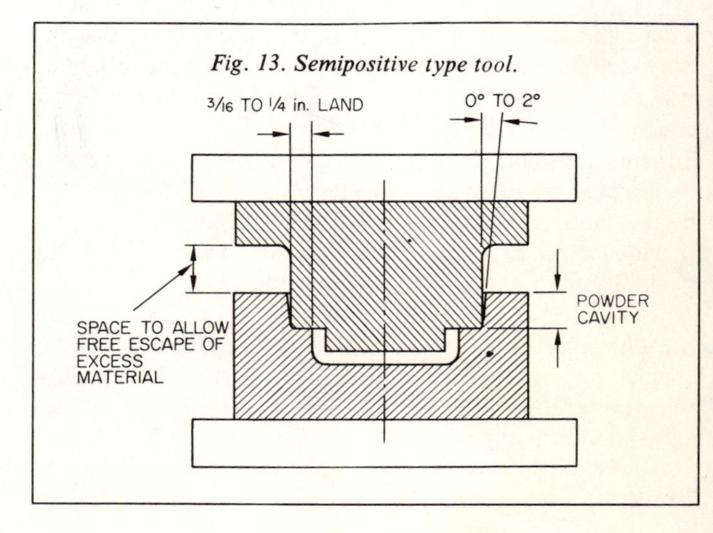

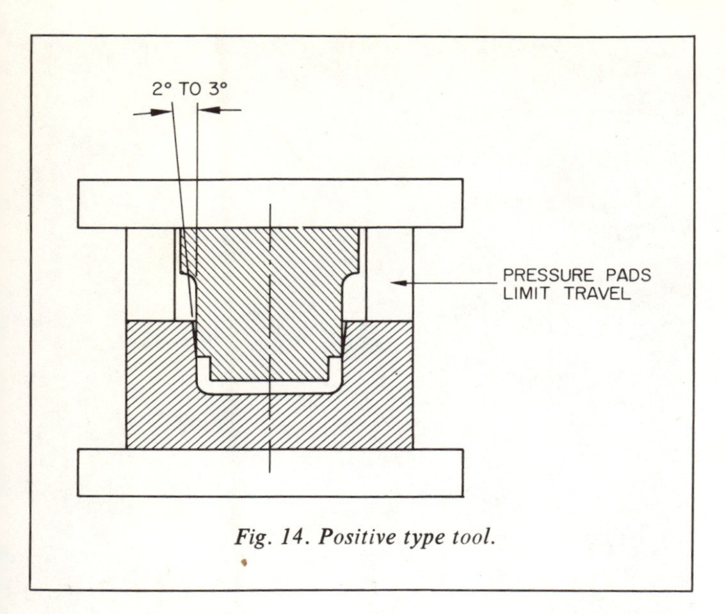

Fig. 14. Positive type tool.

Semipositive mould

As shown in Fig. 13, this most widely used type is similar to the flash mould but with the addition of a powder cavity. The moulding area is bounded by a horizontal land of about 6 mm (0.25 in) which controls the flow of the moulding material and gives a dense component when moulded under properly controlled conditions. An angled clearance is normally provided to allow excess flash to escape easily or, alternatively, vertical grooves of about 0.25 mm (0.01 in) deep may be provided. A further space is allowed between the faces of the male and female dies to allow any flash escaping up the angled or grooved powder land to escape without hindrance. Any overloading with moulding material cannot cause damage as the excess is now free to flow up the walls of the powder cavity without undue restriction.

Materials of a high bulk factor, such as the fabric-filled type, which have a factor of approximately 8:1 are easily loaded into the powder cavity. A suitable powder cavity depth calculation, and one which has been frequently used by the author, is based on the following formula:

$$D = \frac{A - B}{C}$$

Where 'D' is the depth of powder cavity from the top of the moulding face or powder land in inches. 'A' is the total volume of moulding material required in cubic inches. 'B' is the volume of the complete moulding in cubic inches. 'C' is the horizontal area of the moulding plus the horizontal land area in square inches.

The powder land provides a good location between the male and female dies and ensures alignment under all moulding conditions.

The horizontal flash which is produced by the slight opening of the tool when the full press pressure is applied, depends on the grade of material being moulded and also on the number of cavities in the tool. For single cavity tools moulding a general-purpose, wood filled, phenolic material this flash may be reckoned as 0.10 mm (0.004 in) increasing to approximately 0.63 mm (0.025 in) where a multi-impression tool using fabric-filled material is moulded. This flash allowance would be allowed for by the moulding tool designer and need not concern the buyer or component designer to any great extent, except to remember that shrinkages of moulding materials are liable to vary within reasonable limits from one batch of material to another.

Positive mould

This type is not recommended for use with the free-flowing materials which the semipositive type moulds with advantage. They are extremely useful, however, for moulding the dough moulding compounds and the cloth and fabric-filled varieties.

As may be seen from Fig. 14, the tool operation is similar to a plunger working in a piston. Little clearance is allowed between the male die, or plunger, and the female die. This clearance may be reckoned as being about 0.08 mm (0.003 in) all around but is dependent on the area of the component; large mouldings requiring additional clearance.

A vertical flash line is produced which in some circumstances is easily removed by a simple linishing operation.

Fig. 15. A semi-positive tool in production for the base moulding of the 835 Goblin 'Teasmade' at the Streetly Manufacturing Company. (Courtesy of British Industrial Plastics Ltd.)

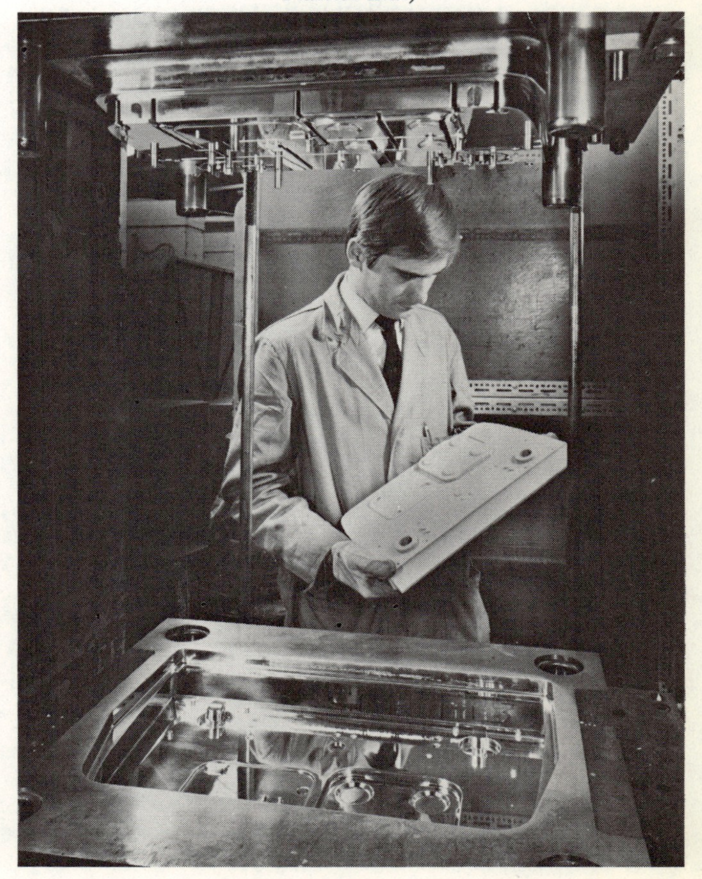

Fig. 16. Mechanical interlock case for a small contactor moulded in a single impression semipositive tool. (Courtesy of Brookhirst Igranic Ltd.)

This finishing move will ensure flat moulding surfaces but, for aesthetic reasons, it should be kept to surfaces not normally visible.

Depth control of the male die into the female die is achieved by the use of hardened steel blocks, usually termed pressure pads, or by having integrally machined areas on the dies. These act as stops to limit die travel and thus control the moulding thickness. Semipositive tools producing large area mouldings often use similar pressure pads but this is to prevent full moulding pressure from collapsing the powder lands.

MOULDING TOOLS

Tools, especially the cavity areas which form the moulded products, have to withstand high working temperatures without distortion, resist the abrasive characteristics of the moulding material, general abuse when handled and have an expected life of probably several million mouldings. For strength, toughness and a good hardwearing surface, the tool cavities should be made from a good grade of tool steel. A typical analysis of this steel would be as follows:

Carbon	0.30%
Manganese	0.50%
Nickel	4.25%
Chromium	1.25%
Molybdenum	0.30%

When correctly hardened at a typical temperature of 800°C and then tempered at 200°C, hardness of Rockwell C 52-58 would be expected. Where extra skin hardness is required,

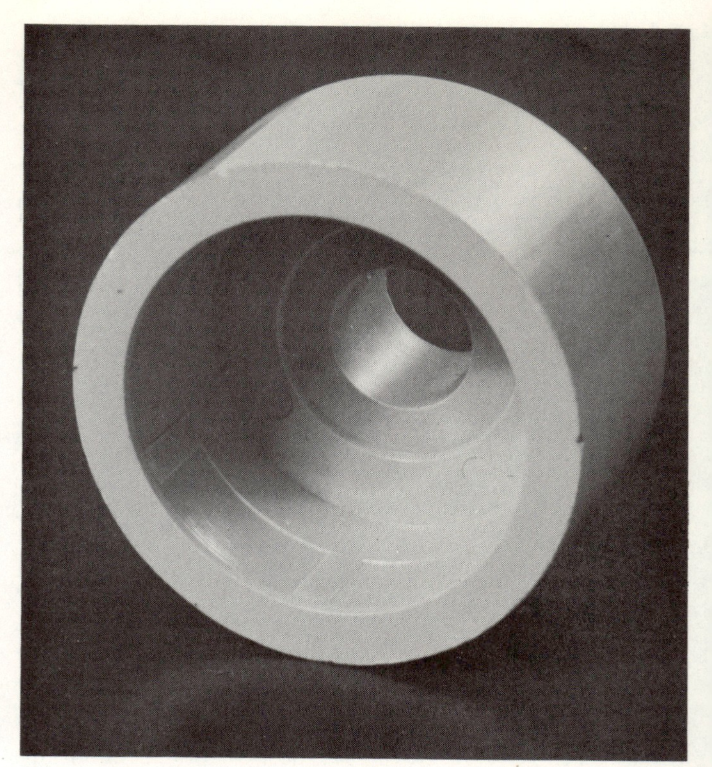

Fig. 17. Insulator moulding in Dough Moulding Compound, moulded in a multi-impression positive tool. (Courtesy of Brookhirst Igranic Ltd.)

as on sliding surfaces of a tool, it is possible to case-harden but any additional grinding, which may be carried out on the final assembly of the tool, will easily remove the surface and expose the softer areas beneath. These are very prone to wear.

Prototype tools, where small quantities of perhaps only 50 mouldings are required for test and evaluation purposes, may be constructed from mild steel or even cast, for instance, in brass. Tool life is limited however and there is no answer other than to order, or manufacture, a production tool which is made from the correct grade of tool steel and hardened. Anything else the author considers as false economy.

It is normal toolroom practice to highly polish the tool cavities and once samples are accepted by the customer, to chromium plate all cavity and powder land surfaces of a tool. Chromium plating, which normally has a thickness of 0.005 mm (0.0002 in) is recommended for the following reasons:

a. A high gloss is imparted from the moulding tool to the moulded component.
b. Ejection and extraction of the moulding are facilitated, especially when moulding deep walled parts.
c. Core pins and thread formers, which are subject to considerable wear, can be replated as soon as any appreciable wear is noticed compared to their original size.

Probably every moulder and toolmaker has his own ideas on construction and build-up of a moulding tool, but the general parts are indicated in Fig. 18. Details will differ from one tool to another according to type, number of cavities and the complexity of the component.

Guide pins, bushes and ejector pins are usually made from standard bought-out items, prove cheaper than making one's own and are easy to rework should they prove too long. Standard die sets of varying designs may also be purchased where only the cavity, together with parts of the ejector system, need be made.

COOLING FIXTURES AND SHRINK BLOCKS

All plastics moulding materials of both the thermosetting and thermoplastic types have a shrinkage. Whilst the former types are generally much more stable it is usual to control warpage and maintain reasonably close tolerances by the means of cooling fixtures of various designs. When a component is removed from a compression moulding press it is still hot and usually handled with gloved hands. Whilst in this state it is often necessary to stabilise the moulding shape until it has fully cooled. Fixtures are usually made from mild steel although laminated plastics, or a combination of both, are used. These are often equipped with quick-release toggle clamps and give a sturdy, easily used rig. Cost is usually small and borne by most moulding companies as a good-will gesture as most fixtures are basically simple and easily made.

Fig. 18. Half sectional view to give the general build-up of a compression tool. In this example a multi-cavity semipositive type tool is shown partly open. A—top plate; B—core pin; C—male die or plunger; D—guide pin; E—push back rods (to return ejection system); F—guide pin bush; G—bottom plate; H—risers or parallels; J—ejector pin; K—ejector or knockout bar; L—female die or die cavity; M—pressure pads.

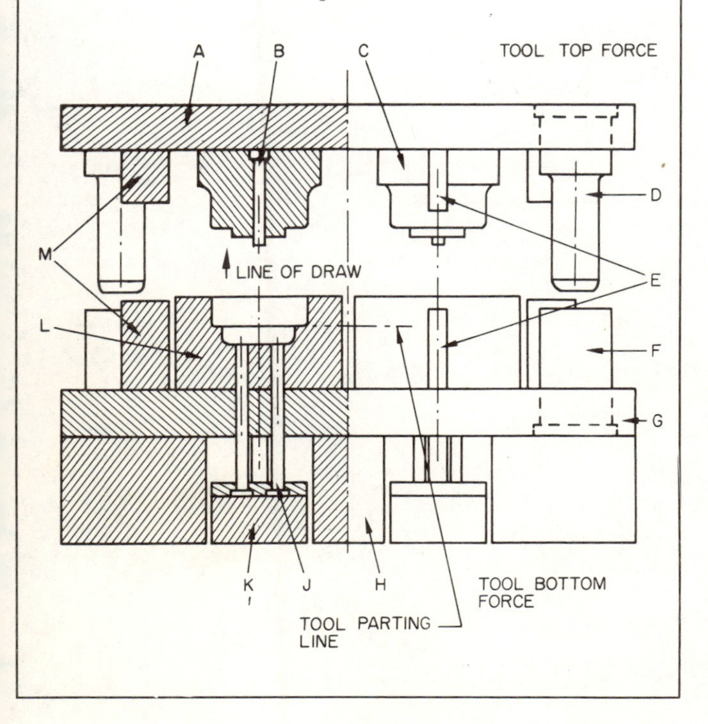

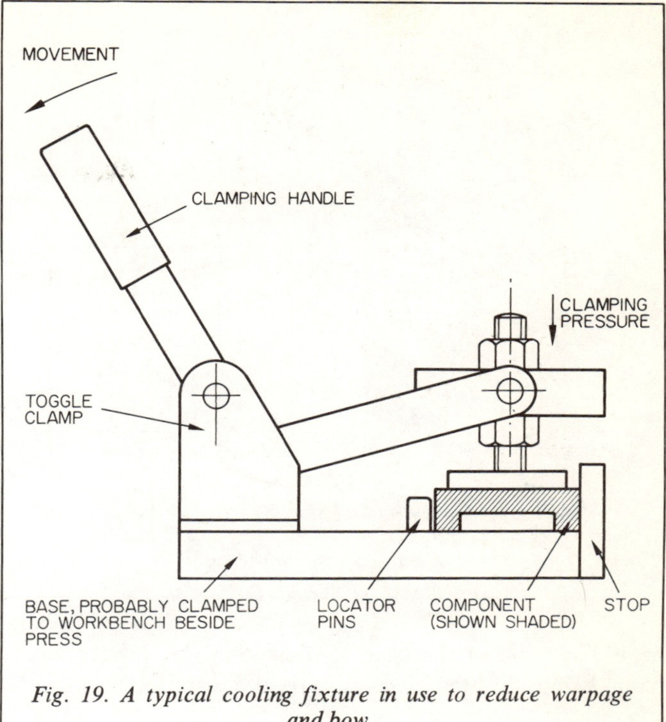

Fig. 19. A typical cooling fixture in use to reduce warpage and bow.

A typical example is shown in Fig. 19 where a toggle clamp is being used to hold a component flat on a base. Location is provided to this example by a stop and locator pins. Toggle clamps are normally purchased as standard items and give adequate pressure with only a small angular movement of their handles. With correct adjustment several hundred pounds of pressure may be exerted.

So that large inserts, especially those with closely toleranced centres, may be quickly cooled to allow the moulding material to shrink to its final dimensions, it is sometimes necessary to cool the fixture with water. This may be by means of sophisticated cooling holes drilled or cut into the fixture but an alternative is to clamp the hot moulding into its fixture and then immerse the complete assembly into a large container of water. Both of these methods using water are considered rather messy and damp but are usually effective as long as the water is not allowed to heat to any appreciable temperature.

Often warpage and distortion can be roughly predicted at the design stages but no ready method of determination is available, due to the many unknown factors such as the changes of shape and size of moulding walls.

Shrink blocks are used to control size and to enable reasonable tolerances to be held. If, for instance, a hole limit must be held to, say, 0.05/0.0 mm (0.002/0.000 in), a plug would be made to the maximum diameter required, or even slightly larger, and fitted into the hot moulding upon extraction from the tool. Whilst the actual dimension of the shrink block is to an extent a matter of trial and error, once ascertained it should give a consistent dimensionally correct

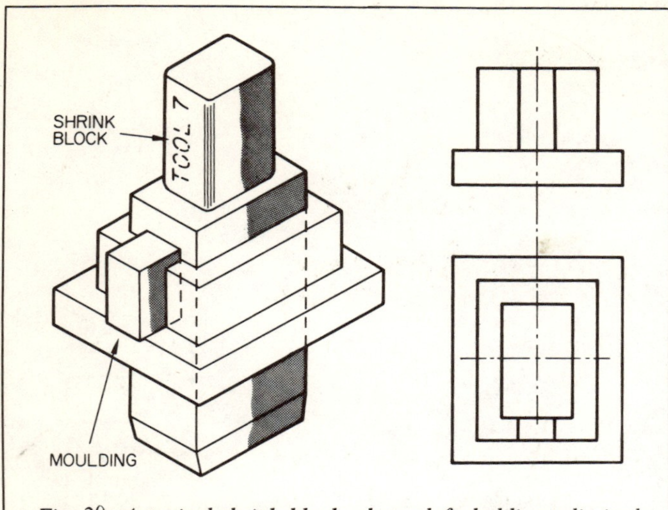

Fig. 20. A typical shrink block, shown left, holding a limited bore tolerance and a closely dimensioned slot. (The moulded component is shown on the right.)

moulding, and in the case of diameters will prevent them from being slightly oval. A typical example is shown in Fig. 20.

FUTURE DEVELOPMENTS

The future of compression moulding for thermosetting materials appears to be swinging mainly towards the requirements of the electrical and automobile industries. Thermosets, this being the diminutive of thermosetting materials, are eminently suitable for electrical components due to their good electrical characteristics and to the fact that they do not melt at high working temperatures. A note of warning must be impressed upon the reader at this point that care must be taken with the choice of materials as the graphite filled materials, for instance, are electrically conductive and not insulants.

Whilst mouldings of the smaller ordering quantities, especially those incorporating moulded-in inserts, are likely to be produced on semi-automatic presses, it is considered that the high quantity orders will increasingly be produced on the automatic compression press or on the relatively newer process of Direct Screw Transfer moulding.

This latter type is not within the scope of this chapter but, briefly, it uses a machine similar in design to that used for the injection moulding of thermoplastics. Material is injected into a heated tool of approximately 160°C in a semi-plasticised state, thus cutting the cycle time and enabling high production rates to be achieved. Output from a single impression tool working 100 hours could well be 6000 mouldings but this will vary according to wall thicknesses, warpage allowances and various finishing moves. Whilst machine cost is high compared to an ordinary compression press, it is possible, in theory at least, to mould fully automatically with low labour costs.

Automatic compression moulding is extremely useful for the moulding of bulk orders and advantageous when multi-impression tools of a generally simple nature can be used. Output will vary from one moulder to another depending on the working hours available, but ideally should be for 24 hours over a complete week. Due to tool damage, wear and press maintainance, this is only theoretically possible, however.

Whilst only the author's private point of view, and one probably not shared by all others, it is expected that thermoplastic materials will gradually ease thermosetting materials, and consequently compression moulding, from their own specialised branches over the next decade. Requirements thereafter will be reduced, but certainly continue to exist and enough to supply the manufacturing capacity of most thermoset moulders for the foreseeable future.

Chapter 7

Transfer Moulding

E D Quiddington
Ariel Pressings Ltd

The word 'transfer' infers movement from one place to another, which is what occurs in the transfer moulding process along with a change of state taking place during the cycle of operation. The transfer of powder to a complete moulding is achieved by placing powder, or a pellet of powder, into a hot chamber and then applying pressure by a close fitting punch entering the chamber, the pressure being applied by an hydraulic system. The material is plasticised in the chamber and forced through the orifice, thus increasing the temperature of the powder. The plasticised material then flows freely and quickly, ensuring that it fills the mould impressions.

EQUIPMENT

Production of mouldings by the transfer process requires the following equipment:

1. A tool containing an 'impression' of the object to be produced.
2. An hydraulic press.
3. A transfer pot and punch.
4. Heating elements for the platens (top and bottom).
5. Thermostat controls to ensure maintenance of correct temperature of platens, during continuous operation.

Ancillary equipment which would be used for the economic production of mouldings in quantity includes:

1. Induction heater for preheating pellets or powder.
2. Pelleting machine.
3. Boards for loading and unloading powder and mouldings respectively.
4. Blasting machine for deflashing mouldings.

The press must at all times remain sound with regard to to the hydraulic system, for loss of pressure during the machine cycle will produce faulty mouldings. The press may be specifically designed for transfer or, alternatively, a straight-forward up-stroking or down-stroking hydraulic press. The tooling for this process can be made in a number

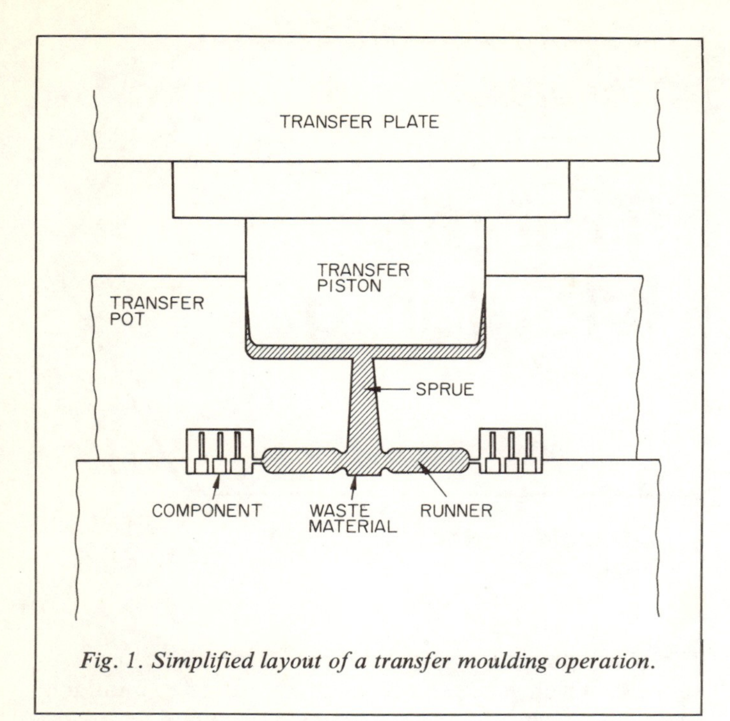

Fig. 1. Simplified layout of a transfer moulding operation.

of ways; for example, it may be simply a top and bottom section if fitted to a transfer press, the press containing an arrangement for mounting satisfactory chambers and punch for the initial material flow operation.

If an ordinary hydraulic press is used, then the tool will have to have the pot built into it and this can be achieved by having a third part to the tool which is hinged and able to swing clear of the mould cavities. This type of tool is heavy for production purposes in that it must be removed from the press at the end of each cycle.

MATERIAL AND DESIGN

The design of products is the controlling factor in the use of the process and it is unusual for one to design a part specifically for transfer moulding. When using the transfer process as described so far, the design of mouldings follows general principles which are related to the use of the product, the type of material required for the product, and the ease of extraction of the moulding from the die cavity after curing. Having taken account of these details when designing the product, the principles which determine the use of the process are the formation of pins or projections within the mould cavity, which, due to their shape, would be weak when considering the pressure applied by the moulding materials being processed.

The materials for transfer moulding are normally thermosets and the range includes phenolics, ureas and melamine. These materials will have resins and fillers combined with them to ensure their free flowing capabilities, the specific types of additives being determined by the rate of flow required to give a correct quality moulding against a particular design, and the cycle time of the operation when considered with other production factors, *ie* manpower, machine capacity.

It has been claimed that there is more dimensional accuracy and stability in parts produced by transfer moulding. In most instances, however, this has not been true, as greater dimensional and warpage problems exist in this particular method of moulding. It is true that delicate inserts, small holes, and side cores may be safely moulded with a minimum of breakage, and in most instances, with the proper mould design, the insert holes can be kept free of material. In many cases the method provides an economic advantage from the standpoint of mould cost, moulding cost, and, in a few instances, finishing cost. Transfer moulding, however, is more complex than compression moulding, and there are many more variables to contend with. A few of the variables which add to the complexity of this method of moulding are listed below:

1. Part design.
2. Type of material.
3. Position of gating.
4. Type of gates.
5. Shape and length of runners.
6. Preheating of the material.
7. Transfer pressure.
8. Clamping pressure.
9. Transfer time to fill the part.
10. Mould temperature.
11. Relief beyond the land area.
12. Air venting.

Any one, or combination of these variables, can be responsible for the following:

Changes in curing cycle.
Variations in shrinkage.
Warpage.

TERMINOLOGY

Various terms are used in moulding and these can be summarised as follows:

Pressure is the pressure applied by the hydraulic system over the moulding area.
Tonnage is the maximum load that the hydraulic system of the machine will apply.
Pot is the chamber or cavity into which the moulding powder is pressed.
Impression is the cavity or shaping for producing one product.
Cycle is the time taken from the start of one sequence of operations through the complete production of the mouldings to the start of the next sequence of operations.
Cure time is the time taken during the cycle to cure the moulding material, that is whilst the powder is under full pressure and at full temperature.

ECONOMICS

The material usage enters very much into the economics of this process in that, to ensure the mould cavity is filled, sufficient material must always be provided in the bottom of the moulding pot and, in consequence, in addition to the sprue and feeder (Fig. 2) there is a disc of cured material in the base of the pot on every cycle. This wastage, for thermoset materials cannot be re-used in the same manner as thermoplastics, is applicable whether producing one article or a number of articles per complete cycle.

Bearing this utilisation of material in mind, it can be seen that from a point of view of economy of use of material, it would be preferable to employ a straightforward compression moulding operation. However, as previously noted, the design of the product will often determine that the transfer process must be used, and really a state of availability of technique becomes priority, rather than economy.

In allowing for these various sprue feeders, runners and gating to components, it will be readily understood that tooling cost must also increase, for each of these necessary features must play their part in allowing the material to flow readily and correctly to the various impressions in the tool.

Account must be taken of the amount of material being wasted in these feeder arms but not, of course, to the detriment of the production of a good quality moulding. Considering these factors it is understandable, therefore, that the tooling costs for this type of process are high. Single impression tools usually have a starting price in the £300–£400 price range, with relative increases for multi-impression tools, and tooling prices in excess of £1 000 are commonplace.

It is essential, therefore, that the production run of the product be of considerable duration and, preferably, an item that is liable to repeat over a time period of 2 to 3 years. This time ensures that the depreciation of the machinery, the running costs, *ie* electric power, manpower, plus the capital cost of the tooling, can be spread over this time period. For example, a single impression tool costing £350 for the production of only 100 products results, before anything else is taken into account, in a cost of 350/100 = £3.50, thus making the product very expensive.

Fig. 2. Typical sprue and runner system for transfer moulding, indicating the high wastage factor.

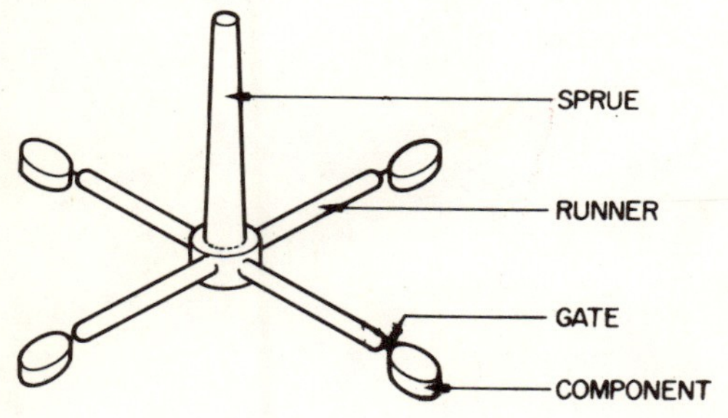

APPLICATIONS

Those products whose designs demand that they are produced by transfer moulding are, to a large extent, components for electrical and electronic equipment. The reason this area is predominant is simply because of the insulation requirements which occur so often between close terminating points or operating parts of circuits through which current is passing, either in a form powerful enough to arc between points or to produce effects which would disrupt the function of the product. Typical applications are the small mouldings as used in plugs and sockets for audio equipment and the bases and sockets of relays and contactors.

DEVELOPMENT IN THE TECHNIQUE—DIRECT SCREW TRANSFER

The discussion so far in this chapter has applied to transfer moulding as a technique which has been used in the Moulding Industry for many years, and the references to the principles of the technique, mechanical equipment used, materials and design, the economics of the process and the design applications, are all relevant. However, during the last few years a total revision of the transfer moulding process (and in fact many compression mouldings) has taken place by the introduction of a series of machines based on the injection moulding principle, but with cylinders, heaters and control equipment which enable modified thermoset materials to be moulded.

The principles of operation for injection moulding thermosets are basically the same as those of the previously described original transfer method. However, with the further sophistications of the machinery, one is able to apply, in addition to the operating pressures and temperatures, additional hydraulic pressures which ensure a clamping of the tool faces, thus avoiding the problem of back pressure splitting the tools and causing excessive flashing as was always a possibility in the transfer method. This new type of machine also allows for semi-automatic or fully automatic running, as distinct from the manual operation of the earlier process.

Equipment

The machinery for injection moulding of thermosets, because of the additional built-in facilities, is considerably more expensive than the conventional hydraulic press. However, advantages occur with the more precise and constant temperature controls and operating cycle which can be achieved with this machine. Many of the machines also have the advantage of being able, by the change of a cylinder and other ancillary equipment, to be used for injection moulding of thermoplastic materials.

In this type of machine there is no transfer pot and punch as with the original process, for the powder is fed from a storage hopper through to the cylinder in which it is to be kept in a semi-plasticised state within a very precise temperature range for a limited period.

This process also eliminates the need for induction heating of pellets and powder, pelleting machines and boards or containers for manually loading correct quantities of powder. The screw in the cylinder ensures that sufficient powder is passed through to fill the mould impressions, runners and sprue within the capacity for which the machine was originally designed.

Material and design

As discussed earlier, the design of the product has been the controlling factor in the use of the original transfer technique. With the injection process, which uses the same method of processing the material through to the impressions, the problems of small cross-sections and shapes for which pins are used are naturally catered for in the same manner as applied in the original transfer technique. The controlling factor in the injection technique is the design criteria, however, these points of design are normally the same as those for thermoplastic injection mouldings and thermosetting compression mouldings, namely:

1. Sections should be as uniform as possible.
2. Suitable radii should be allowed to ensure flow of material.
3. Sections should be sufficient size to allow for flow of material.
4. It should be remembered that the thickness of sections dictates the length of cure time for the moulding.
5. No inserts should be included in the moulding, but provision should be made for any necessary insertion as a secondary operation.
6. Care should be taken to allow a suitable point on the product where the feeder gate may be positioned.

The process does cause heavy wear on the tooling gates due to the fast flow and abrasive nature of the materials. Consequently, one should endeavour to design the tools with replaceable gates, and also ensure that the tools are hardened and able to resist wear on the actual impressions of the mouldings.

The biggest problems encountered are with regard to the strength and positioning of the gating, which must be placed such as to avoid damage to the finished product and yet be in a position which will allow full and easy flow, thus producing full impressions.

The materials employed are similar to the general thermosetting materials but modified particularly for use with this process, and raw materials are frequently marked with their references suffixed with the letters '*IG*', indicating an Injection Grade material. These materials have been designed to give trouble-free automatic moulding, taking into account the following features:

1. Fast cycle time.
2. High gloss finish.
3. Quality moulding.
4. Trouble-free operation.
5. Consistency.

These materials are available with fillers of wood flour, paper, glass, natural fibres, cotton, nylon and in many grades and colours, thus ensuring that design is not restricted by availability of materials.

The fault problems of burning/blistering still occur in this process as with normal transfer and compression but, with the greater sophistication of control, they are more readily eliminated. With the faster cycling, and the manner in which the tools are manufactured, there is an increase in the problem of gassing (*ie* the trapping of gasses and air in small cavities, stopping the material filling or causing a very hot area which generates a burn on the material), and the need for more careful venting is created.

Economics

The injection of thermoset materials, on the type of machine described, produces mouldings in a far more economical manner than the transfer process described earlier. Whilst the injection technique still produces waste elements in the form of a sprue and runners, the waste element is reduced in that no disc is formed at the bottom of the pot. The largest benefit from this process, however, comes with the reduction of the over-all cycle time, and this, dependent upon the product part, can give a time saving of between 40% and 60% when compared with normal transfer or compression moulding of equivalent parts. For example, a moulding produced with a time cycle of 4 minutes by transfer moulding can be produced in 2¼ minutes by the injection technique. The moulding is also of superior quality and throughout the production run retains a 'flatness' which is essential in the application of the product.

Whilst the overhead element of cost incurred due to the higher purchase price of the machine and tooling must be considered, the over-all saving in labour time and the better quality of mouldings ensures that the economies of the process are considerable.

Chapter 8

Injection Moulding

V E Moore
Invicta Plastics Limited

The use of injection moulding is well known and therefore it is only intended to deal briefly with the characteristics of the process here. Its products are many and varied, typical applications being housewares, crates, tote boxes, closures, automotive and marine components, toys, machine parts, *etc*. The high costs of the moulds limits the process to components requiring relatively long production runs.

Rapid production rates can be achieved with little limitation on size and shape. Cycle times may be as low as 10 seconds for small components, being dependent on the time required to fill the mould and the cooling time. Filling time is proportional to the pressure drop across the nozzle raised to a power dependent on the type of plastics being injected, and cooling time is roughly proportional to the square of the wall thickness of the moulding.

A high degree of dimensional accuracy can be attained providing the correct allowance is made for the mould shrinkage of the material used. It should be noted that differential contraction on cooling can lead to residual strains in the component, and the directed flow of material frequently causes the components produced to exhibit marked anisotropy. Mouldings generally have good surface finish.

The major advantages of the injection moulding process are, therefore, its suitability to a wide range of products, its fast cycle times, its moulding accuracy and the attractive economics of large scale production runs. The techniques and equipment have undergone, and continue to undergo, various modifications often associated with the development of new materials.

THE PROCESS

Basically, the process involves the injection under pressure of a predetermined quantity of heated and plasticised material into a relatively cold mould (Fig. 1). After the material solidifies it is allowed a further interval to cool

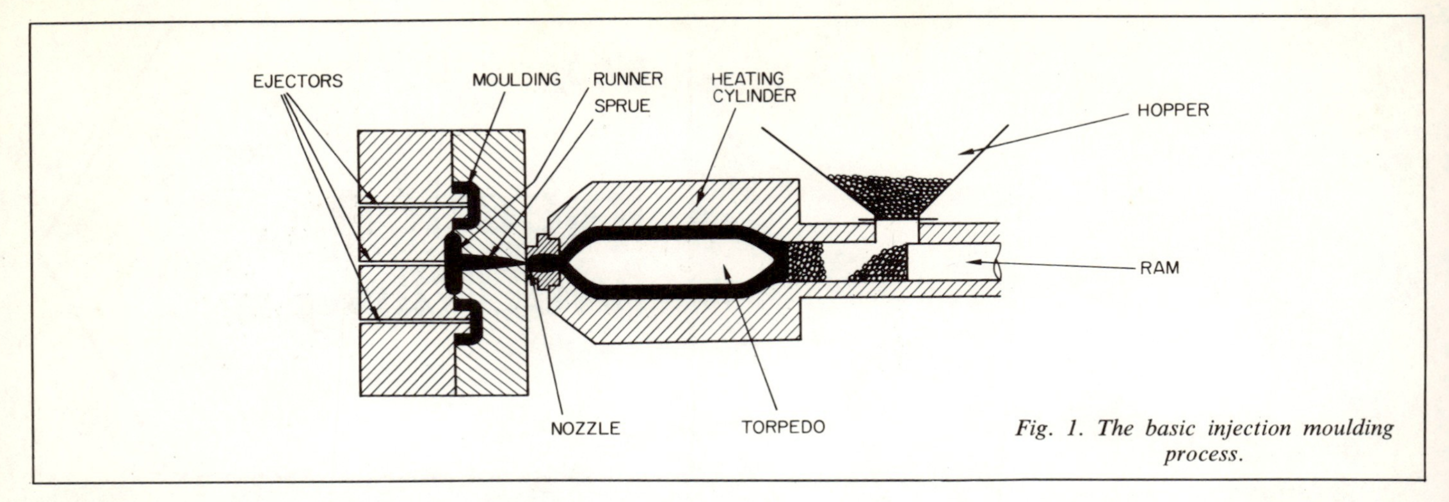

Fig. 1. The basic injection moulding process.

before the mould is opened and the product removed. In some respects, the basic process is very similar to pressure diecasting. The operations which make up a cycle of this basic process can be listed as follows:

(a) The measuring and feeding of a predetermined quantity of plastics, *ie* sufficient to fill the mould cavity or cavities, runners, *etc*, from the hopper to the heating or injection cylinder.
(b) The injection of this material (shot) by the use of a plunger or ram into the heating cylinder, thus displacing the previous shot, now heated and softened, through a nozzle and into the mould cavity via sprue and runners.
(c) Continued pressure of the ram during initial cooling.
(d) Retraction of the ram and final cooling.
(e) Opening of the mould and removing the product.

Developments of the process

A problem with the basic process is the contradictory requirements of uniform heating and high pressure at the nozzle. Uniform heating calls for internal channels within the heating cylinder that are as small in cross section and as long as possible in order to plasticise the material effectively. Conversely, these channels should be as large and as short as possible in order to minimise pressure drop. This problem has, however, been overcome by preplasticising systems.

In-line screw system. The in-line screw system, or screw ram system as it is sometimes called, is the most important development in injection moulding. In this system the plastics is softened by passing it through a heated cylinder by the use of an extruder type screw, the softened plastics building-up in front of the screw. The build up of plastics in front of the screw causes the screw to move axially backwards. The shot size can therefore be easily controlled by stopping the screw rotation at a predetermined position. When the preplasticising operation is complete, the screw is moved forward under pressure pushing the plastics into the mould (Fig. 2).

Preplasticising units. Further modifications of the basic process include the use of preplasticising units. In these systems the plastics is melted in a separate chamber and then passed to a pressure chamber for injection. This separate melting chamber is called a preplasticiser and it can be either a conventional heating chamber with ram similar to that used in the basic process or a screw system can be used. The advantage of the screw system is that, due to the action of the screw, lower temperatures can be employed making it suitable for moulding heat sensitive materials, such as unplasticised PVC.

EQUIPMENT

The type of injection moulding machine most widely used is the horizontal machine in which the injection units and clamping equipment employed to open and close the mould operate in the same plane. Vertical machines are manufactured and are particularly useful where inserts are required, these machines are however largely manual in operation.

Most injection moulding machines currently available can be operated either manually, semi-automatically or automatically. The source of power is invariably electricity, but drives are based on mechanical, hydraulic or even compressed air modes of operation.

Moulds

Both single moulds and multiple moulds are employed. Single moulds have certain advantages which include the mouldings produced being completely identical which is very difficult to achieve on multiple moulds, furthermore single moulds are comparatively inexpensive and therefore suitable for short runs. Multiple moulds are often employed in an endeavour to use the capacity of an available machine for the production of products which are small in relation to shot capacity. However, with multiple moulds, quality is difficult to control due to runner length. When material in the gates is insufficiently solidified to withstand internal mould pressure there will be a backflow from the mould to the runners and internal mould pressure will drop to the level of resistance of the material in the gate.

Standardised moulds with interchangeable inserts are used wherever possible in order to reduce costs. These are known

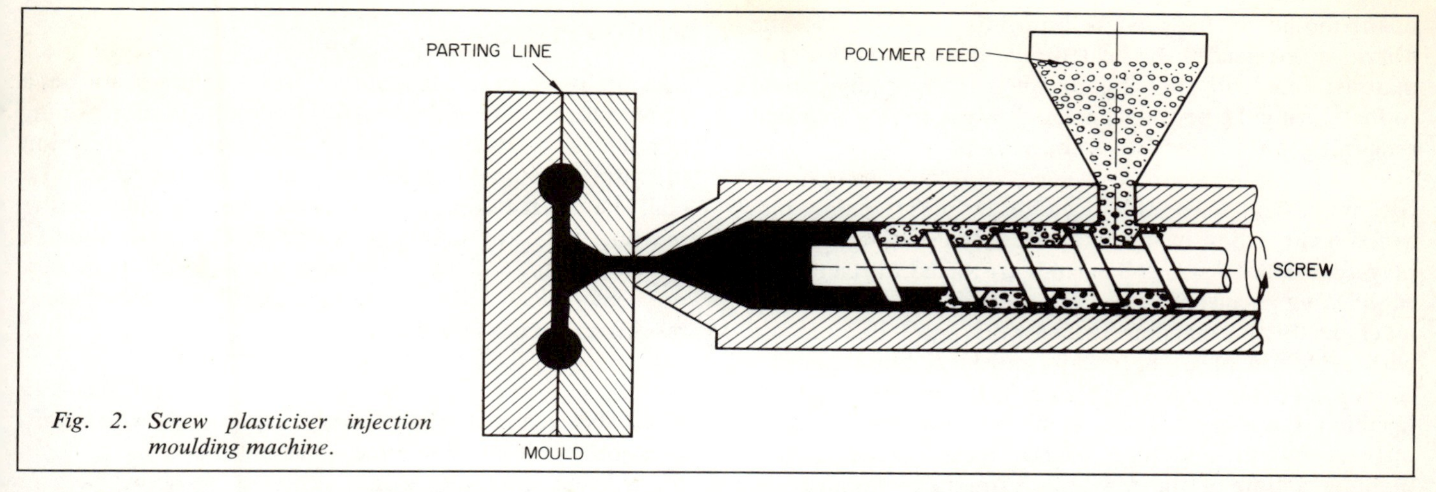

Fig. 2. Screw plasticiser injection moulding machine.

as standard or basic moulds and are usually available for either rectangular or circular inserts. Most companies have developed their own range of basic moulds and new products are often considered from the point of view of an insert in a suitable size of basic mould.

The temperature of the mould is controlled. For some materials, *eg* polystyrene and polyethylene, it is usually provided with cooling facilities to achieve rapid cycle times. For other, *eg* nylons and acetals, it is heated in order to achieve the desired degree of crystallinity and thus optimise the properties of the moulded article, while for thermosetting materials it must be hot enough for the resin to cure.

Barrel construction

Barrels and screws for injection moulding processes are generally similar to those used in extrusion, but because of more severe conditions there are several features which call for greater attention. For example, the barrel of an extruder is frequently designed to withstand a pressure of 1500 bar (20 000 lbf/in^2) although it very infrequently will experience this pressure in use. However, an injection moulding machine cylinder often does experience pressure of this order and frequently higher. This means that the design and selection of the barrel and particularly the liner is quite critical.

Similarly, in the case of a screw that is also used as a ram, the strength must be adequate to withstand arduous service condition, especially when it is considered that the screw might be required to start and stop from full speed several times a minute during the plasticising cycles. This also, of course, applies to the thrust system employed.

Heating

Electric resistance heating is employed almost exclusively in the heating cylinders. Zoned temperature regulation is one method which has been favoured and involves the use of several heaters along the plasticising cylinder, each heater being controlled by separate temperature control equipment. It is thus possible to heat the cylinder in stages to match the flow of material and its stage of plasticisation and therefore to ensure a homogeneous melt by approximating the optimum temperature in each zone. The material is also protected against over-heating.

Since a plasticiser screw normally operates discontinuously and for economy its cycle must match the time the mould takes to cool, attempts are frequently made to increase the melting efficiency of the unit by raising the temperature of the extruder cylinder. When this is the case, the temperature of the injection cylinder should be maintained below that of the plasticising cylinder so that the melt forced into the injection cylinder will not degrade during the stationary period.

MATERIALS

Thermoplastic moulding materials are usually produced commercially in granular form. The shape and size of the individual granules differ according to the type of material. Practically all thermoplastics can be injection moulded, with varying degrees of ease. The particular form of granulate is dictated primarily by processing requirements, but may also be dependent upon production techniques used during manufacture. The most important requirement for processing is a clean, free running granulate without any tendency to agglomerate.

Thermoplastic materials have very low heat conductivity and poor thermal stability which are disadvantages in effecting efficient melting, a variety of design modifications have been developed to overcome this difficulty. There is always a danger of overheating the layer in contact with the heating surface, before the inner layers are softened. The screw extrusion type of plasticising unit, which has now become popular, overcomes this problem. The change in consistency accompanied by the intensive kneading of the granulate offered by this system, whilst it is passing through the heating chamber, gives considerable advantages due to the fact that all layers of the material are continuously forced towards the heated cylinder wall, resulting in a much more consistant melt temperature.

Thermosetting materials may also be injection moulded, but great care is necessary to ensure proper heating of the moulding powder without premature setting in the passage of the mould. In the process known as 'jet moulding' the nozzle is arranged to be heated only during the time that material is actually flowing, and it is then rapidly cooled so that during the period when no flow occurs, the material remaining in the nozzle does not harden.

New processes have been evolved for the injection moulding of foamed plastics, particularly the polyolefins, toughened polystyrene and ABS. The plastics are mixed with a chemical blowing agent before feeding to the injection moulding machine. In the machine the blowing agent decomposes with evolution of a gas (usually nitrogen) but expansion cannot take place in the constrained conditions of the injection moulding machine. A metered amount is injected into the mould, and filling of the mould occurs by the foaming action of the plastics/gas mixture. Because the pressures developed in the mould are comparatively low it is possible to use low cost aluminium moulds at anything from 25% to 60% of the cost of a fully machined steel mould. Mould costs can, therefore, be amortised over shorter runs than in conventional injection moulding. Production costs, on the other hand, are generally higher because the cooling cycles are longer. An important factor affecting economics is the fact that the use of a foamed structure gives greater rigidity for the same material weight, or equivalent rigidity for less material.

ECONOMICS

Essentially, injection moulding still continues to be a process for moulding thermoplastics despite its increasingly popular use for the moulding of thermosets. Injection moulding equipment is generally more expensive than an equivalent sized compression moulding press, the smaller machines are approximately twice as expensive and the larger sizes are often five or six times as much. However, pre-heaters and pelleting machines are not required thus resulting in some savings.

Cycle times for injection moulding of thermosets can often be reduced dramatically compared with compression moulding, sometimes to as much as a sixth. Times for transfer moulding would probably be somewhere in between.

However, mould costs are greater for injection moulding than for compression moulding, by a factor of about four. For large and intricate items such as a dairy crate, the tooling cost is in the region of £10,000. With tooling costs of this order, it is important to have long production runs. The high cost of moulds is dictated by the fact that they have to withstand high pressures. These high pressures are necessary to cause the very viscous liquids to flow.

Chapter 9

Extrusion

K J Braun *BSc CEng MIMechE*
Bone Cravens Ltd

Recently published figures for the United Kingdom show that an average of about 50% of the main thermoplastics consumed (polyolefins, PVC and polystyrenes) is finally extruded. These figures do not take into account the very important application of extrusion machines in the manufacture of polymers where extruders are used as reactors, melt pumps, compounders, devolatilisers and so on. The large majority of all thermoplastics materials now produced will have been extruded at some stage of manufacture. It should also be borne in mind that other plastics processes, not usually associated with extrusion, employ extruders for auxiliary functions. For example, extruders are used in both blow moulding and injection moulding to plastify the polymer.

TYPES OF EXTRUDERS

Single screw extruders. By far the most common type of extruder used in the plastics industry is the single screw machine (Fig. 1). The principal features of a single screw extruder are illustrated in Fig. 2. In its most simple form the machine consists of the screw and barrel with a feed pocket, some means of revolving the screw and of resisting the thrust developed by the screw in forcing the melt through the various passages leading to the final die orifice, and finally some means of controlled heating (often also cooling) of the barrel.

Most modern extruders are fitted with steplessly variable speed drives, such as a.c. commutator or thyristor controlled d.c. motors, to allow precise selection of the optimum screw speed for the particular application.

The extruder barrel is generally heated by electric resistance heaters although many other forms of heating may be found, *eg* steam, oil, mains frequency electric induction, *etc.* Where cooling is used, air or water are the most usual heat transfer media. Temperature regulation of the extruder barrel is generally fully automatic. For this purpose, the barrel is divided into a number of zones to allow variation of conditions along the length of the barrel.

Fig. 1. Single screw extruder. (Courtesy of Bone Cravens Ltd.)

Twin screw extruders. Twin screw extruders find their main application in the extrusion of unplasticised PVC, particularly large diameter piping produced from powder, and also in a number of more difficult compounding applications.

The main difference in the method of operation between single and twin screw extruders is the way in which the material is conveyed within the machine; the single screw relies on friction between the material processed and the cylinder to move the material, whereas the configuration of two screws can produce positive displacement. This means that, for example in the case of unplasticised PVC, the formulation that can be processed on a twin screw machine is not as critical as for a single screw extruder, since the quantity and type of lubricant does not affect the material feed. Furthermore, it also means that less work may need to be done (by virtue of less frictional heat being generated) on the polymer and consequently degradation may be avoided.

Conversely, the positive displacement feature of the twin screw machine permits generation of shear stress in excess of that which can be achieved on single screw machines and hence a twin screw extruder can perform some compounding operations on which the single screw may fail. However, mechanically the twin screw extruder is more complex and typical cost per unit output is 2–3 times that of the single screw machine. Also, the proximity of the two screws in a twin screw machine introduces problems of torque and thrust transmission which do not exist on the single screw machine and which may prove to be the limiting factors in some demanding applications. Similarly, the single screw machine offers infinite scope for variation of screw profile which is not the case with an intermeshing twin screw extruder.

Extruders incorporating more than two screws in parallel are not in general use in the plastics industry. Occasionally 2-stage twin screw machines are referred to as four screw extruders (see Fig. 3).

Fig. 2. Single screw extruder — main features.

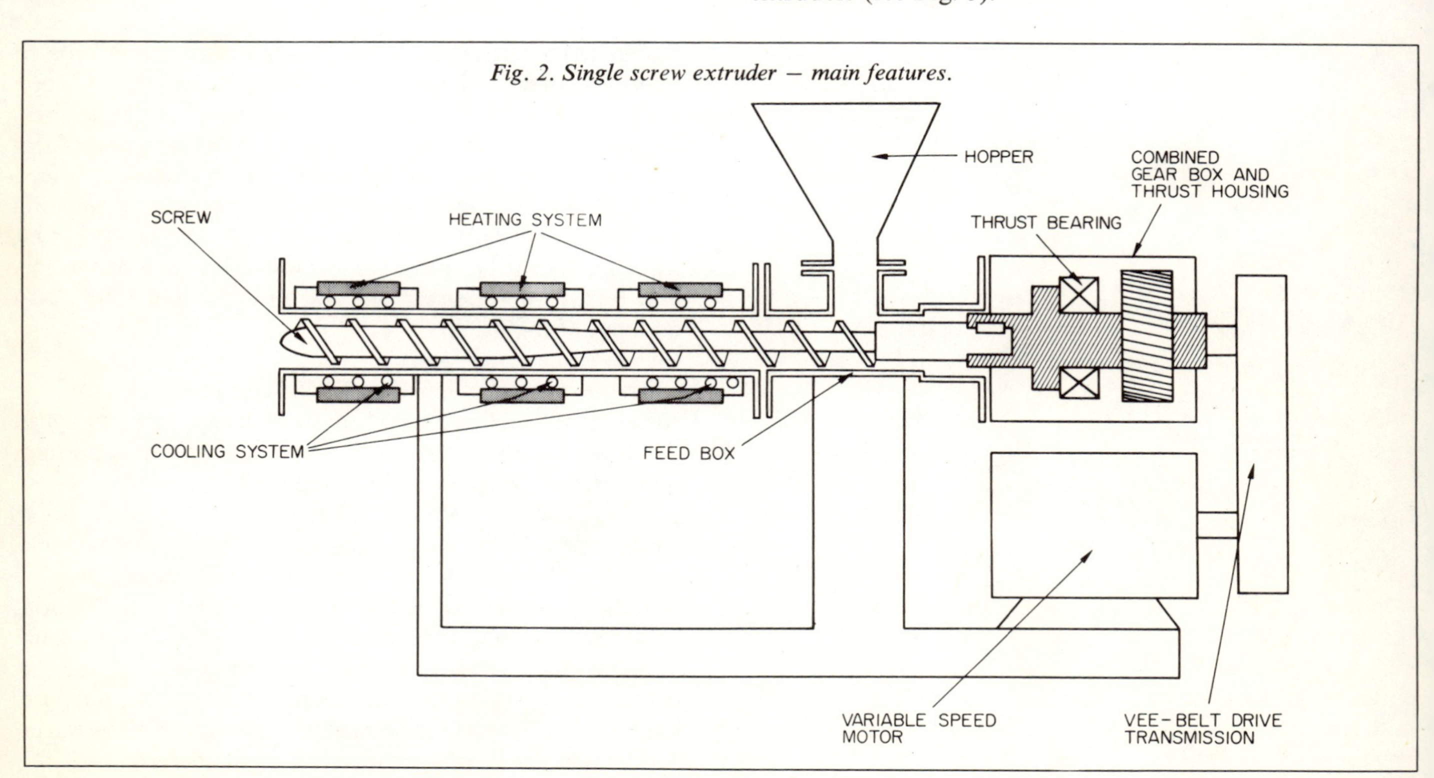

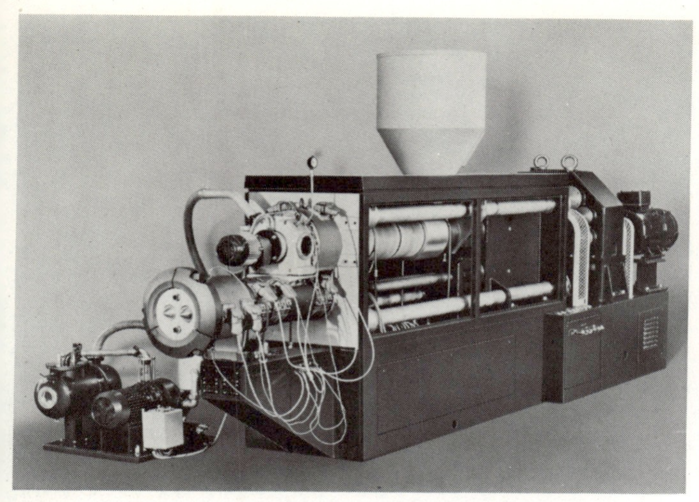

Fig. 3. 2-stage twin screw (4-screw) extruder. (Courtesy of Anger GmbH.)

THE EXTRUSION PROCESS

The basic functions of an extruder used in a finishing application are illustrated in Fig. 4. Thus, very simply, the extruder has to convert the polymer feed stock to an homogeneous melt which is pumped to the die at a uniform rate. To achieve this simple aim, the screw has to convey the solid material, melt it, and deliver it at a uniform temperature and pressure.

Whether a single screw will feed or not depends on the balance of frictional forces acting on the material. The mechanism of conveying relies on the material adhering to the barrel and slipping on the screw. The frictional behavior of the plastics depends to some extent on the surface finish of the material in contact and to a very large degree on the metal and polymer temperatures.

The frictional behaviour of the feed stock material also has a considerable influence on the rate of melting that can be achieved and this in turn dictates the design of the feed section of the screw. The frictional forces necessary to convey the material must inevitably generate heat which contributes to and, in some cases, can be largely responsible for, melting of the polymer. Heat transfer to the polymer by conduction from the barrel can be very limited, particularly at high rates of throughput; and considerable improvement in extruder performance can be achieved by correct design of the feed section of the screw and selection of appropriate operating conditions.

Reverting to Fig. 4, the last function of the screw is to deliver the melt to the die at a uniform temperature and pressure. In a conventional single screw (Fig. 5a), the metering section has the task of ensuring uniformity of output and adequate mixing to give an homogeneous melt. The mixing requirements in turn can be subdivided into dispersive and distributive mixing. The dispersive mixing is the action of de-agglomeration—breaking down of solid granules, which may not have melted, breaking down of gels, *etc.* Distributive mixing may be defined as the action which produces random spatial orientation. In a single screw machine, screws are limited in their ability to adequately mix highly viscous melts, which requires a wide range of velocity vectors within the screw channel. To some extent this can be achieved by increasing the discharge pressure at the inevitable expense of higher melt temperature which is generally undesirable. Alternatively, some form of multi-stage mixing screw may be used to obtain the required degree of homogeneity at a lower level of energy input. The configuration of one such screw is illustrated in Fig. 5b. Broadly speaking, a multi-stage screw separates the metering function from mixing and the two types of mixing required (*ie* dispersive and distributive) are separately catered for by sections of the screw specifically designed for the purpose.

TEMPERATURE CONTROL

Accurate temperature control plays an important part in the extrusion process. The temperature regulation of the process within the extruder, is generally confined to the barrel although provision is often made for cooling of the feed pocket and the screw. Cooling of the feed pocket may prevent the feed stock from sticking or bridging due to heat conducted from further down the barrel and, in the case of the single screw extruder, cooling of the screw may be used either to improve material feed, if applied in the feed section only, or to alter the effective screw characteristic, if the length of cooled section of the screw is extended into the melt region. Screw cooling does not directly influence the melt temperature in either case but in the latter, broadly speaking, the result will be a lower rate of throughput but better homogeneity.

Fig. 4. Basic functions of a plastifying extruder.

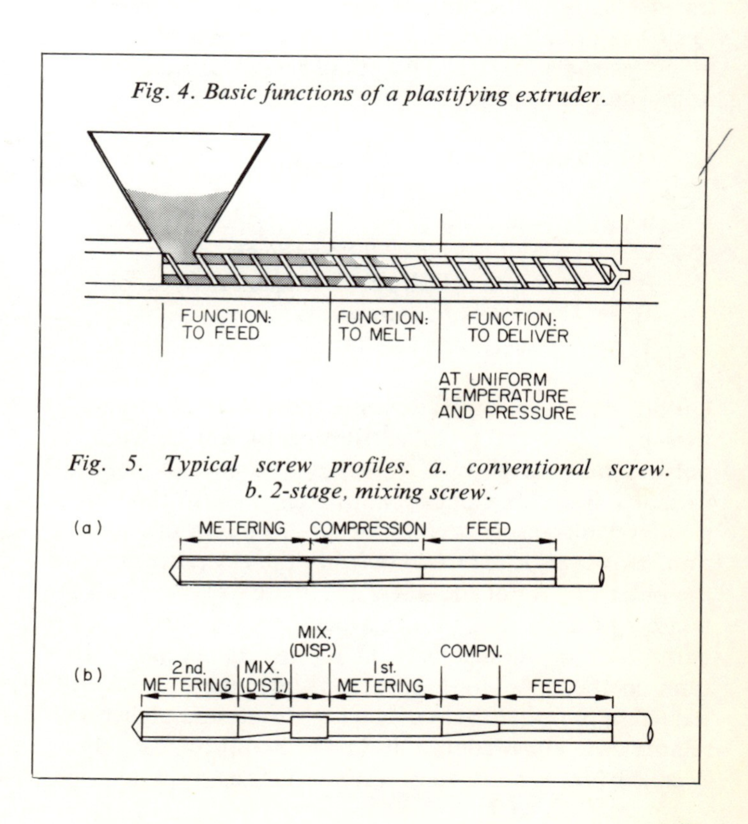

Fig. 5. Typical screw profiles. a. conventional screw. b. 2-stage, mixing screw.

The principal aim in developing extruder screw design is to increase productivity and maintain quality for the same size of extruder at the same capital cost. In many extrusion processes this requirement is coupled with the need for lowering melt temperature.

A number of advantages are to be gained by extrusion of plastics at relatively low melt temperatures; for example, a melt at a low temperature is more viscous and therefore retains its shape more easily, shorter time and distance are necessary for subsequent cooling because of the small heat content in the melt, and the risk of thermal degradation is reduced.

The continuing trend towards higher output from extruders of a given size has meant that extruder cooling is progressively playing a more important part in the extrusion process. In many applications the main difficulty arises from the limited heat removal capacity of the barrel cooling system coupled with inadequate mixing action of the screw. An effective barrel cooling system coupled with inadequate screw design usually increases variations of melt temperature.

Complete screw design must of necessity take into account the overall energy balance of the extruder and therefore cooling must also be considered. There are probably as many (if not more) different variations of barrel cooling systems as there are extruder manufacturers, but the majority of them are based on the use of either air or water or a combination of the two. To assess the relative merits of the different cooling systems available, it is necessary to consider two basic aspects, namely their heat transfer capacity and their controllability. The question of heat transfer capacity is more easily considered by reference to a basic equation governing heat transfer in a system such as an extruder barrel. This may be represented by the following relationship:

$$Q = UA(T_p - T_c)$$

where Q = heat transferred
U = Overall heat transfer coefficient
A = area over which the heat transfer takes place
T_p = Mean polymer temperature
T_c = Mean temperature of the cooling medium.

Unfortunately, the major factor in the total resistance to the flow of heat is the so called film coefficient between the polymer and barrel wall. Although a number of empirical formulae have been put forward from time to time, there is an ever increasing need for accurate data in this connection. However, it is recognised that the clearance between the outer surface of the screw and the barrel bore and the number of times that a flight wipes the inner surface of the barrel are the fundamental factors affecting the plastics film coefficient. Thus, a worn screw may show little reduction in output but could easily result in a higher melt temperature under conditions in which appreciable cooling is called for. A multi-start screw which gives more wipes per revolution will create more favourable conditions for heat transfer.

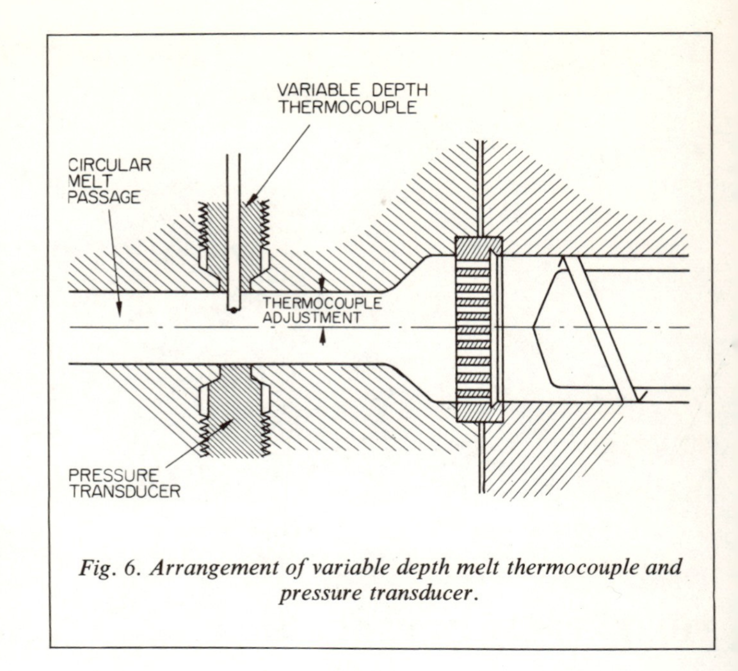

Fig. 6. Arrangement of variable depth melt thermocouple and pressure transducer.

Reference to the equation for heat transfer does show that for any given system the heat removal capacity is directly related to the overall temperature difference, *ie* the difference between the polymer and the temperature of the primary cooling medium. From the point of view of heat removal capacity, direct water satisfies two important requirements:

i. High value of specific heat
ii. Excellent heat transfer properties, and ease of handling by pumps, control valves, *etc*, due to the low viscosity.

Fig. 7. The tubular film process.

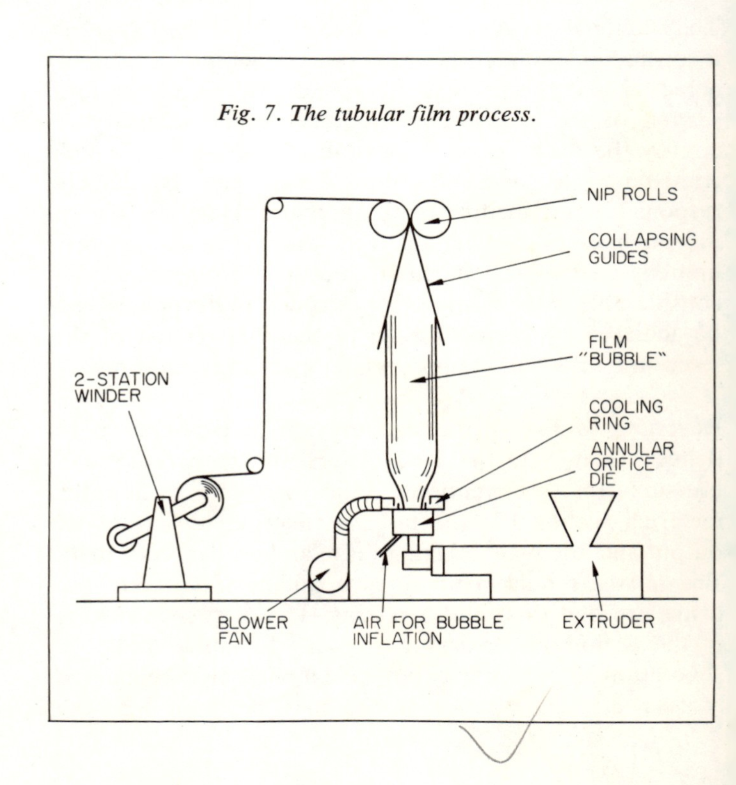

Direct water raises control problems because of its relatively low boiling point coupled with high latent heat vapourisation, but recent advances in instrumentation have largely overcome these limitations. Furthermore, with cooling systems using water, there is always a potential danger of scaling, sludging or corrosion, although these problems are usually not difficult to overcome. If the present trend of ever increasing outputs from a given size of extruder continues then undoubtedly direct water cooling systems with their high heat removal capacity will continue to gain application.

EVALUATION OF EXTRUDER PERFORMANCE

One of the problems of evaluating the performance of an extruder is that of the two fundamental parameters, namely output and quality, only output can be measured in direct terms. However, there are at least two fundamental measurable variables which can be used to define the stability of the extrusion process and the quality of the final product. These are melt temperature and melt pressure as measured at the end of the extruder screw.

Pressure variations offer a quantitative indication of the fluctuation level of the extruder throughput, whilst positional and time dependent temperature variations are a very reliable pointer to the quality of the extrudate. Melt temperature variation is generally attributed to either a lack of residence within the extruder or inadequate mixing. Discharge pressure fluctuation can be caused by an actual variation in rate from the extruder or by melt temperature variation of material passing through the die.

Melt temperature variations may be measured by variable depth thermocouple and a pressure transducer, or a grease-filled pressure gauge may be used to determine the variation of pressure. Fig. 6 illustrates an arrangement for measuring the two fundamental variables. Measurements of melt temperature and pressure may also be used to establish optimum operating conditions for the extruder.

FINISHED PRODUCT EXTRUSION PROCESSES

Extrusion processes in which a finished product is made may be variously divided into catagories; one such division would be: film, sheet, pipe, monofilament and profile (or section).

The precise definition of these products is somewhat arbitrary. For example, thick film becomes sheet and there is no exact limiting thickness in either case. Generally, plastics web below 0.010 in (0.254 mm) is referred to as film and above this thickness as sheet. The term foil is also used sometimes to describe thin sheet in the thickness range of (say) 0.010 to 0.060 in (0.254–1.524 mm). Pipe or tubing is described as such so long as the section remains round in the finished product. Tubular film is sometimes referred to as lay flat tubing.

Fig. 8. Tubular film plant for the production of heavy gauge polyethylene film. (Courtesy of Bone Cravens Ltd.)

Monofilaments are strands of plastics materials ranging in diameter from about 0.005 in to about 0.0625 in (0.127–1.6 mm). Profile or section is a product that does not fall into any of the above four categories and includes such items as curtain rails, refrigerator sealing sections, window frame sections, *etc.*

Film

The majority of plastics films made today are produced by the tubular process, although for certain applications the 'flat film' casting process is used. The so-called water quench system in which film is extruded directly into water is now not in general use.

The principle of the tubular film process is illustrated in Fig. 7. The important aspects of the process variables which

Fig. 9. Principle of the cast film process.

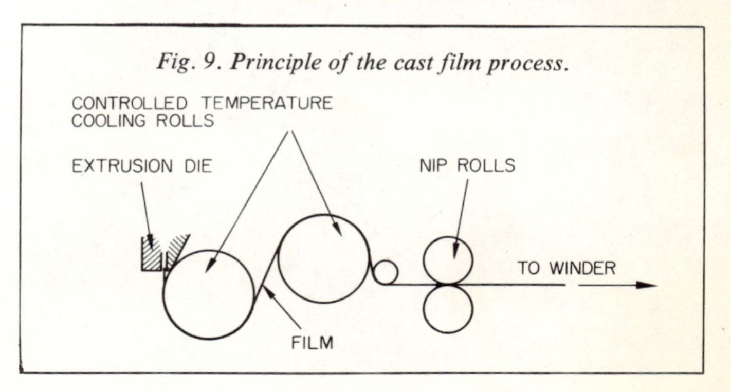

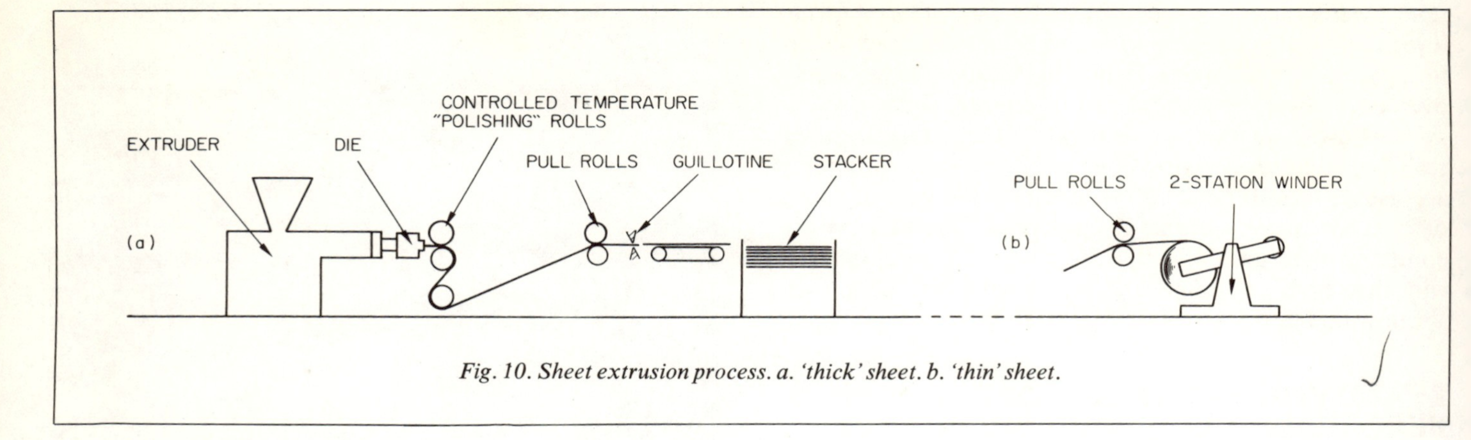

Fig. 10. Sheet extrusion process. a. 'thick' sheet. b. 'thin' sheet.

determine the quality of the film are the uniformity of melt temperature, the flow distribution in the die, and the cooling ring design, which not only controls the rate of cooling but can affect the thickness distribution in the film and the stability of the bubble.

Since the rate of film cooling is the usual output limitation, the level of melt temperature at the die orifice and the cooling ring play important parts in determining the maximum possible output for a process. Good air cooling ring design will ensure high velocity air flow directed along the bubble contour and the ring may improve the process stability by producing a venturi effect at the bubble.

The tension characteristic of the winder plays an important part in the production of the reel. It is generally recognised that a reducing or 'taper' rather than constant tension is desirable during the build up of a reel to eliminate excessive pressure on the inner layers.

As a further aid towards the production of good reels the extrusion die is often rotated in relation to the winder; this

Fig. 11. Sheeting line based on a 6 inch extruder. (Courtesy of David Bridge & Co Ltd.)

may be done by rotation of the die only, or the complete extruder with die, or the winder and haul-off together with the extruder and die stationary. By this rotation small irregularities in the film are distributed on the reel. The rotation does not, however, affect the film itself in any way.

The principle of the cast film process is illustrated in Fig. 9. Film is extruded through a straight slit and cooled by contact with two or more rolls whose temperature is controlled by recirculation of water. An air knife is sometimes used to improve the contact of films with the first roll.

The film made by the casting process, generally, has better optical properties than blown film but has no orientation in the transverse direction which produces weakness in this direction. Edge trim is usually necessary and this increases the conversion cost but on the other hand line speeds are not limited by the rate of cooling to the same extent that tubular film is.

The 'flat' die offers better facility for gauge adjustment but this is offset by the inability to distribute thickness variations on the reel by complete rotation of the die. Capital cost of cast film plant per unit output is substantially higher than tubular film plant.

An important development in the extrusion of film is the so called co-extrusion technique, where two or more layers of film are extruded simultaneously to produce a laminate with improved barrier, appearance or mechanical properties. Each layer is fed from a separate extruder. Both cast and blown film techniques are used; the composite tubular film may be rotated in the same way as in simple lay extrusion whilst the flat film die allows individual adjustment of the melt streams.

Low density polyethylene is the most common film material in use today, although the new development of paper-like high density polyethylene films promises to be an area of large expansion. Films are also extruded from polypropylene (mainly biaxially oriented), PVC, nylon and polyester. The main application for polyethylene film is in packaging but considerable quantities of film are also used in building, industrial sacks and many other applications.

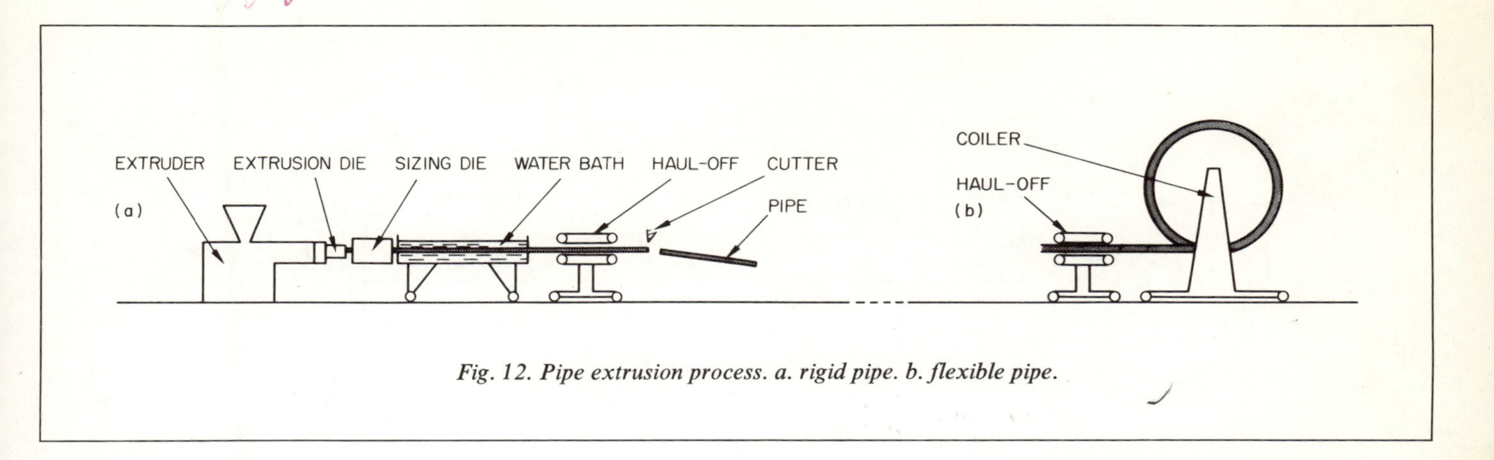

Fig. 12. Pipe extrusion process. a. rigid pipe. b. flexible pipe.

Sheet

The process is illustrated in Fig. 10. In principle this process is similar to cast film but a polishing roll is added to give good surface finish on both sides of the sheet. A slight degree of calendering does occur in the nip between the first and second rolls giving improvement of sheet gauge. The surface finish on one side of the sheet may be further improved by lamination of foil or flame treatment of the sheet surface. Die design which gives good flow distribution before the final die jaws is very important in eliminating uneven stresses in the finished sheet as is temperature control and minimum variations across the width of the polishing rolls. Since two of the main materials from which sheet is made (high impact polystyrene and ABS) tend to absorb moisture, vented extruders are finding application in this process.

High impact polystyrene is the most important sheet material. PVC, ABS and high density polyethylene are also extruded into sheet. A large percentage of all extruded sheet is subsequently thermoformed and this is sometimes done in-line. The thermoformed articles made from sheet, range from disposable cups and containers to refrigerator liners. ABS is beginning to be used for car bodies; decorative panelling and corrugated roofing is made from PVC.

Pipe

Pipe is extruded horizontally (Fig. 12) and generally sized on the outside diameter, using either pressure or vacuum. Where pressure is used the pipe is inflated against the walls of the sizing die and the pressure is maintained either by folding over the free end of the pipe or, if the pipe is rigid, a floating bung attached to the inner mandrel of the die is used. The newest and now most popular method of sizing pipe up to about 4 in (100 mm) in diameter is the use of the vacuum tank. Both methods are illustrated diagrammatically in Figs. 13 and 14. In the United States a significant percentage of plastics pipe is sized internally using a cooled die mandrel extension.

The main plastics used for pipe are polyethylene and PVC. Unplasticised PVC piping finds application in irrigation, guttering, soil pipes, electric conduit, *etc*, whilst soft plasticised PVC is extensively used for flexible tubing in a variety of applications. Domestic cold water piping is often made from low density polyethylene, whilst large sewage pipes are often made from high density polyethylene. Nylon tubing is used extensively in hydraulic and pneumatic applications.

Monofilaments

Monofilaments extruded through a multi-hole die are usually quenched in water maintained at a controlled temperature and then stretched between two sets of so called 'godet' rolls to produce orientation in the machine direction. The stretching stage may be followed by an optional annealing stage. After stretching (or annealing if used) each strand is individually wound-up. The process is illustrated diagrammatically in Fig. 15. Monofilaments are made from nylon, polypropylene, polyethylene, polystyrene and PVC.

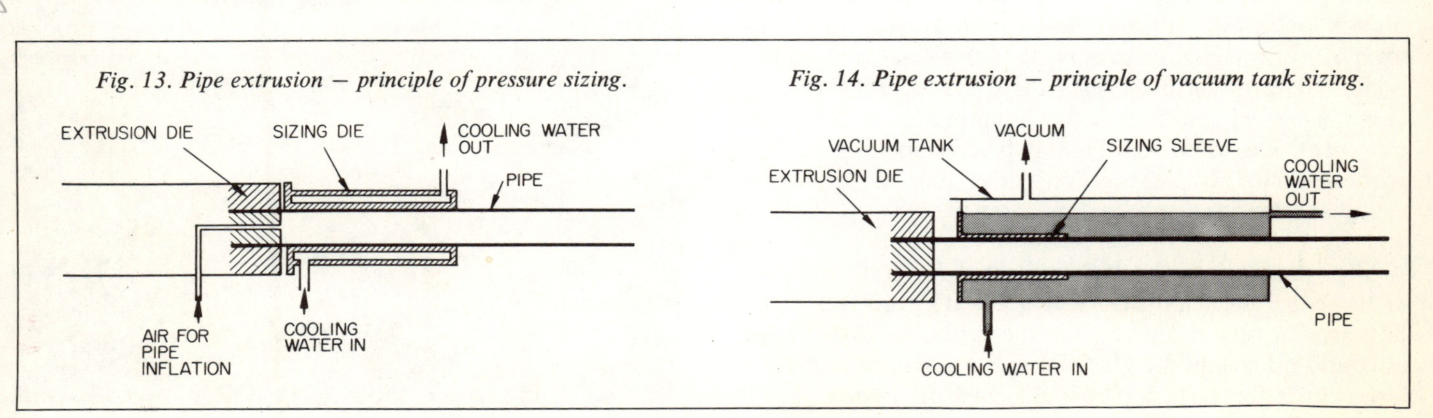
Fig. 13. Pipe extrusion — principle of pressure sizing.

Fig. 14. Pipe extrusion — principle of vacuum tank sizing.

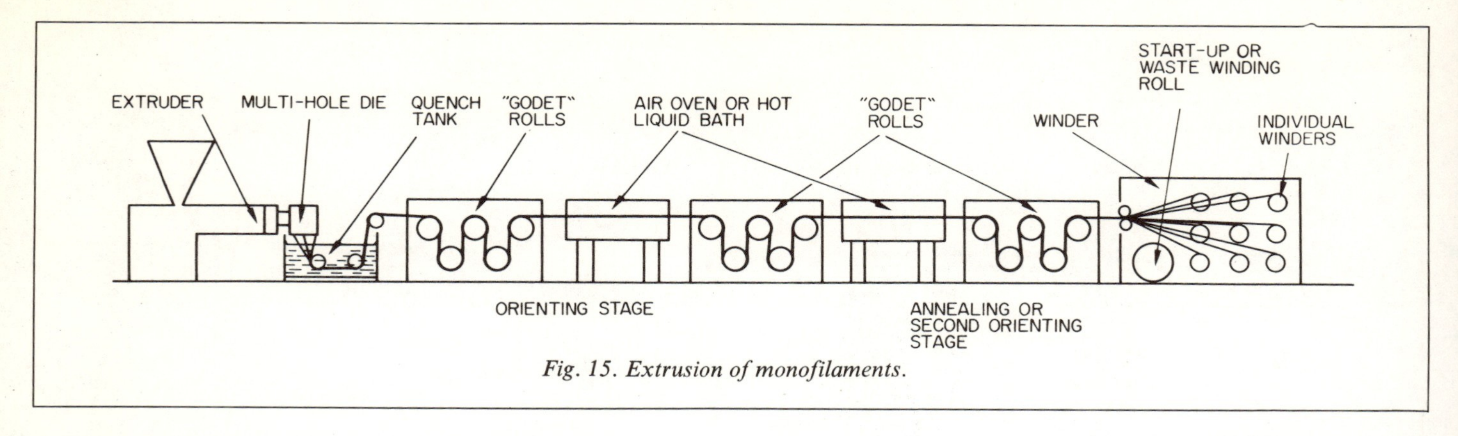

Fig. 15. Extrusion of monofilaments.

A comparatively recent development in the technique of monofilament production is based on extrusion of film which is subsequently slit and then processed in the same way as the conventional monofilaments. Tapes produced from slit film are used in many applications ranging from strapping, ropes and twines, to woven sacks and carpet backing. The main material here is polypropylene with some of the lighter tapes made in high density polyethylene.

Profile

The equipment used in extrusion of various profiles is generally similar to that used for pipe or tubing. The main difference is in the die, whose shape corresponds to the contour of the particular profile. The shape of the die determines the ultimate shape of the section but the two are not the same because of the complex flow within the die. Many of the problems associated with the die design for profile can and have been solved theoretically, but a majority of dies made still rely on practical experience and trial and error to achieve the exact shape required. The most important material used for miscellaneous sections is PVC both plasticised and unplasticised.

THE COATING PROCESSES

Two very important plastics extrusion processes in which molten polymer is used to coat another material are extrusion coating and cable covering.

The term extrusion coating (Fig. 16) is used in reference to a process in which molten plastics film, usually polyethylene, is applied to a continuous sheet (web) of flexible material such as paper, fabric, aluminium foil or another plastics film. No adhesives are used but in many applications some form of priming is necessary to achieve a satisfactory bond. The priming may be in the form of preheating by flame or heated roll, chemical coating or corona discharge treatment. Extrusion coating may be used to form a sandwich laminate with the polyethylene film in the centre acting as adhesive.

A large proportion of extrusion coated materials are used in the paper sack industry where a waterproof layer can be provided very cheaply and in the carton for distributing milk and other liquids. The packaging of frozen foods is a large outlet for extrusion coated board. Extrusion coated

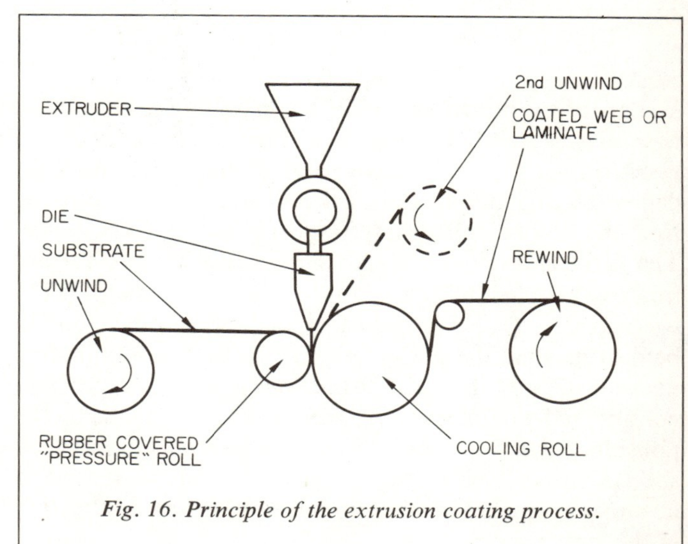

Fig. 16. Principle of the extrusion coating process.

Fig. 17. Extrusion coating process. (Courtesy of Bone Cravens Ltd.)

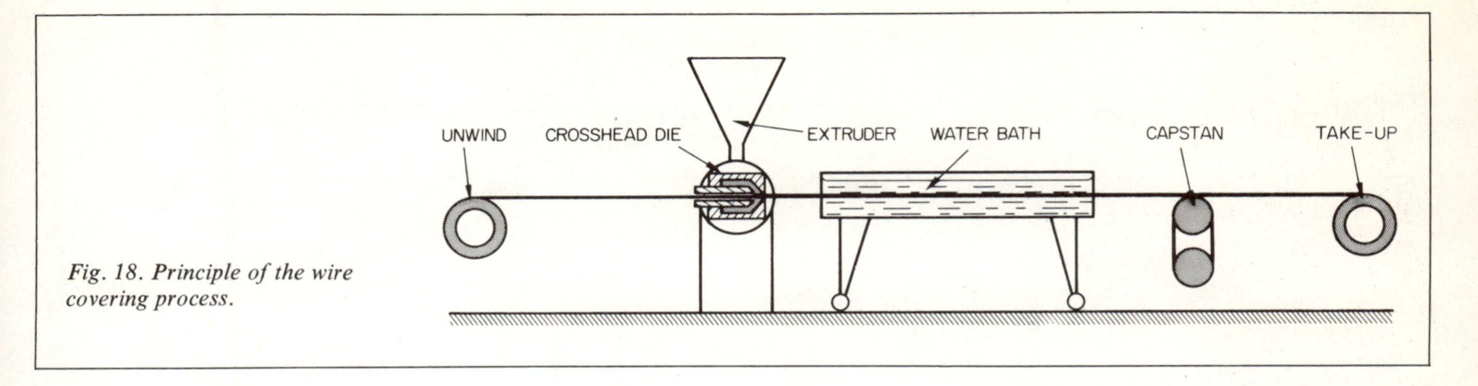

Fig. 18. Principle of the wire covering process.

packaging materials include polyethylene coated hessian, polyethylene coated woven high density polyethylene sacks, and so on.

In the wire covering process the conductor is passed through the extrusion cross-head and coated by the plastics melt either with the die or immediately after. Fig. 18 illustrates the principle of the process. The main plastics used for wire insulation are PVC and polyethylene and almost any wire can be coated by the extrusion process.

Fig. 19. Extrusion line for insulating household wire. (Courtesy of Francis Shaw Ltd.)

ECONOMICS OF EXTRUSION

In most extrusion processes the cost of material represents the major proportion of total operating cost and a typical breakdown of production costs for an extrusion process would be as follows:

Raw material	— 80%
Labour and overheads	— 14%
Depreciation	— 4%
Power	— 2%

It is evident from the above that any equipment feature which either increases the rate of production or reduces the amount of waste are of primary importance in the operation of an extrusion plant.

Good screw and die design combined with the correct choice of operating conditions can be used to achieve significant improvements in output and quality. Better quality of extrudate not only results in an improved product but in many cases can also reduce the usage of raw material. For example, in the manufacture of film a certain minimum thickness may be required to achieve the necessary mechanical strength and, therefore, with close gauge tolerance costly wastage can be reduced.

APPENDIX

Forming techniques known to be normally employed, or capable of being used, for each plastics

TYPE OF PLASTICS	Compression moulding	Transfer moulding	Injection moulding	Extrusion	Blow moulding	Thermoforming (vacuum forming)	Calendering	Rotational moulding	Dip moulding	Powder coating	Sheet for forming	Casting	Laminating	Expanding (foaming)	Encapsulation	Sintering	Glass fibre reinforcing
THERMOSETS																	
Alkyds (DAP and DAIP)	×	×	×									×			×		×
Aminos (urea and melamine formaldehyde)	×	×	×[1]										×	×[7]			
Epoxides	×									×		×	×		×		×
Phenolics	×	×	×	×								×	×	×	×		×
Polyesters	×										×	×	×	×	×		×
Polyimides	×															×	×
Polyurethanes	×			×								×	×	×			
Silicones	×	×										×	×	×	×		×
THERMOPLASTICS																	
Acrylics			×	×	×	×		×[3]			×	×	×	×[5]			
Acrylonitrile butadiene styrene (ABS)			×	×	×	×	×				×			×[3]			×
Cellulose acetate	×		×	×		×					×						
Cellulose acetate butyrate (CAB)	×		×	×		×	×	×		×	×						
Cellulose propionate	×		×	×		×				×	×						
Chlorinated polyether			×	×	×	×				×							
Chlorinated polyethylene			×	×	×	×		×			×			×[3]			
Ethylene vinyl acetate (EVA)			×	×	×												
Fluorocarbons: PTFE				×[2]							×		×			×	
PCTFE	×	×	×[2]	×[2]													
FEP			×	×		×					×						×
Ionomers			×	×	×	×											
Nylons			×	×	×[1]			×[4]		×		×				×	×
Polyacetal			×	×	×												×
Polycarbonate			×	×	×	×		×[3]									×
Polyethylenes			×	×	×	×		×			×			×			×
Polyethylene terephthalate			×	×													×
Polyphenylene oxide (PPO)			×	×	×												
Polyphenylene sulphide	×		×		×					×							
Polypropylene			×	×	×	×				×	×			×[6]			×
Polystyrenes			×	×	×	×		×			×			×			×
Polysulphones			×	×	×	×					×						×
Polyvinyl chloride (PVC)	×	×	×	×	×	×	×	×	×	×	×		×	×			×
Styrene acrilonitrile (SAN)			×	×	×												×
TPX (methylpentene polymer)			×	×	×	×											
Polyurethane (thermoplastic)			×	×			×										
Phenoxy			×	×	×												

NOTES: 1. Some grades. 2. With difficulty. 3. Not known to have been exploited commercially at this time. 4. Nylons 11 and 12. 5. Produced in Japan. 6. Produced in USA. 7. Urea formaldehyde.

Chapter 10

Sheet and Vacuum Forming

J C Evans
Telcon Plastics Ltd

Thermoplastic sheeting is one of the major basic material forms upon which the plastics industry has built up its every increasing potential over the past several decades. The very earliest practical thermoplastic sheeting materials available were gutta percha and subsequently celluloid, and from the knowledge that such materials could be heated, softened, shaped and reshaped, there has developed a comprehensive and sophisticated range of thermoplastic sheeting materials to cover a very wide range of applications and service conditions.

There are of course many applications in which plastics sheet is used in its original flat form, but the fact that it can be heated and shaped makes it of considerable additional interest and this chapter sets out to describe the main thermoplastic sheeting materials available, the facility with which they can be vacuum formed, the nature of the vacuum forming process and the equipment currently in use.

Before proceeding, however, it will be of value to have an appreciation of the essential characteristics of thermoplastic sheet materials, namely that when they are heated to just below melting point they become rubbery or elastic in nature to an extent which enables them to be stretched out rather like a balloon. They thus display a good 'hot melt strength' and it is this feature which enables various types of sheet material to be successfully formed.

Manufacture

Plastics sheet is manufactured by the main processes of press moulding, casting or by extrusion. In general, the very thick gauges of sheet continue to be press moulded, but by far the greatest proportion is produced by the continuous extrusion process (Fig. 1), and by this means sheet thicknesses ranging from 0.010 to 0.5 in (0.25 to 12.5 mm) can be economically produced, and thicknesses up to 0.75 in (22.8 mm) are also possible by extrusion.

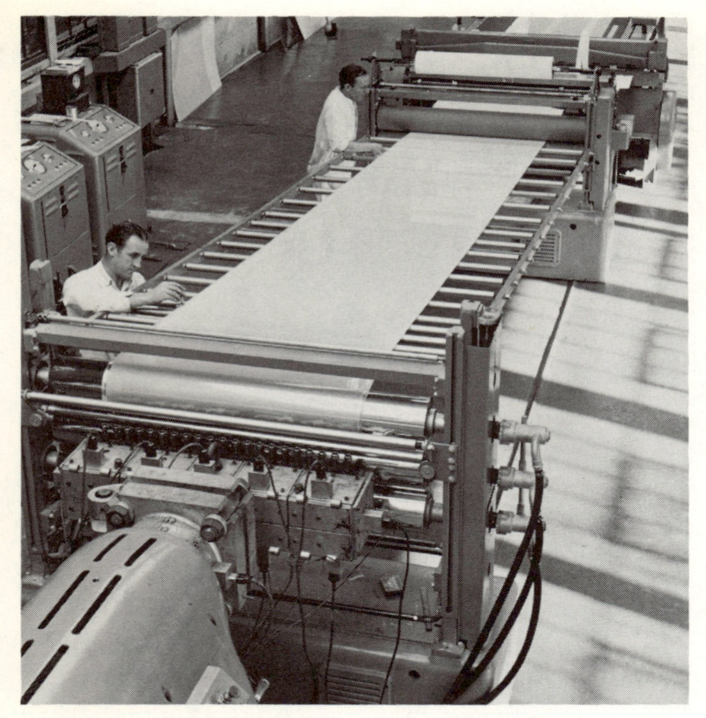

Fig. 1. Continuous sheet extrusion. (Courtesy of Telcon Plastics Ltd.)

Maximum widths have for some years been limited to 48 or 56 in (1219 or 1422 mm) but extruded sheet is now available up to 80 in (2030 mm) wide, and new extrusion plant will shortly be in operation capable of producing sheet up to 10 ft (3048 mm) wide.

VACUUM FORMING

Vacuum forming equipment in its most simple form consists of a vacuum box with air outlet and clamping frame, a heating panel, air compressor and mould.

The mould, which is partially hollow underneath and drilled through with a number of fine holes, is placed over the air outlet. The plastics sheet to be formed is placed over the open top of the box and clamped down by the frame, thus sealing off the box and making it an airtight compartment (Fig. 2a). The heater panel is then placed over the plastics sheet, at a distance of 5–6 in (127–152 mm) in order to heat the sheet as uniformly as possible (Fig. 2b).

When the sheet has been raised to a temperature a little below its melting point, the heater is withdrawn and air is evacuated by means of the vacuum pump. This causes the plastics sheet to be sucked down into and over the mould and to form an accurate reproduction of the mould contours (Fig. 2c). The mould can be either male or female or a combination of both.

When the sheet has cooled and hardened the clamping frame is then released, the formed sheet removed from the mould and the surplus material trimmed off. A basic machine is shown in Fig. 3.

Fig. 3. Basic vacuum forming machine. (Courtesy of M L Shelley & Partners Ltd.)

Fig. 2. The vacuum forming process.

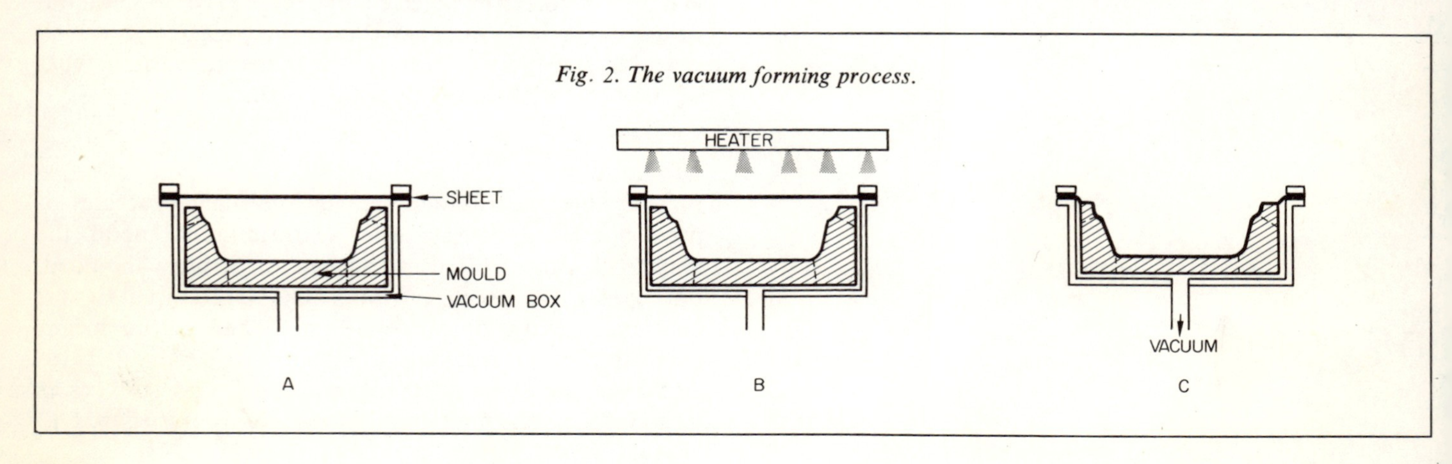

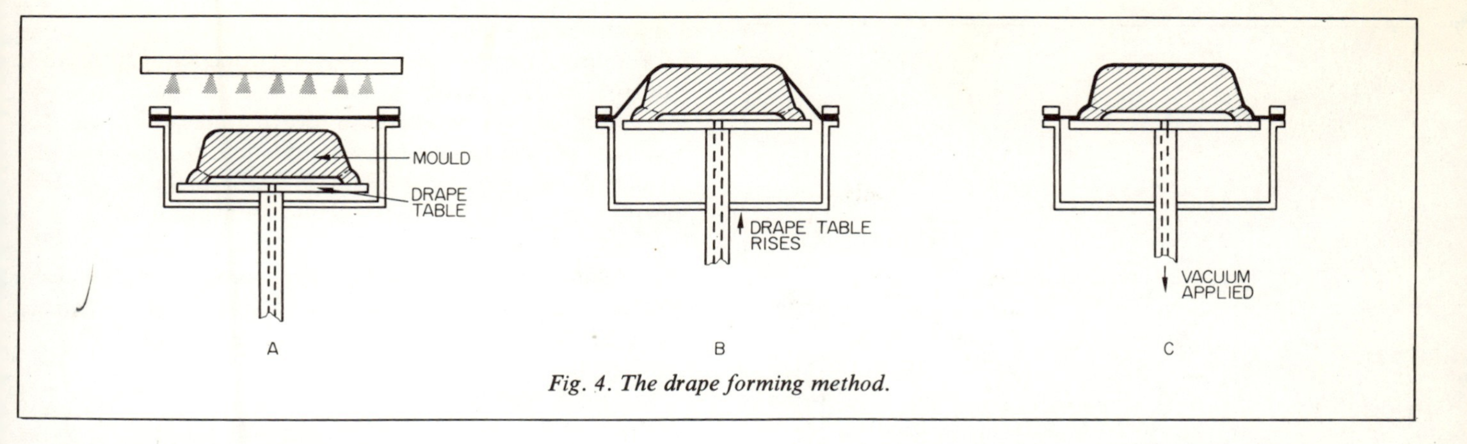

Fig. 4. The drape forming method.

PRODUCTION MACHINES

A custom made vacuum forming machine for continuous production will have a number of additional refinements, according to purchase price, as follows.

Frame

In the more simple machines hand operated toggle clamps are used, but with the more expensive equipment the frame will be hydraulically operated, the clamping margin around the edge of the sheet is usually a minimum of 1 in (25.4 mm).

Drape table

An hydraulic ram enables the mould mounting platform to rise within the vacuum box and push the mould into the softened sheet immediately prior to the vacuum being applied (see Fig. 4). This allows a certain degree of pre-forming to take place and also facilitates subsequent mould withdrawal.

The depth of the vacuum box, and hence the extent of travel of the drape table, is an important feature of a vacuum forming machine, since it determines the depth to which a product can be formed. Thus, a machine specified as having a '12 in (305 mm) drape' would be capable of producing products to approximately this depth.

Bubble or air assist

A valve switching system enables air to be pumped into the vacuum box to pre-stretch the sheet into a bubble prior to the mould table rising (Fig. 5). This facility lessens the possibility of the hot sheet thinning out at the sharper or higher parts of male moulds in particular, and makes for a more uniform cross section. This arrangement is particularly useful when forming some of the more difficult materials such as polypropylene, as further illustrated in Figs. 6 and 7.

Care must be taken to ensure that too large a bubble is not formed as this will result in a surplus of sheet material when forming. A photo electric cell and scanning beam can be fitted at a suitable height above the sheet so that when the bubble breaks the beam the air is automatically shut off (Fig. 5).

Fig. 5. Pre-stretching by the bubble assist method.

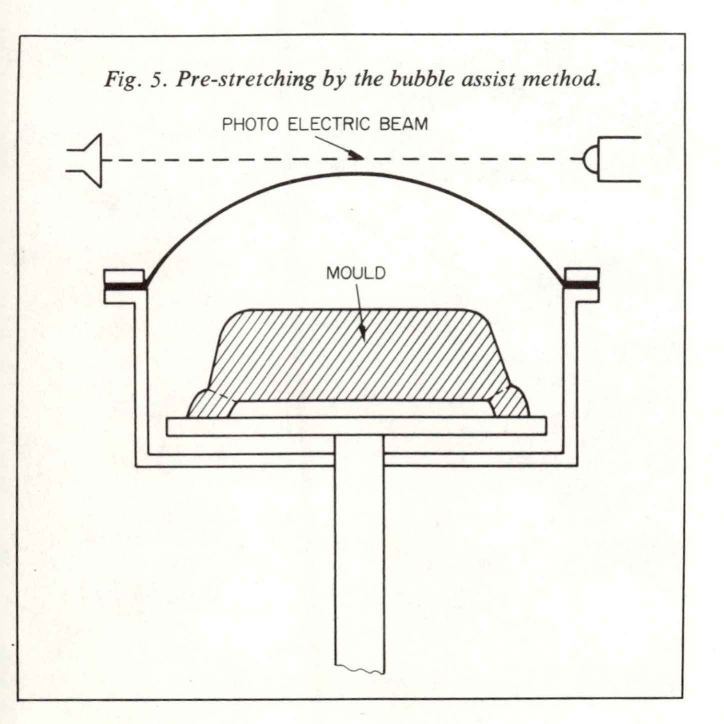

Fig. 6. Bubble forming polypropylene sheet. (Courtesy of Component Plastics Ltd.)

Fig. 7. Bubble forming completed. (Courtesy of Component Plastics Ltd.)

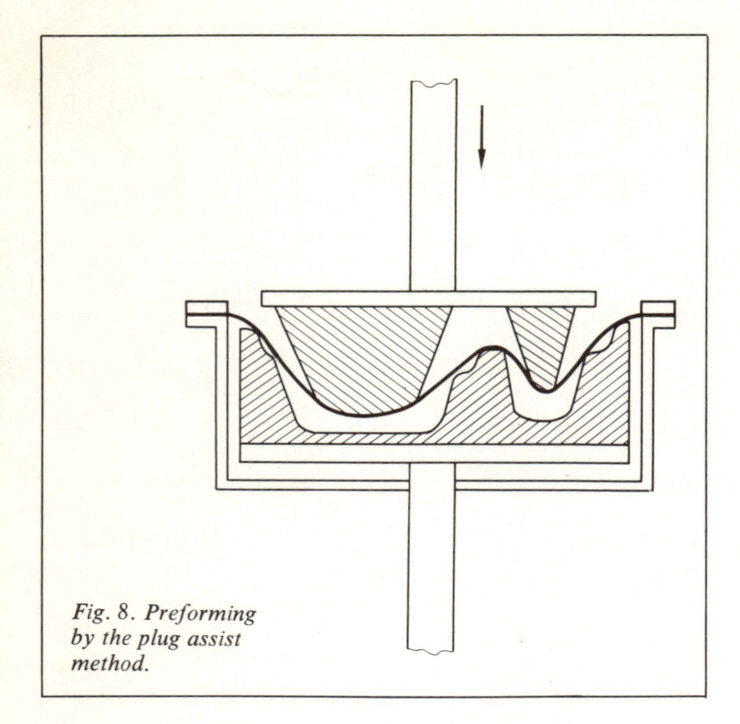

Fig. 8. Preforming by the plug assist method.

This blowing feature can also be used for assisting removal of the formed sheet by blowing it off the mould when sufficiently cooled and hardened.

Plug assist

This consists of another hydraulic ram which enables a rough preform mould plug to be pushed into the top of the sheet, again immediately prior to vacuum being applied. This is of particular value with deep formings on a female mould (Fig. 8).

Heaters

Heaters can be of several types but the most widely employed is the infra-red ceramic type of about 9 ... 3 in (228 ... 76 mm) with a wire element sealed into it. A series of these units are assembled to make up heater panels of various sizes, and can be wired into zones with individual controllers, thus enabling more heat to be switched to certain parts of the sheet if required.

The heater is arranged to slide over the sheet by hand or by hydraulic operation. Single heaters are widely used and will form sheet up to about $\frac{1}{8}$ in (3 mm) dependent on the material. Double or 'sandwich' heaters (*ie* on both sides of the sheet) assist greatly in speeding up the heating time and will certainly be essential for the forming of polypropylene and HD polyethylene in excess of $\frac{1}{8}$ in (3 mm).

The heating time may be anything from $\frac{1}{2}$ to 5 minutes according to material, thickness and type and size of equipment. As an approximate guide a time of 1 second per 0.001 in thickness of material would be required for materials such as polystyrene and ABS, heater loading being about $1\frac{1}{2}$ kW/ft^2 plus about 1 kW/ft^2 for the bottom heater where sandwich heating is used, and about twice this figure for polypropylene.

Automatic operation

The more expensive machines are fitted with automatic controls which enable a completely automatic cycle of operation of any or all the above features following the loading of the plastics sheet. Standard machines are available but most manufacturers will make up special purpose machines to suit particular applications.

Fig. 9 shows an automatic machine, complete with process timers, that enables the following sequence of operations to be carried out once the sheet is loaded:

(a) Push start button.
(b) Heater comes forward for set time.
(c) Heater retracts.
(d) Air injected to form bubble—if required.
(e) Drape table rises or plug assist descends.
(f) Vacuum applied.
(g) Cooling air or water mist spray applied to sheet surface.
(h) Vacuum and air off, plug assist (if used) withdrawn.
(i) Air pressure on to release sheet from mould.
(j) Air pressure off.
(k) Drape descends or plug assist ascends.
(l) Operator removes forming and reloads.

Fig. 9. Automatic forming machine with double heat and plug assist. (Courtesy of M L Shelley & Partners Ltd.)

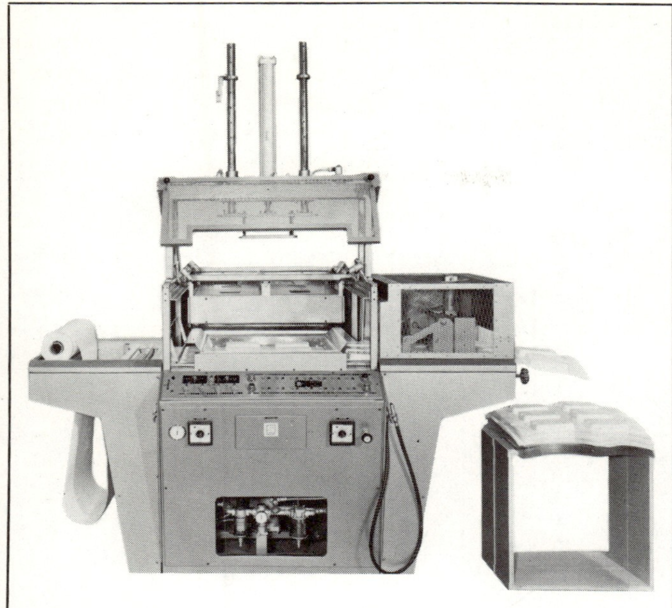

Fig. 10. Automatic vacuum forming machine with roll feed and cutter. (Courtesy of M L Shelley & Partners Ltd.)

Fig. 11. Roller trimming press. A demonstration composite cutting board is shown and such presses can be supplied to trim formings up to a depth of 15 in (380 mm). (Courtesy of Ridat Engineering Co.)

Further automation can be provided with thinner sheet formings by continuous roll feed as shown in Fig. 10.

Moulds

For forming prototypes and small batch quantities, wood or plaster moulds may be used. Hard wood is frequently employed but for extended production runs epoxy resin or aluminium are to be preferred.

Male moulds are easier to construct and hence cheaper and are generally used for deeper formings. The sides of the mould should be slightly angled or tapered to facilitate easy removal of the formed sheet, particularly with a male mould. This taper is not quite so important with a female mould as the formed sheet will contract upon cooling and release freely from the mould. Whilst it is possible to form undercuts, this complicates the mould by necessitating removable parts and lengthens the forming cycle. Vacuum holes through the mould must be kept small otherwise the sheet will be drawn into them, a maximum hole size of 1/32 in (0·7 mm) is recommended and they should be concentrated particularly around the more detailed part of the mould.

The main vacuum outlet hole in the base plate must not be obstructed and it is usual to mount the mould onto a sheet of wire gauze on top of the base plate in order to ensure uniform suction.

Trimming

Trimming equipment can vary from an inexpensive knife to a sophisticated cutting press. For small to medium sized products a roller press (Fig. 11) is very popular and high production rates can be achieved. For the larger items a power driven router operating with a suitable cutting jig is most effective and such equipment is usually set up from local resources.

MATERIALS

Although there are a few thermoplastic materials which are not entirely suitable for vacuum forming, there is a sufficiently comprehensive range of sheeting available to enable the best combination of characteristics to be chosen for both the end product application and the production facility with which the article can be vacuum formed.

The materials most generally used for forming may be listed in descending order of processing facility as follows:

Polystyrene
ABS (acrylonitrile butadiene styrene)
PVC (polyvinyl chloride)
Acrylic
CAB (cellulose acetate butyrate)
Polycarbonate
Polyethylene (high density)
Polypropylene
Polyethylene (low density)

Each material has its own properties and characteristics and in many cases there are several different grades of each material. The general physical characteristics are detailed in Table 1.

Polystyrene

Polystyrene in its basic form is a clear material, however, it has a low impact strength and there has been developed the so-called 'high impact' (HI) or rubber toughened polystyrene. This is an opaque material, excellent for vacuum forming and the cheapest and easiest material to form.

Although HI polystyrene still does not have a particularly high impact strength compared with other sheet materials,

Table 1. General guide to the properties of thermoplastic sheeting.

	High impact polystyrene	ABS	PVC	Acrylic	CAB	Polycarbonate	Polypropylene	High density polyethylene	Low density polyethylene
Specific gravity	1.1	1.05	1.4 approx	1.18	1.2	1.2	0.90–0.91	0.94–0.96	0.92–0.93
Tensile strength—lbf/in²	3000–4000	5000	6000	7000–10000	3000–4000	9000–9500	4000–5000	3000–5000	1000–2300
kgf/m²	210– 280	350	420	740	210– 280	650	280– 350	210– 350	70– 160
Elongation %	approx 20	20–140	approx 20	2–10	50	90	200–700	50–1000	100–600
Impact strength Izod ft lb/in. notch	2	3–12	5–10	0.5	Up to 5	16	Up to 6	Up to 20	Does not break
Hardness—Shore	—	—	D.80	—	—	—	D.90	D.65	D.45
Rockwell	R.50–100	R.100	—	R.99	R.100	R.120	—	—	—
Vicat softening point (°C) approx.	80–90	85	75	115	70	160/170	150	120	90
Cold bend temperature (°C)	—	−40	approx −10	—	−40	−90	approx 0	below −75	below −75
Dielectric strength (volts/mil)	550	400	350	320	400	400	1000	1000	1000
Volume resistivity (ohm cm)	10^{16}	10^{16}	10^{16}	10^{16}	10^{15}	10^{15}	10^{17}	10^{17}	10^{18}
Flammability	Burns	Burns slowly	Self-extinguishing	Burns slowly	Burns slowly	Self-extinguishing	Burns slowly	Burns slowly	Burns slowly
Outdoor weathering	Fair	Good	Good	Excellent	Good	Excellent	Moderate—but 2% carbon black pigmentation and/or U/V stabilisers provide protection		
Clarity	Opaque (sheet) Clear (film)	Opaque	Clear to opaque	Clear	Clear	Clear	Translucent	Translucent	Translucent
Formability	Excellent	Excellent	Good	Good	Good	Good	Fair	Good	Moderate
Approx order of price comparison per square foot	1 (cheapest)	5	5	5	6	7	2	4	3
Resistance to chemicals	Fair	Good	Very good	Good	Moderate	Good	Excellent	Excellent	Excellent

Fig. 12. Catamaran style hull and inner liner formed from ABS sheet. (Courtesy of Marbon Europe NV.)

Fig. 13. Mini-front replacement panel in ABS. (Courtesy of Motoline Ltd.)

it is sufficiently robust to be widely used for the less critical applications. The thinner sheeting is extensively used for pre-packed foods. Clear film is also available and this is produced by a biaxially orientated process which effectively toughens it, the resultant film being widely used for blister packs.

ABS
This material forms very readily and there are various grades available varying from very high to medium impact strengths. The high impact grades are very tough and the material is of particular interest for larger area formings such as boats, caravan parts, car bodies, containers, *etc* (see Figs. 12 and 13).

ABS is hygroscopic and must be kept dry prior to forming, the moisture absorbtion is relative however, and does not in any way affect its end performance in service.

PVC
Has the feature of being self extinguishing, can be obtained as a clear as well as an opaque sheet. A vacuum forming grade must be specified. The thinner sheet gauges form well and are also used for food packaging and blister packs, but thicker sheet may need to be pressure formed.

Acrylic
Although acrylic has for many years been traditionally shaped by pressure forming, the advent of extruded acrylic sheet has made vacuum forming a more economic process as extruded sheet is cheaper and more readily formable. A tough attractive material in clear or opaque with excellent weathering properties. The low impact strength figure given in Table 1 for acrylic may appear somewhat misleading, the explanation is that Izod is a notch impact test, and acrylic is a notch sensitive material hence sharp corners in design should be avoided.

CAB
This is also a good forming material similar to acrylic but with a higher impact strength and requiring a relatively high forming temperature. Whitening in the formed product indicates insufficient heating. CAB like ABS and acrylic tends to be hygroscopic and must be kept dry prior to forming.

Polycarbonate
A clear material, vacuum forms well and has an exceptionally high impact strength and remains tough at considerable extremes of temperature. Its use has so far been limited to specialist applications due to high cost, but the material is likely to become more competitive in price.

Polypropylene
A rigid tough translucent material but more difficult to form (unless the right equipment is available) largely due to its high and narrow temperature softening range and the fact that the sheet tends to distort into hills and valleys when being heated. This can be overcome to a certain extent by preheating the sheet, also excessive sheet sag can be reduced by using high wattage heaters 3–4 kW/ft^2 which enable the sheet to be heated in the shortest possible time. The forming temperature is very near to melting point and this can be clearly seen by the sheet becoming almost transparent immediately prior to forming.

Sheet up to 0.1 in (2.54 mm) will form with reasonable facility on single heater machines but beyond this double heaters are required. Polypropylene is the lightest thermoplastic sheet now available.

HD polyethylene
Many of the recommendations made for forming polypropylene can usefully be applied to HD polyethylene except that the material has a somewhat lower melting point and heats more uniformly.

It is a rigid tough translucent sterile material and useful for a range of vacuum forming applications, as it is economically priced and capable of being formed into deeper drawn articles such as tanks (Fig. 14) which can be used over a wide operating temperature range.

Fig. 14. 100 gallon container formed from HD polyethylene. (Courtesy of Arun Plastics Ltd.)

LD polyethylene

A tough flexible material not particularly suitable for vacuum forming, it will, however, form quite well into shallow draw articles providing there are no sharp corners.

Finish

Sheeting is normally supplied with a plain surface finish, but embossed surface patterns are available and decorative film can be laminated to the base sheet in the case of polystyrene, ABS and PVC. All the above sheeting can be obtained in a range of colours.

ECONOMICS

The particular attractions of vacuum forming are the low tool costs and the relatively low capital cost of vacuum forming plant when compared with the cost of injection moulding machines to produce similar sized articles. Indeed many of the larger formings are quite beyond the practical physical capabilities of injection moulding equipment. Vacuum forming moulds can be readily modified and quickly changed. Whereas an injection moulding machine will only become a viable proposition with an initial order of many thousands, the vacuum forming process is sufficiently flexible to deal economically and profitably with initial orders of a few hundred.

The operational time cycle for vacuum forming will of course vary according to the complexity of the product and the nature and thickness of the sheet material. Smaller items can be produced on multi-impression moulds and fully automatic special purpose machines are available which will produce from thin sheet food pack containers trimmed, content filled and sealed at a rate of many thousands per hour.

With thicker larger articles the cycle time will take several minutes, but even in the case of a boat or caravan shell a time of little more than ten minutes may well suffice.

It is of particular economic advantage if the trimming operation is adjacent to, and can be combined with, the the vacuum forming operation. In this way with a machine having an automatic or semi-automatic operation, one

Table 2. Relative costs of vacuum forming machines.

Frame size	Drape (Vacuum box)	Heater	Addition for plug assist	Approximate price		
				Manual	Semi-automatic	Fully automatic
19 × 13 in (483 × 330 mm)	4 in (102 mm)	Single	—	£280	—	—
18 × 18 in (457 × 457 mm)	4 in (102 mm)	Single	—	£450	—	—
26 × 26 in (660 × 660 mm)	8 in (203 mm)	Single	—	£525	£1000	£1570
26 × 26 in (660 × 660 mm)	8 in (203 mm)	Double	—	—	£2300	£2850
36 × 30 in (914 × 762 mm)	15 in (380 mm)	Single	£420	—	£2550	£3150
36 × 30 in (914 × 762 mm)	15 in (380 mm)	Double	£420	—	£3500	£4120
50 × 48 in (1270 × 1220 mm)	15 in (380 mm)	Single	£560	—	£4100	£4750
50 × 48 in (1270 × 1220 mm)	15 in (380 mm)	Double	£560	—	£4750	£5500
80 × 48 in (2032 × 1220 mm)	24 in (610 mm)	Double	£1650	—	—	£9100

Considerably larger machines are available, generally custom built, at corresponding higher costs; frame sizes of 15 × 8 ft (4570 × 2438 mm) and up to 20 × 10 ft (6096 × 3048 mm) are currently in commission.

Fig. 15. Model car bodies formed in ABS. (Courtesy of M L Shelley & Partners Ltd.)

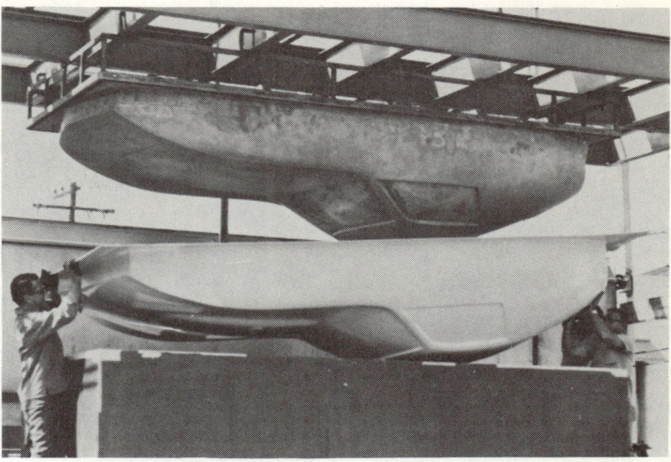

Fig. 16. Formacar in ABS manufactured by Centaur Engineering Co of the USA. (Courtesy of Marbon Europe NV.)

operator having loaded a sheet into the frame and started the forming cycle, can in many cases effectively trim the previous forming and maintain a combined form/trim operation.

The choice of vacuum forming machine will depend upon how extensive a range of products are to be produced, but if it is decided to invest in a size of machine somewhat larger than the one immediately required, then many machines have adjustable frames and zoned heaters to enable smaller items to be accommodated or one can, with better advantage, plan to produce on a two or more impression mould. Table 2 provides a general guide to relative costs of vacuum forming machines and additional facilities.

APPLICATIONS

Applications for vacuum formed products are legion and many articles can be formed from several types of material depending on the service and quality required. The following list provides a general application guide relative as far as possible to the different types of sheet material.

Polystyrene. Advertising models and signs (internal), display cards, packaging trays and containers, egg boxes, tote boxes, refrigerator liners, plant holders.

PVC. Packaging, tote boxes, containers, wall panelling, lighting panels, machine covers and cases.

ABS. Boats, caravans, trailers, cars, lorry and tractor cabs, vehicle body parts, luggage, refrigerator liners, tote boxes, materials handling containers, prams, machine covers and cases, baths, sinks, shower trays, surf boards, shelters (bus, workmen, telephone), wall panelling and external facing.

Acrylic. Lighting fittings and panels, advertising signs, roof lights, aircraft domes, baths, shower screens and machine guards.

CAB. Similar applications to acrylic but with higher impact strength. Also packaging applications for thin film.

Polycarbonate. Similar applications to acrylic and CAB but more expensive. Used where very high impact and clarity is required and higher temperature service application.

Polypropylene. Food containers and packaging, non-toxic containers that can be boiled and used in micro-wave ovens, television and radio back covers, engine covers, higher temperature service applications.

Polyethylene. (A)High density—water tanks, industrial containers, tote boxes, fish boxes, boats, gravel and salt holders (road side), mud guards, deep freeze containers. (B) Low density—flexible shallow formings, medical face masks.

DEVELOPMENTS

With regard to items formed from thinner sheet such as packaging and food containers, development has been largely concentrated on obtaining faster production cycles by improved machine design providing more efficient automatic facilities, faster heating and removal from mould facilities, built-in trimming devices, *etc.*

Fig. 17. Forming process of a caravan body section by Centaur Engineering Co of the USA. (Courtesy of Marbon Europe NV.)

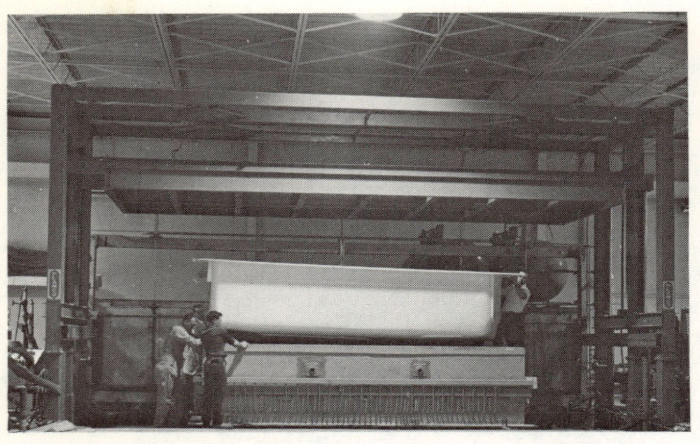

Fig. 18. Completed caravan body section formed from ABS sheet by Centaur Engineering Co of the USA. (Courtesy of Marbon Europe NV.)

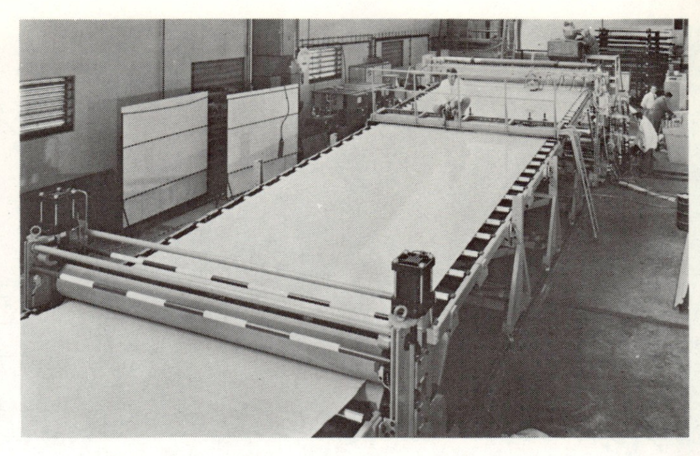

Fig. 19. 3 400 mm wide sheet extrusion plant now installed by Telcon Plastics Ltd.

The major developments over recent years, however, have been centred around the ever increasing size of formings and hence the wider range of major applications to which they may be applied. The toy car bodies shown under manufacture in Fig. 15 are typical of the size and nature of formed articles widely produced throughout the industry. The possibility, however, of forming products of the very considerably larger dimensions of the vehicle body shown in Fig. 16 presents a new field of particular interest to the designer and applications engineer. Machines are now becoming available capable of forming sheet up to 10 ft wide by 25 ft long (3 × 7.5 m), and although the limitation has been largely that of obtaining sheet of such dimensions, new sheet extrusion plant will shortly be in production to adequately satisfy this demand (Fig. 19).

Fig. 20. Exhibition building externally clad in ABS vacuum formed panels. (Courtesy of Marbon Europe NV.)

It will be appreciated that formings of this size require not only a material that can be readily formed, but also one that is particularly tough and robust to withstand the more rugged service conditions that such large components would be subjected to. The material that predominantly fulfils these requirements is ABS and this is amply illustrated by Figs. 17 and 18 which show the forming and completion of what is to date the largest vacuum forming in the world. This is a caravan body produced in the the USA on a four stage rotary machine and formed from sheet 87 in (2210 mm) wide by 192 in (4850 mm) long. The forming could equally well be a large storage tank, a shipping container or a one piece formed bath room unit.

Although sheet of a fairly substantial thickness, from about $\frac{1}{4}$ in (6 mm) upwards, would be required for such depth of formings, a sandwich construction comprising inner and outer formed sheets filled with injected polyurethane foam will provide an extremely rigid tough yet lightweight construction which can if necessary have a cross section of several inches.

Vacuum formed products can thus be very large but they can also be very extensive and versatile in application, an example of this is the building shown in Fig. 20 externally clad in ABS decorative panels. This type of development is extending considerably in Europe although building regulations in the UK continue to present certain restrictions. However, much work is being done on further development of fire retardent materials and it is to be hoped that more progressive application of plastics sheet to the building construction industry will become increasingly possible.

It will thus be seen that the versatility of thermoplastic sheet and the techniques of vacuum forming present a field of application which is not only intriguing to the production and design engineer but also demands particular attention for the commercial viability and the widening marketing prospects which its end products will increasingly command.

Chapter 11

Calendering

K J Hardman *CEng MIMechE MIPlantE AMCT*
Imperial Chemical Industries Ltd

The actual calendering of plastics is the final act of production after a lot of effort has been put into the preparation of the material at the right temperature and texture for introducing to the calender. It is therefore proposed to take the process of producing PVC sheeting step-by-step and discuss the most important or difficult points.

Raw material for plastics production is often loosely referred to as a resin, a term which covers all PVC polymers and copolymers. Resins, along with the various fillers, which are usually inert mineral compounds added to adjust the cost and modify the physical properties, are supplied in standard 25 kg bags. These are formulated along with colour pigment, lubricant to assist processing, stabilisers to delay deterioration, and, if soft sheet is required, a large percentage of plasticiser oil such as DAP or TXP.

The formula is first blended to give complete dispersion then subjected to violent agitation and heat to cause the plastics to gel. Further blending and straining or filtering follow before the material is calendered, cooled, measured and wound up into rolls. Fig. 1 shows a typical layout of the various items of plant used.

PRE-CALENDERING PROCESS

The various dry powders in the formulation are loaded into a ribbon type blender with the exception of the colour pigment (see Fig. 2), and agitated long enough to produce an even dispersion, the time being found by experiment and test. If oil is to be added to produce a soft mixing, this is sprayed into the powder during the initial period of blending (*ie* the first quarter of the cycle). When the dispersion is satisfactory, the blend is removed via a valve in the base of the mixing chamber (see Fig. 1) and weighed-out into batch tins which can be stored on a conveyor ready for the next process. The colour pigment is added to each separate batch at this stage. This is usually pre-weighed into PVC plastics bags for easy handling.

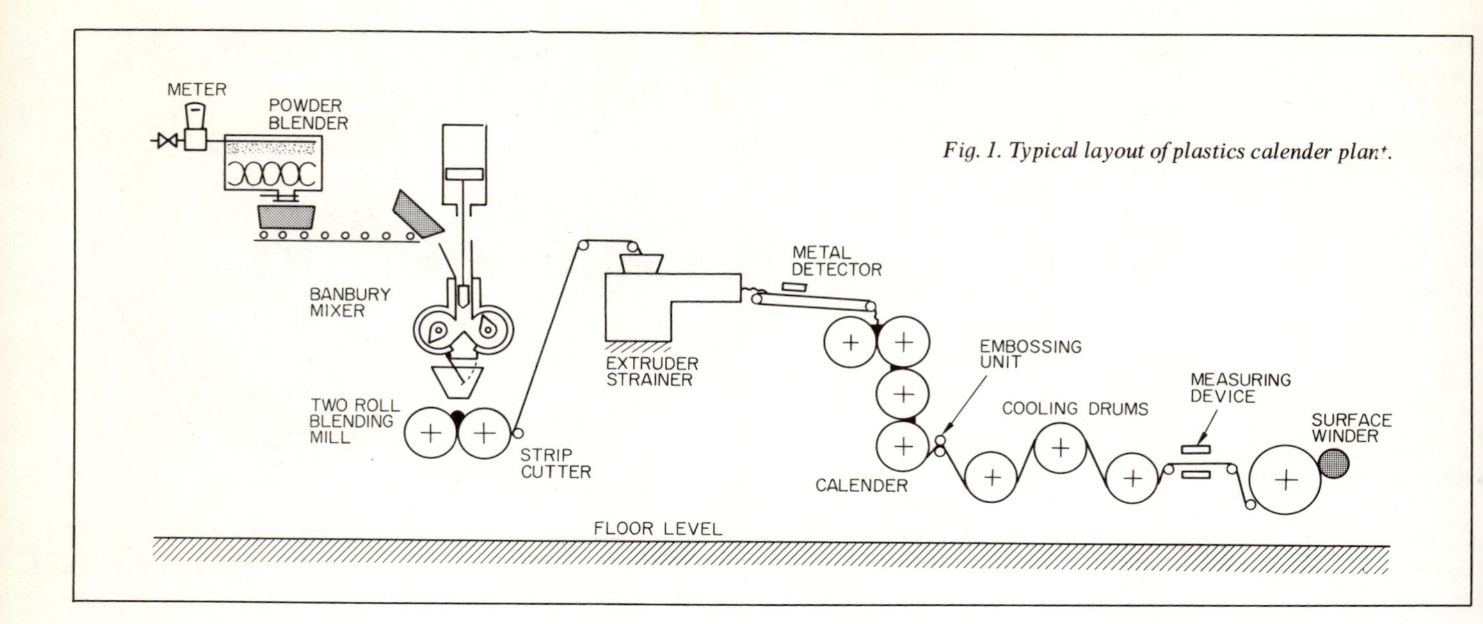

Fig. 1. Typical layout of plastics calender plant.

Mixing and 'gelling'

Up to now the polymer and powders have not been subjected to any heat and only a mild form of agitation. They are therefore still in their original form and can be stored for considerable periods if desired, provided they are covered to exclude unwanted dust and impurities.

The batch tins are now emptied into an enclosed type of primary mixer (indicated in Fig. 1 as a Banbury mixer). The advantages of an enclosed machine for mixing powders will be obvious, especially if the batch contains a minimum of plasticisers. The agitation and beating action of the machine cause powder to become airborne, and the enclosure thus confines most of it and ensures that it ends up where it is wanted.

The Banbury mixer, named after the inventor, is a heavy duty machine consisting basically of two hollow shaped rotors, each of which has four 'wings' cast in, which, when rotated in opposite directions, impart a kneading action to the plastics mix. The rotors are provided with glands to make them suitable for steam heating and the outer casing is arranged for either heating or cooling. The batch is loaded by lifting the 'floating weight' (see Fig. 1) with the air cylinder provided and then by applying air pressure in the reverse direction, a downward load is applied to the batch which assists greatly in the mixing process.

The Shaw Intermix is a similar machine in outward appearance but differs in its mixing action. The rotors 'intermix' with a friction ratio of approximately 1.5:1, and the mixing action of this machine tends to take place more between the rotors. The downward pressure on the floating weight is controlled to suit different mixing formulations and the outer casing arranged for either heating or cooling.

Sealing the holes in the end casings through which the rotors pass is very important. The mixing in the casing must not be lost but it must not get trapped in the seal where it can degrade and contaminate subsequent mixings. Lubrication is solved by the use of a high pressure pump using an acceptable plasticiser oil. In this way any excess oil supplied to the seals is taken up by the mixing without causing any trouble. The standard method of releasing the mixes is by means of a hinged door which normally forms the base of the mixing chamber.

A high speed type of primary mixer known as a Gelimat is sometimes used (see Fig. 3) which takes up much less floor space and will partly 'gel' very dry mixings, which are either very difficult or not possible on other types of machines. This machine has a single solid shaft on which are welded, at right angles to the axis, a number of rectangular flights with the tips alternately twisted to the left and right. The end flights are, however, set at a slight angle to the axis and are given a minimum end running clearance; these end flights keep the material moving towards the middle of the mixing chamber. Because of its high speed the material is thrown by centrifugal force to the outer diameter of the casing and, as a result, any escape of

Fig. 2. View of interior of ribbon type powder mixer in operation (Courtesy of ICI (Hyde) Ltd.)

powder from the shaft gland is minimal. The mixing shaft rotates at speeds from 750 to 1 500 rev/min in a simple cylindrical casing which can be opened along the centreline for easy cleaning. Heating or cooling of the rotor or casing are not necessary.

Yet another type of mixer, quite different from either kind mentioned, is known as a Buss Ko-Kneader. It consists of a cylindrical centre mixing element which has three rows of parallel teeth, and operates with a combined rotary and reciprocating action. An outer casing also cut with shaped teeth is arranged to split open to assist with cleaning. This machine is good for long sustained runs of one quality, but when short runs and quality changes are required, cleaning time is a major consideration. Edge trimmings cannot be returned directly to the feed hopper of the Buss Ko-Kneader but must first be granulated. The use of a blending mill is necessary as with other types of primary mixer.

The choice of a primary mixer is vital to the success of the calendering plant and before deciding which mixer best suits the requirements many questions must be answered, some of the more important of which are listed below:

1. Which type will efficiently mix the range of formulae required at the correct rate?
2. Which machine fits best into the layout and building?
3. Which type is easiest to keep clean and show a minimum of contamination when colour changes are made?
4. What are the relative capital and running costs?

Blending mill

A standard mill (see Fig. 4) of the type in general use in the plastics industry, has two horizontal rolls geared together to run in opposite directions, with the surface speed of one roll 10% faster than the other.

The partly 'gelled' mix from the primary mixer above will pass through the 'nip' formed by the rollers and, with assistance from the operator, form a continuous sheet round the front roller. The excess material in the mill will form a bank which must be made to circulate by partly cutting away the continuous sheet formed on the front roller and allowing it to be replaced from the bank above the nip, thus leaving room for the excess to pass through the nip. This action of rolling the material in the nip will blend and even out the temperature of the whole mass to that of the surfaces of the rolls thus carrying the gelling process a stage further.

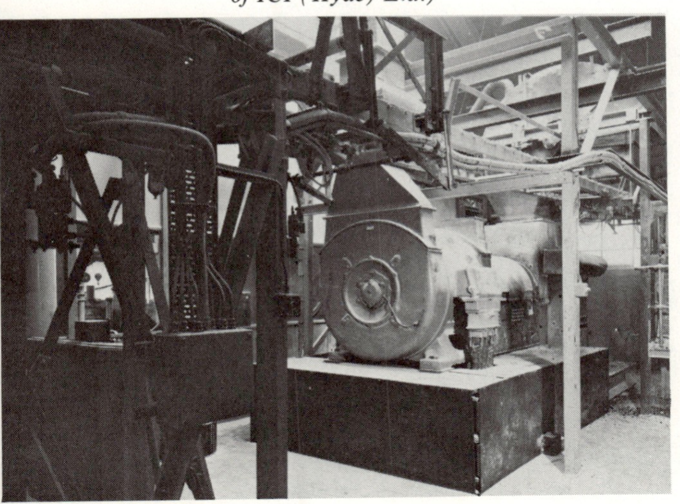

Fig. 3. General view of a Gelimat mixer and drive. (Courtesy of ICI (Hyde) Ltd.)

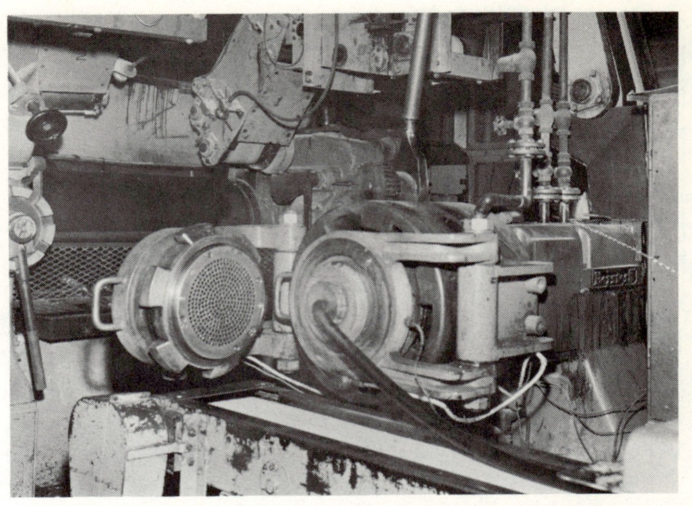

Fig. 4. Front view of an extruder/strainer with double swing head and 2 roll mill for feeding. (Courtesy of ICI (Hyde) Ltd.)

The power consumed by the mill is, however, once again, converted to frictional heat in the nip. The temperature control system for this mill must first heat and stabilise the roll temperature at the correct preset figure (say 150°C for 'soft' and 170°C for 'hard' mixes), and secondly remove the excess friction heat.

The plastics material has now reached a 'gelled' state at an even temperature but may still contain impurities and 'lumps' of unmixed powder or resin. When proposing to make good quality sheet and especially thin sheet (say 0.1 mm thick) one must ensure that all coarse particles are removed.

Straining

The strainer shown in the Fig. 1 layout consists of a single screw extruder of about 5:1 ratio (*ie* the screw being five diameters long). Hot strip cut from the front roll of the mill is transmitted by conveyor to the hopper of the strainer. Inside the strainer and immediately in front of the screw is a filter screen consisting of a very fine stainless steel woven wire mesh, backed up by a medium and coarse mesh for mechanical support. The thrust of the pressure generated by the extruder, which is necessary to force the plastics through the fine screen, is however taken by a thick perforated strainer plate of large diameter. The material having passed through the strainer is now in a refined condition and is extruded by a polished cone-shaped die into a fast moving small section ribbon direct to the calender.

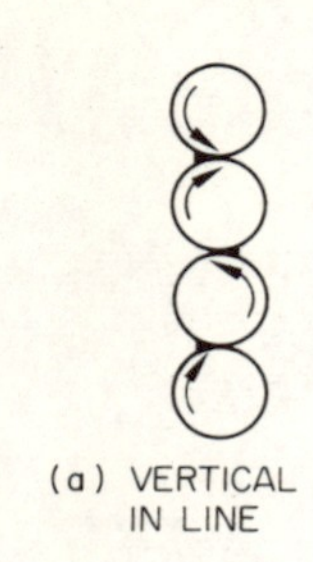

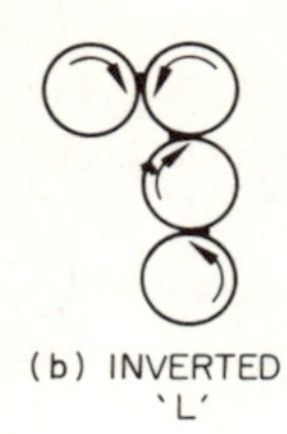

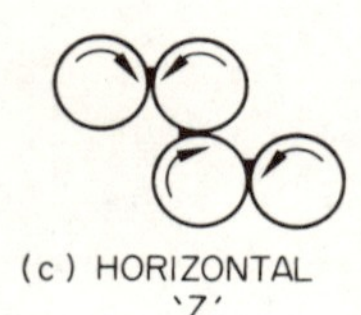

Fig. 5. Various roll arrangements for calenders. (Arrows indicate the direction of rotation.)

In order to sustain a long run, it is necessary to keep up a steady supply of refined material to the calender and, since the screens will need changing regularly, it is advisable to make provision for doing this changing quickly. Fig. 4 shows an arrangement whereby a double strainer head is used, one held in readiness whilst one is in use. The output from the strainer should pass through a metal detector arranged to stop the conveyor and interrupt the feed to the calender if metal is suspected. This device avoids steel particles of broken screen in the strainer pack, passing to the calender which could damage very expensive rolls, and gives a warning of the need to change the screens.

CALENDER

A modern plastics calender is a very expensive and sophisticated piece of equipment and needs a very skilled operator in order to avoid equally expensive damage to its rolls. The rolls can be arranged in a variety of forms as shown in Fig. 5. The 'in line' type as shown in Fig. 5a has the advantage of taking up a minimum of floor space and offers the designer a more simple problem for the strength of the side frames and gives good visibility. The types shown in Fig. 5c and d, which have alternate rolls at right angles and are known as 'Z' calenders, have the advantage that 'bending' of the rolls caused by pressure of material in the nip has no effect on the succeeding nip and therefore tends to improve thickness accuracy. The shape of the side frames for the 'Z' formation does, however, give the designer a much more difficult problem in order to provide adequate strength and it also increases weight, machining and material costs, as well as taking up more floor space and giving visibility problems. The inverted 'L', as shown in Fig. 5b, offers a good compromise since the first pair of horizontal rolls hold the input material easily and the succeeding two nips can be arranged to give good control of gauge.

The calender rolls, or 'bowls' as they are sometimes called, are the most important feature of the machine and call for very high precision in manufacture. They are made of high quality cast iron and cast in a 'chill' to provide a hard outer surface which will take a high degree of finish and resist abrasion and other mechanical damage. They work under very arduous conditions, *ie* continual heating and cooling, and are subject to very high bending stresses—difficult duty judged by any standards.

The rolls are provided with a heating medium such as oil or high pressure water carried in a series of holes drilled across the roll from end-to-end in the softer part of the 'chill' effect (see Fig. 6). Temperature accuracy of ± 1°C can be obtained over the whole of the working face by this method. The final grinding accuracy of the roll face relative to the journal faces should give figures for 'run out' better than 0.010 mm, measured hot (*ie* at an average working temperature). Continental roll makers grind the journals with the roll hot and held in fixed steadies on the roll face.

On the calender, the plastics passes for the first time through a fine friction nip and the gelling of the material is completed. The material must be kept continually on the move in the banks formed at the nips of the calender, cooling-off cannot be tolerated as poor quality material will be produced. The second and final nip should ideally have a small rolling bank of material present which should be just big enough to prevent nip starvation and therefore avoid holes in the sheet. With individual d.c. motors and geared drives to each roll, friction ratio between rolls can be set and this will cause the nip banks to roll and improve quality.

The temperature of the last roll is usually set lower than the others so that the material can be pulled away more easily by the take-off roller. The thickness is now set and the sheet should be cooled as quickly as possible, measured and wound up into batches.

Thickness control

Passage of the material through the nip exerts a considerable force on the rolls and produces a slight deflection at their centre, thereby tending to make the sheet thicker in the middle. When it is required to produce thin sheet (0.1 mm to 1.0 mm) to an accuracy of $\pm 2\%$ or better, this effect must be countered. The rolls can be made with a slight increase in their centre diameter, known as 'camber', so that when deflected they will produce flat sheet. Finding a sufficiently good average camber to suit a range of qualities, thicknesses and temperatures would be expensive and very limiting on quality. Rolls are nevertheless always equipped with a slight camber (say 0.025 mm difference in diameter).

Because of the high nip pressures and temperatures involved the best bearings on calenders consist of plain

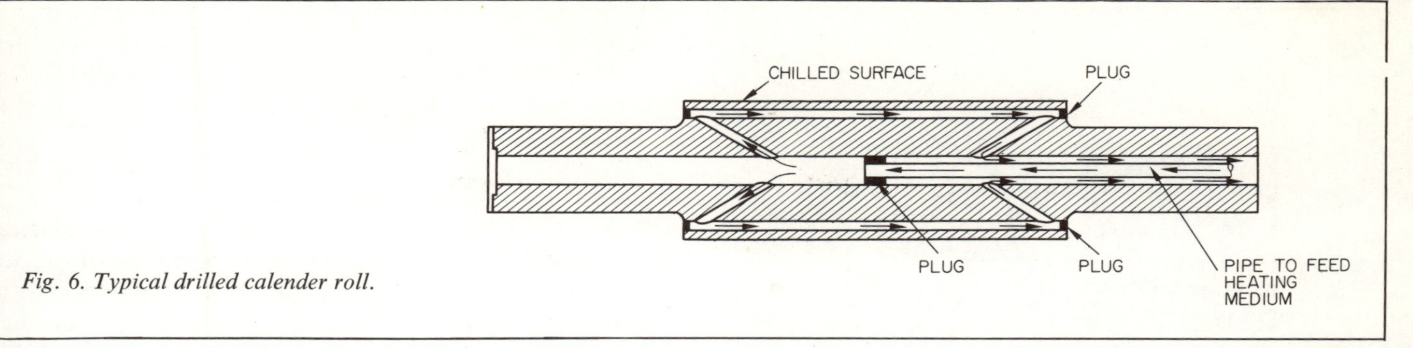

Fig. 6. Typical drilled calender roll.

journals running in brass bushes and flood lubricated with heated and pressurised oil. Since the running clearance of this type of bearing will be comparable with the thickness of sheet, the roll must not be allowed to 'float' around at will within these clearances. Hydraulic rams operating through another set of bearings force the main journal to occupy a controlled position in its bush. This feature is popularly known as 'pull back' although it more often than not 'pushes back' and produces the second method of thickness control.

A third method of thickness control is achieved by 'roll bending'. Extra bearings set on extensions to the roll ends in order to provide leverage are forced apart by hydraulic rams situated between them. By using the main journal as a fulcrum (see Fig. 7) it is possible to bend the roll in opposition to the deflection caused by the material and thus neutralise it. If a roll bending system is fitted it will do the 'pull back' operation as well.

A fourth method, sometimes used, makes use of the 'cross axis' principle. Consider the vertical centreline of the calender as the fixed reference point. The roll journals can be moved by an equal amount in opposite directions and put one roll out-of-line or 'cross the axis'. A pair of rolls with crossed axes will have greater clearance at the edges than at the centre and a very small angle of cross will counter all the bend likely to be produced by the material.

Cooling

Sustained high temperature working of PVC is very bad for the material, causing discolouration and degradation, so the time during which it is held at working heat (say 170°C) should be kept to a minimum. The cooling drums should be fitted as close to the calender as possible and, if they are to work efficiently, be machined to an accuracy comparable with that of the calender roll. Surface accuracy ensures a maximum amount of contact is obtained between the sheet and the cooling surface.

Thickness gauging

There is little point in giving a calender operator first class equipment to make plastics sheet, unless some method is provided to accurately and continuously measure the material produced. The thickness and dimensions which matter are those which the customer receives and should therefore be made just before batching up. This is a point where the material is cold and firm but is unfortunately a long way, in terms of path length, round the cooling drums, from the calender and any calender adjustments take a long time before they can be assessed. Delicate measuring instruments fitted near to the calender or hot plastics sheet are subjected to heat variations as a result of radiation and, if the sensing device has to contact the sheet, the hot plastics could be spoilt.

The Beta-Ray equipment gives an average reading of thickness over an area. Fig. 8 shows the transmission system of measurement in which radiation from an isotope such as Thallium 204 is passed through the sheet and collected in an ionisation chamber. The radiation received is inversely proportional to the weight per unit area of the material being gauged[1]. Finally, Fig. 9 shows a general view of the controls of a calender including cooling drums.

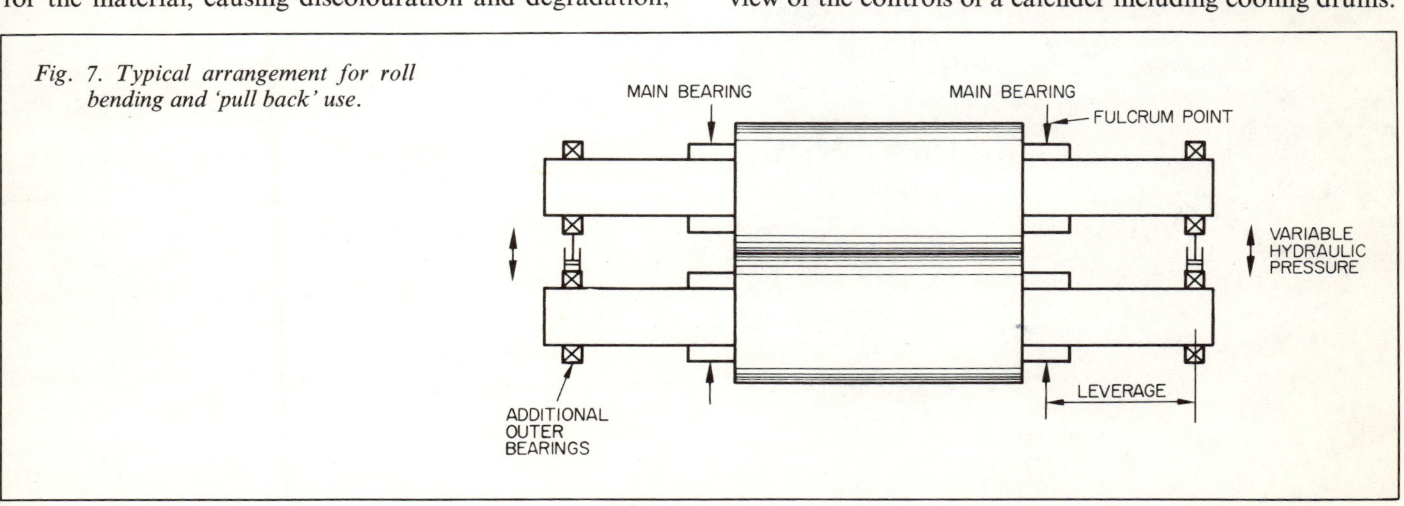

Fig. 7. Typical arrangement for roll bending and 'pull back' use.

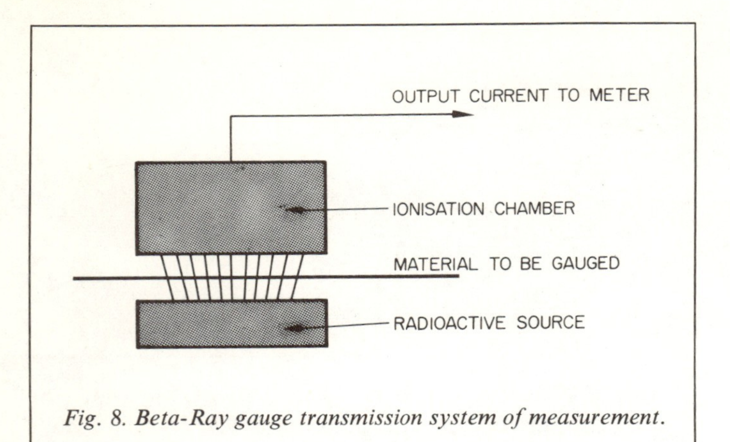

Fig. 8. Beta-Ray gauge transmission system of measurement.

Wastage

Approximately 10% of the width of sheet is trimmed-off before winding up. This is returned to the blending mill with a minimum of delay and re-used in the same run. Unusable material made at the start and run out of a batch is granulated and stored for use in subsequent runs of the same quality. The loss is therefore minimal and not more than 2%.

Fig. 9. A general view of the control side of a modern calender. (Courtesy of ICI (Hyde) Ltd.)

Materials for calendering

Most of the products calendered at present are either PVC or ABS or a mixture of both. A wide range of grades are available today, each with their own particular physical properties.

Calender utilisation time

If long runs of one quality can be made, operating efficiencies of 95%+ can be attained; if, on the other hand, the quality, thickness and colour are continually changed for short runs, this figure may drop as low as 75%. With such an expensive set-up of equipment as a modern calender plant it must be used to maximum efficiency.

FUTURE DEVELOPMENTS

The direct calendering of a controlled layer of dry powder would have very considerable advantages if a satisfactory method could be found. The time during which the plastics is held at a process temperature would be minimised and the high cost of preparation machinery considerably reduced.

Doubling-up thicknesses of material satisfactorily would take on some thick sheet work which is now extruded for roofing and vacuum forming.

The d.c. drive system for a modern calender is very expensive and, with all the necessary gearboxes, bulky, and makes close access difficult. A system of hydraulic drive using direct acting motors on each calender roll and item of take-off drive would have many advantages and probably reduce cost.

Plenty of scope in these possibilities to interest anyone in calender operation and design.

REFERENCE

1. J T Laing, Web thickness control by Beta-Ray. *International Plastics Engineering*, vol. **3**, No. 9, Sept. 1963.

Chapter 12

Rotational Moulding

T Russell
Thermo Plastics Ltd

Of the many processes for producing complex designs from thermoplastic materials, rotational moulding is probably the most simple. The process requires relatively inexpensive equipment with low power inputs and exerts only small pressures on the material being formed.

In plastics forming processes the basic polymer is heated until it flows, moulded into shape under pressure and finally cooled. In rotational moulding the whole of the process takes place inside a female mould. The process exploits the principle that finely divided thermoplastic material becomes molten in contact with a hot metal surface, and will take up the shape of that surface. If the polymer is then cooled while still in contact with the metal a solid copy of the surface will be produced.

In production, a measured quantity of powdered polymer is placed inside a female mould. The mould is then heated and rotated in such a way as to allow the polymer to flow over all parts of the mould. This part of the process is critical if an even or controlled wall thickness is required. When all the polymer has melted and is adhering to the mould the cooling cycle is started. The mould continues to rotate until it is cooled down sufficiently to enable the product to be taken out without distortion, after which the next moulding cycle can start. It will be noted that as the complete heating and cooling cycle takes place within the mould the process must necessarily be much slower than processes where the polymer is transferred from a heating section to a cooled mould, as in extrusion or injection moulding.

Rotational moulding machines are designed to rotate the mould biaxially during the heating and cooling cycles according to a programme established for that particular mould. The machines themselves do not come into contact with the moulding material, nor do they exert any pressure upon it, so that the machine design is inherently simple. As only a female mould is used, and the only pressures exerted are those induced by gravity and centrifugal force,

Fig. 1. A typical 'rock and roll' type moulding machine.

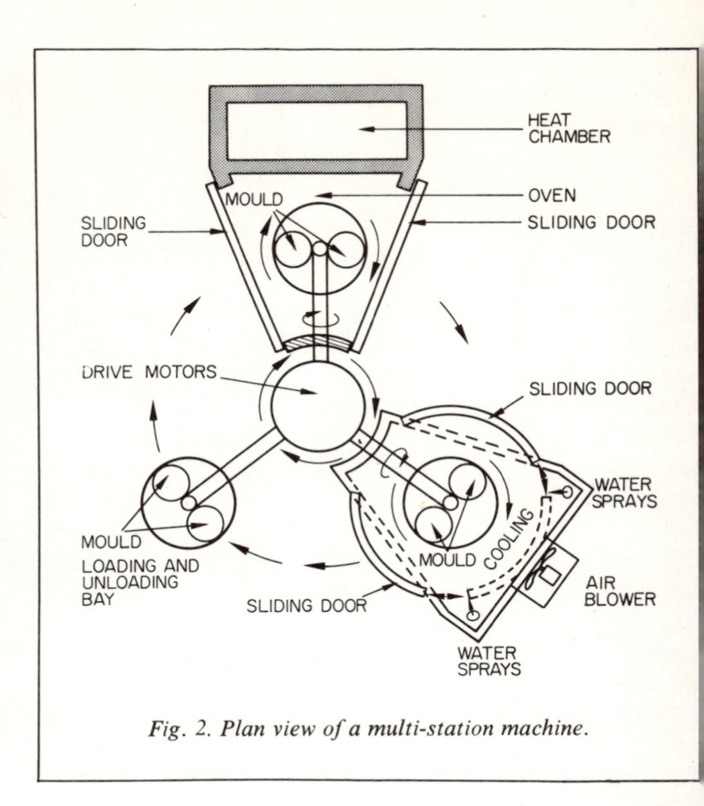

Fig. 2. Plan view of a multi-station machine.

mould design and manufacture can be comparatively inexpensive.

Rotationally moulded products were almost entirely made from plasticised PVC in the early years and these were mainly of small size, *ie* dolls, balls and containers, but the use of low density polyethylene in the early 1960s increased the potential for the process enormously. The ability to produce rigid mouldings, particularly large tanks and containers, was largely responsible for the growth of the process, and a variety of machines using different heating and rotating principles have been designed. Currently, many different polymers are moulded at good production rates with sophisticated equipment, although the process itself retains its simplicity. Mouldings are produced without the use of high pressures, thus without 'moulded-in stress' which is a feature of most other processes.

EQUIPMENT

Rotational moulding machines must be designed not only to produce technically suitable mouldings but they should also be capable of earning a reasonable return on their investment. There are a variety of machines available which meet this criteria, from the primitive, slow and cheap machine to the sophisticated, fast and expensive. The basic differences between these machines are the method of heating the mould and the means of reducing the non-productive parts of the moulding cycle. The moulding cycle is as follows:

(*i*) Load the mould with powder
(*ii*) Heat and rotate the mould
(*iii*) Cool and rotate the mould
(*iv*) Unload the moulding from the mould

In the most simple single-station machine the heating and cooling processes must take place sequentially so that their utilisation must be quite low, whilst in the more complex machines the heating and cooling process is continuous, the moulds being moved between stations automatically.

Rock and roll type machine

The 'rock and roll' type of machine is one of the least complicated versions (Fig. 1). Each mould for the machine is built between two parallel hoops. When the mould is loaded on to the machine the hoops rest on a bed consisting of four rollers which, when driven, cause the mould to rotate. During rotation the bed rocks from side-to-side at right angles to the plane of rotation. The mould is heated during this movement by gas jets carefully placed to produce the heat distribution which will give the required material dispersion within the mould.

When the heating cycle is complete the mould is either cooled *in situ* with water sprays or moved by block and tackle to a second rocking bed for cooling so that the first machine can immediately be re-loaded with a second mould. The operator can control the direction and speed of rotation of the mould in order to achieve optimum results, and several experimental cycles are often necessary before a satisfactory product can be made. The wall thickness of the product tends to be greatest in the hottest areas of the mould. It may even be necessary to attach an insulating material such as glass fibre to certain outer areas of the mould to achieve the desired heat distribution. Similarly it may be necessary to preheat certain detailed areas of the mould, such as rims and valve areas, with a gas flame prior to moulding to obtain a sufficient thickness.

Multi-station machines

Where high production rates are required the utilisation of the heating and cooling cycles has to be increased and the downtime between stages and cycles has to be reduced. One way of achieving this is to have a machine with three arms mounted on a central pillar which can rotate as required (Fig. 2). One or more moulds are mounted on the end of each arm, which can be up to about 1.5 m (5 ft) in length. There is gearing at the end of each of the arms to impart biaxial rotation to the moulds. Situated around the central pillar is the heating oven, the cooling area and the loading bay. During production there is one mould being heated and one mould being cooled, consequently each processing stage is continuous and the moulds can be moved rapidly between stations.

The mould heating oven may be heated by direct gas flame, hot air, infra red, molten salts or fuel oil. The moulds are usually cooled by water immersion or sprays, and occasionally by cold air. Cooling is not critical but one must be aware of the effect of rapid cooling on the relative crystallinity of polymers.

Loading the polymer can also be accomplished at high speed by having pre-weighed material ready to be piped into the mould as soon as it becomes available.

Jacket mould machines

So far we have considered single skin female moulds being directly heated during rotation by a variety of means. The correct heat distribution, especially for moulds of complex shape, can be difficult to attain and can mean loss of production while gas jets or heaters are re-positioned between jobs.

A method which claims to overcome some of these difficulties uses a double skin or jacketed mould. During the heating cycle hot oil, at a controlled temperature, is circulated through the jacket, thus heating the mould. The mould is subsequently cooled with oil at a much lower controlled temperature. The mould is supported by a double L frame during rotation through which hot and cold oil is alternately pumped into the mould from reservoirs. Heat transfer is rapid with this system, enabling short cycling times to be obtained, but it is a single station machine, the heating and cooling media being brought, in turn, to one mould station.

Fig. 4. A double skin (jacketed) mould mounted on a Thermovox machine.

This type of machine (see Figs. 4 and 5), developed by Thermovox of Germany, has an advantage of requiring less floor space but mould costs are high compared with other methods and any increase in production will require further moulds and associated plant.

Fig. 3. An air heated rotational multi-station moulding machine. (Courtesy of John Orme Ltd.)

Fig. 5. A Thermovox oil heated machine for the double skin type of mould.

Fig. 6. A 1600 gallon rotationally moulded tank.

TOOLS

Tools for rotational moulding need only be of light construction as high pressures are not used. The moulds are rotated only sufficiently to let the polymer tumble round the internal surfaces and centrifugal forces are not significant.

Moulds are generally manufactured from sheet steel or light metal castings, dependent upon the requirements of the finished product. Sheet steel (14–16 swg) is preferred where surface finish is not critical and for the larger moulds, though interior welding must be neat and ground-off to a smooth finish. For smaller and more detailed products castings are normally used, though occasionally electroformed copper nickel tools are made for the very intricate surfaces, especially in doll making. Stress relieving is vital with all moulds as the heat cycling they will be subjected to during production may cause distortion.

Where moulds are split to enable the moulding to be removed, accurate matching is important. There should be no gaps which may allow molten polymer to leak out or air to enter the mould. Mould splitting should be made rapid and easy, using quick release clamps or compressed cylinders. Undercuts are only permissible if the product is flexible enough to be collapsed out of the mould or if the moulding will shrink out of the undercut. Moulds for the jacketed system are obviously more expensive to manufacture as the oil circulation has to surround completely both halves of the mould ensuring an adequate and rapid heat exchange.

Although tools are light and relatively simple the method of heating and rotation must be carefully considered during the tool design stage to reduce to a minimum the number of process changes that have to be made between set-ups for different tools. Effective mould design can considerably reduce production cycle times and much experience has been gained in this field.

MATERIALS

Almost all thermoplastic polymers and some thermosetting materials can be rotationally moulded in small quantities under laboratory conditions, but commercially the range of materials is quite small, the most typical being those given below:

PVC — suspended in a plasticiser and known as plastisol. Was the first material used for rotational moulding and still widely used.

Polyethylene — the most popular material today. Low to medium density grades.

Polystyrene — very good surface finish and most suitable for decoration. Special grades developed to overcome flow difficulties and embrittlement due to high temperature moulding.

CAB — (cellulose acetate butyrate) very stable and used for transparent applications.

Work has also been carried out on acrylic, polycarbonate and nylon, but few commercial products are currently available. Fillers and pigments can be used in the process but they do affect the physical properties of the material and should only be used after consultation with the material suppliers.

DESIGNS

Technically, the nearer the product approaches a sphere, then the easier it is to mould as the process relies only on centrifugal force to pull the molten material into intimate contact with the internal surfaces of the mould. It becomes more difficult to mould products with a uniform wall thickness when the length of the moulding becomes large in relation to the diameter and for practical purposes the length should never exceed four times the diameter.

Sharp radii or corners are difficult to fill evenly whether concave or convex unless a very fine powdered polymer is used, but even then they should be avoided where possible.

With experience, thin sections, or areas left completely open, can be moulded by insulating external areas of the mould with asbestos or glass fibre. Similarly, inserts such as hinges, valve fittings, *etc*, can be inserted into the mould prior to moulding to produce a finished assembly.

The inherent simplicity and versatility of rotational moulding can often enable difficult design problems to be overcome. For instance, a large tank, subjected to large hydraulic pressures from a highly corrosive liquid would require a large wall thickness of expensive material, a combination which could probably price the product out of the market. The tank could be made much more cheaply by processing two or more materials successively in the mould to produce a laminated structure. The outer skin, processed first, could be a carbon black loaded polyethylene to give protection from ultra violet degradation. The next layer could be a cheap or foamed material, giving the required thickness to counteract the hydraulic pressures. The final inner layer is of the best material to withstand the corrosive attack of the contained chemicals. The wall of the tank has thus been constructed to exploit the qualities of three different materials.

A more simple and common example of this technique is the lamination of natural polyethylene with a thin outer skin of ultra violet resistant material. If an insert is placed inside the tool during the first outer skin moulding stage, and removed when the second inner material is moulded, the inner material will mould through the outer skin to the surface of the finished moulding. This enables tanks to be constructed with a built in sight glass. The black outer skin is moulded with a strip insert in position which is removed when a natural inner material is moulded. The liquid level can then be seen through the translucent strip in the finished tank.

Fig. 8. This rotationally moulded bollard is currently in production. (Courtesy of John Orme Ltd.)

PRODUCTION ECONOMICS

One of the highest single costs in plastics processing is the cost of the material used. In most processes considerable material wastage takes place during changes of tools, colours of material or material types. Production runs must, therefore, be long to reduce the wastage to a minimum. Rotational moulding processes use only the amount of material required for one product. Changes of tool, material or colour do not incur any material losses.

The capital cost of the machine and tools is low in comparison with other forming equipment, but output rates are also lower. Cycle times on modern rotational moulding machines are rarely below ten minutes, except for the very small mouldings, and are often as high as twenty minutes. Injection and blow moulding rates tend to be in the one to three minutes range for larger mouldings.

The choice of process for a particular product must be influenced by the number required and the rate at which they will be required. Generally, assuming the product is technically suitable, short runs of large mouldings are more economical on rotational moulding equipment than on blow or injection moulding machines.

It is not uncommon to tool a product for rotational moulding in the first instance to test market acceptance. The success of this initial market investigation will then determine the process which will ultimately be used for major production.

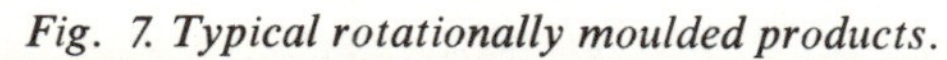

Fig. 7. Typical rotationally moulded products.

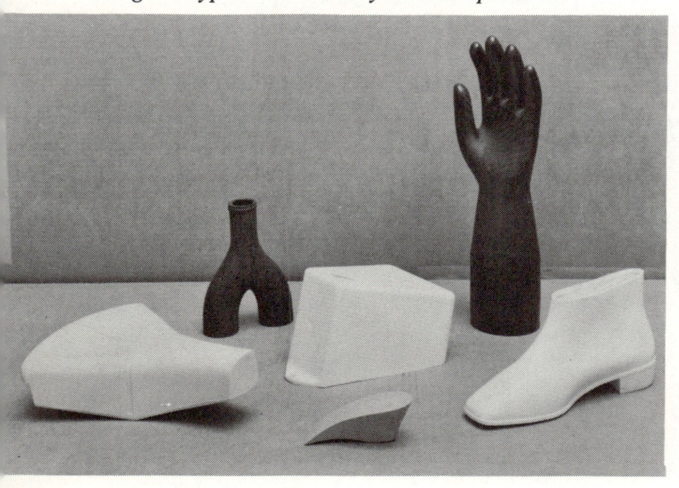

Higher outputs can be achieved with the multi-rotation machines using several tools for each product. As many as six tools are used on some moulding machines, thus reducing the cycle time per item to around two to three minutes, though mould costs are also proportionately increased.

FUTURE DEVELOPMENTS

One of the main advantages of the process is that it can produce enclosed or hollow products, and it is in these areas that it is most likely to grow. Fuel tanks for domestic use and for the motor car are currently being evaluated. The automotive industry may also provide a substantial market for glove compartments, heater ducting and other partially enclosed fittings.

The development of materials other than polyethylene and plasticised PVC will open up further market possibilities. Rigid mouldings should find acceptance in the furniture industry where large hollow modules are required, and the cost of wood is rapidly increasing.

Rotational moulding is still developing, with cycle times being reduced and the number of available materials increasing. Whilst it is doubtful that it will ever compete with injection moulding, it will always have a place in the areas already discussed which are technically unsuitable for other processes.

ACKNOWLEDGMENTS

The author wishes to acknowledge the co-operation of the following:

John Orme Ltd., Engineers, Wellingborough, Northants
P. B. Stewart Ltd., (U.K. Agents for Thermovox), Henley-on-Thames, Oxon
G.P.G. Holdings Ltd., (Tanks Division), Leamington Spa, Warwickshire.

Chapter 13

Coating Techniques

G E Barrett *BSc PhD*
Plastic Coatings Ltd

Plastics coatings were originally applied to metals to protect them against corrosion. This is still one of the major uses, but the field has now widened to embrace cushion coatings to prevent damage to delicate parts, coatings for electrical insulation purposes and coatings for decorative finishes. In the chemical industry, engineers are making increasing use of specific plastics for corrosion protection against an ever increasing combination of compounds which attack mild steel and very often stainless steel. The automotive industry has a very large number of carrying-jigs coated with a resilient grade of PVC to prevent chromium plated components from being scratched, and instruments from being jolted during in-plant movements. Tubular steel furniture is being coated in PVC, epoxy and, in particular, nylon to provide an exceptionally durable, non-chip, decorative finish for both indoor and outdoor use.

Coating process

The technique of plastics coating consists, in essence, of heating a metal article in an oven to a temperature such that when it is dipped into a thermoplastic polymer, the polymer which adheres to the hot metal melts, flows and fuses into a coherent coating. This process may be carried out manually (Fig. 1) or on fully automatic machines with ovens designed to heat saturate the metal.

In practice, it is usual to put the coated article back into the oven after dipping to complete the melting or sintering process. Which particular polymer should be used depends upon the properties required of the final finish and also on the precise nature of the metallic article being coated.

Plastics coatings may be applied as liquids or as solids. The liquid system applies only to PVC and is confined to a straightforward dip process. Solid or powder processes apply to polyethylene, nylons, PVC, chlorinated polyether, epoxide, polyurethane and polyesters. These coatings, which do not involve solvent evaporation, are of the order of 0.01 in (0.25 mm) and upwards rising to 0.25 in (6.35 mm) for PVC cushion coatings.

A variety of methods of application are possible; namely spraying by flock gun, flame gun, electrostatic gun and other electrostatic techniques, as well as fluidised bed dipping, vacuum coating and cast lining. These will be described later when discussing the coating materials in detail.

In addition to the previously mentioned materials, there is the family of fluorocarbons (PTFE, FEP, *etc*) which are applied from dispersion.

PROCEDURE

There are four major stages in the dip coating operation:

1. Metal preparation
2. Pre-heating
3. Dipping
4. Sintering or curing

It cannot be over emphasised that plastics coating should be considered at the design stage of any article or structure.

Fig. 1. A vertical indexing, pre-heating oven requiring manual loading, unloading and dipping. However, a 60% saving in floor space over an equivalent linear oven is achieved. (Courtesy of Plastic Coatings Ltd.)

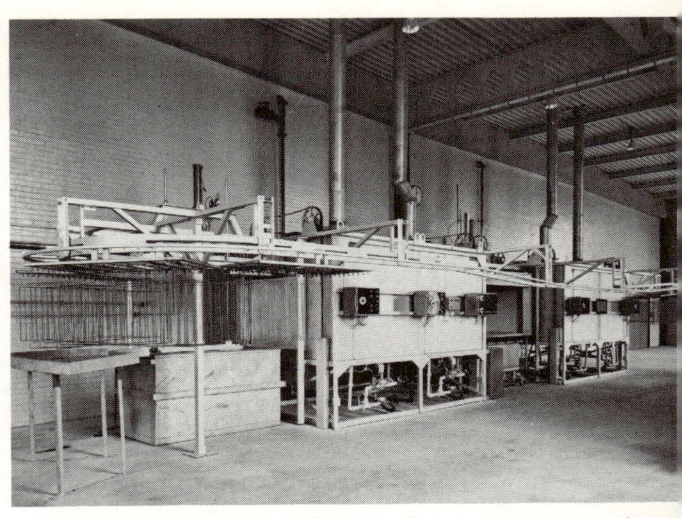

Fig. 2. An automatic, in-line powder coating system, the Autocoat, based on the fluidised bed technique. It is capable of repetitively coating 15,000 articles per 8 hr. shift in polyethylene, PVC or nylon powder. (Courtesy of Plastic Coatings Ltd.)

Whilst many of the difficulties which are encountered due to this lack of foresight can be overcome by modifications to coating procedures, it is inevitable that such changes will increase the cost of the coating. The requirements, which will enable the coating procedure to be simple and, therefore, as economic as possible, will be apparent from the ensuing description of the coating process.

Metal preparation

Metal cleaning. The first stage in metal preparation is the removal of all forms of oxidation and contamination. This is carried out either by the conventional means of shot-blasting, or by chemical removal of rust, scale, grease and paint.

Priming. The second stage of preparation is the priming of the cleaned article. Adhesive primers which promote a satisfactory bond between the substrate and the coating are applied. The primers require to be heated after application in order to develop their adhesive properties. Since the article has to be heated to pick-up the coating, this operation also serves to cure the primer. The primers vary for different coating materials, but they can all be rendered ineffective by over-heating. Therefore, they constitute the first limitation to the coating process in that the time and temperature of pre-heat are restricted by the nature of the primer.

Pre-heating

The cleaned, primed article is pre-heated at a temperature and for a period of time, both of which are best determined by experiment, so that when the article is dipped, it contains sufficient heat to gelate (in the case of liquid PVC) the PVC onto the metal, or, in the case of a fluidised powder, to melt the particles so that they will flow and form a coherent coating. It is obvious that the article to be coated must be capable of withstanding the temperature of the process. High output coating equipment (Fig. 2) may run at very high temperatures and it is essential to ascertain the

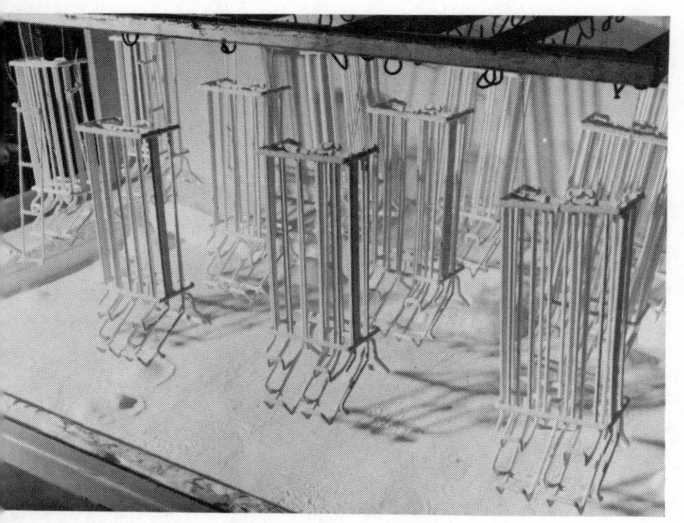

Fig. 3. A fluidised bed of polyethylene powder showing wire display racks prior to being sintered. The cropped wire ends are completely covered showing a saving in fettling costs. (Courtesy of Plastic Coatings Ltd.)

naximum to see that the item will not melt, crack or distort. 'illers are usually unreliable and tend to bubble during the ıeating. Soft solder and galvanising will often be quite mpracticable. Certain fluxes and other deposits may cause rouble during the heating and cause blisters in the final oating, although the normal pre-treatment method will ;enerally remove them.

The amount of coating that is picked up is a function of the ›re-heat conditions, since the coating thickness depends ıpon the amount of heat available above the melting or ;elling point of the particular material. This heat depends ıpon the temperature of the article, its specific heat and its ross-section. For a given material and temperature, one vould expect the heavier sections to attract a thicker oating than the thinner sections and this is what, in ›ractice, will occur, although the method of heating can lo much to minimise this effect.

t is pertinent to note here that the designer must make dequate allowance for the thickness of coating particularly when a number of coated parts have to be assembled. Thus, a 0.02 in (0.50 mm) coating on a number of roof modules will lead to an error of several inches on a large roof expanse causing failure to register with support beams.

There must be no porosity – this applies particularly to castings – although it does not imply that every hole in the surface will cause faulty coating, but rather that where the hole is nearly closed over at the top so that it is scarcely visible, there is a danger of the coating bubbling at this point. The same effect can be found where welding is not continuous between two plates. Where the weld does not penetrate completely, there may well be a considerable air space. If this is not enclosed by a complete run of weld, then expansion of the air within will undoubtedly affect the coating during processing, and the same problem applies where tubular components are welded into a structure. Unless the ends of the tubes are completely sealed, there is usually trouble, although an aperture large enough to prevent the plastics bridging across will relieve the air pressure and so avoid trouble. The same difficulty may be encountered with spot welded parts where the metal between the welds spreads slightly, allowing an air gap to form. These kinds of hazards can often be overcome by the experienced coater, but usually at the expense of rate of production, and hence cost. It should be understood that simple welds need no special treatment and may be left unground for plastics coating. This is due to the fact that the thickness of coating tends to even out surface irregularities of the substrate. Indeed, a major reason for plastics coating wirework is the fact that the article can be left in a relatively rough state as far as the wirework manufacturer is concerned, thus leading to overall cost economies (Fig. 3).

The source of heat used for pre-heating can vary considerably; recirculating ovens are generally found to be most efficient, since an even temperature can be more easily maintained throughout. Radiant heaters are also used, but these can cause trouble where a variety of metal surfaces are encountered, since reflective surfaces may be impossible to heat sufficiently, or may take up costly time and fuel in attaining the required temperature.

Fig. 4. Flow plan of plant for the continuous plastics coating of wire using PVC.

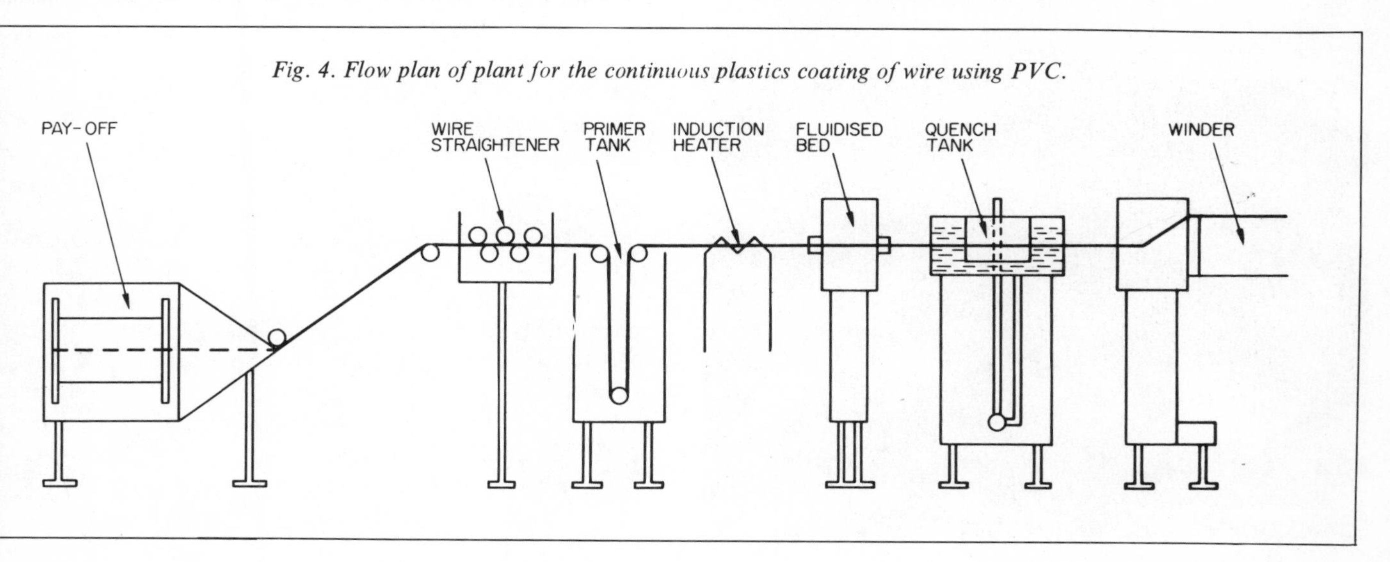

Induction heating, resistance heating and direct flame heating can also be used, but the first two methods impose limitations on the shapes of acceptable articles and the third is difficult to control. However, the continuous coating of long cylindrical sections, *eg* steel tubes and wire, using induction heating techniques is carried out commercially incorporating the fluidised bed method for applying the powder coating (Fig. 4).

Dipping

Drips and runs are the most common drawbacks to an acceptable appearance on PVC plastisol dipped articles. Slow withdrawal from thixotropic plastisol and the use of electrostatic de-tearers can reduce the occurrence of drips but they are seldom completely absent. Fluidised bed coatings using powdered polymers (discussed below) avoid the problem of drips completely.

PVC is the only material applied from the liquid form. Other polymers, such as nylon, chlorinated polyether, and polypropylene could be applied by solvent dipping, but the cost would be excessive because of the solvent loss or expenditure on recovery plant, and the many precautions needed for factory handling. It should be emphasised that PVC plastisol is not a solvent system. In a PVC plastisol, resin is suspended in a plasticiser and the action of heat brings about an homogeneous mixture of the resin and plasticiser; no material is lost during this process (Fig. 5).

All dip coated articles require to be suspended for the dipping operation and where the article is specified for complete coverage it is necessary to seal the suspension point by a subsequent operation. It often occurs, however, that the product does not require an enveloping coat and in consequence the coater can take advantage of this for suspension purposes (Fig. 6).

Sintering or curing

When the dipping operation is complete, the coating, be it plastisol or powder, usually requires a further heat treatment in order that the maximum physical properties of the polymer are achieved. In the case of a plastisol coating, this is generally referred to as curing and, in the case of a powder coating, as sintering.

An uncured plastisol coating is generally weak, crumbly and matt in appearance. A cured coating is tough and glossy. Before curing the plasticiser/resin mixture is not entirely homogeneous, whereas after curing homogeneity is complete. In the case of powder coating, sintering ensures that adherent unfused particles are fused yielding a uniform coherent coating. After sintering or curing, the articles are generally water quenched to speed up handling.

POWDER COATING

As previously mentioned the problem of drips from dipping into liquid systems can be avoided by the use of powder coating. There are several techniques available for the application of powder coatings, namely:

Fluidised bed dipping
Spraying – Flame spraying
Flock spraying
Electrostatic spraying
Dispersion spraying
Vacuum coating
Rotational lining (similar to cast lining)
Electrostatic – Fluidised bed and cloud chamber.

Fluidised bed

Dipping a hot metal article into a box of powder is, unfortunately, not a viable proposition, as wherever the article is dipped a hole in the powder results and successive articles will not pick-up any powder. This limitation has been overcome by the use of a fluidised bed – which, as its name implies, is a method of making a powder behave like a fluid.

Basically, the principle of the system is a gas passing upwards through a vertical column containing finely divided

Fig. 5. Fume extractor hoods protected inside and out by a single dipping in PVC plastisol. A thickness of $\frac{1}{8}$ in (3.175 mm) on 10 swg sheet is usual. (Courtesy of Plastic Coatings Ltd.)

Fig. 6. Mating faces of these pruner handles are free from coating and provide a surface for holding whilst dipping. (Courtesy of Wilkinson Sword Ltd.)

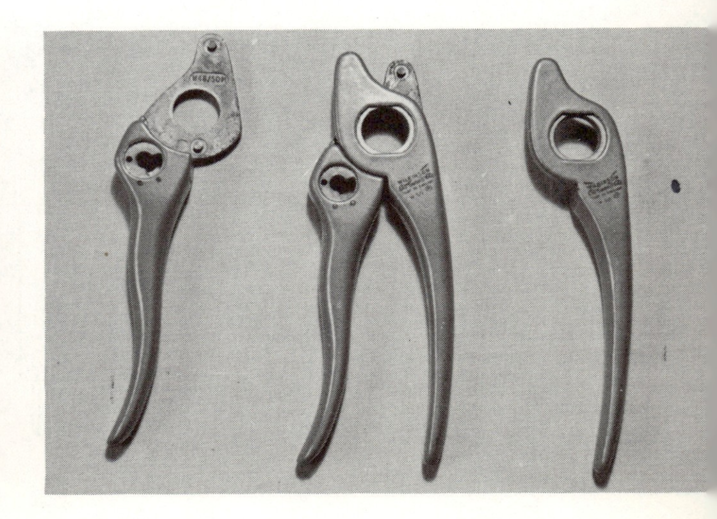

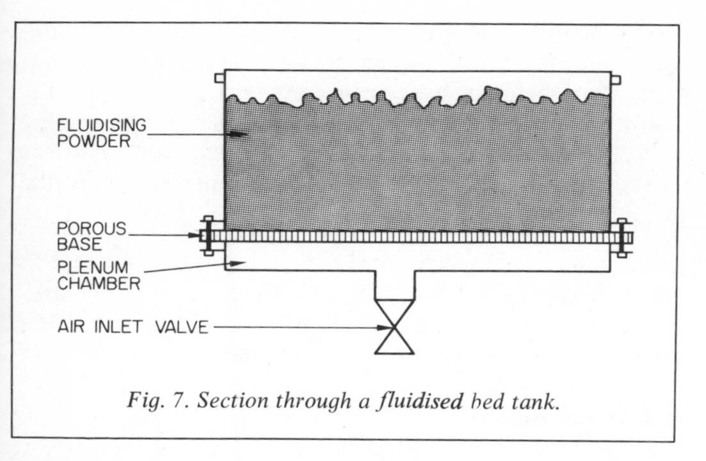

Fig. 7. Section through a fluidised bed tank.

particles such that the gas flow is uniformly distributed over the entire cross-section of the column. At some particular gas velocity, the weight of the particles will be slightly exceeded by the 'buoyant' force of the gas. At this point, the entire system will assume the apparent properties of a boiling liquid.

The ability to attain smooth fluidisation over reasonably broad bed expanses depends on a proper balance of such parameters as particle size, particle size distribution, particle shape, flow character of the powder, gas velocity, humidity, ratio of height of bed to diameter, and pore size and pore volume of distributor plate.

As far as the actual technique is concerned a fluidising container consists of a box with a false bottom with a fine microporous substance (*eg* ceramic tile) dividing the false bottom from the top part of the box. A large volume of air at low pressure is passed into the false bottom and this is then evenly diffused into the box proper by the microporous material (Fig. 7). This air then lifts the powder in the box, diffuses through it causing it to bubble-up and fall back in different places. The whole effect is very similar to simmering water, and successive articles dipped into the fluidising powder will pick up a uniform coating.

Electrostatic techniques

Powder coatings may also be applied by electrostatic spraying or in electrostatic fluidised beds. For spraying, a high voltage is applied to the nozzle of a powder spray gun and the powder spray, produced at a low velocity, acquires a strong negative charge. The object to be coated is earthed and so attracts the charged powder (Fig. 8). As all parts of the object are earthed the powder will envelop it completely, and not just coat the visible surfaces as in conventional spraying. The coated object is then heated to fuse the coating powder into a continuous film.

Normally, film thicknesses of 0.003–0.008 in (0.076–0.20 mm) are achieved by this method since mutual repulsion between the powder particles sets a finite limit. The film thickness can be increased by lowering the resistivity of the powder by the use of antistatic agents or by preheating the metal article. However, the powder only sticks to the object by virtue of its charge, so that it will tend to drop-off if too much antistatic character is built into it. It follows from this that particular characteristics are needed in a powder to ensure that it may be sprayed electrostatically to yield a useful coating. Among these requirements are suitable dielectric constant, electrical conductivity, self electrification, particle size and shape, and specific gravity. So far this technique has only been carried out on a commercial scale using certain epoxy resins, although some special suitable PVC and nylon powders are becoming available.

In the case of the electrostatic fluidised bed, a combination of the electrostatic spray technique and the fluidised bed technique is used. The apparatus consists of an array of electrostatic elements on the bottom of a shallow bed. A small quantity of powder is placed on the elements and fluidised by the normal method. The elements impart a negative charge to the powder with the result that a cloud of charged powder is obtained. An earthed object held in the cloud collects a layer of powder by electrostatic rather than fluidised bed principles. Unfortunately, the density of the cloud varies considerably from top to bottom and is sensibly uniform over only 4 in. This is offset somewhat by the fact that an electrostatic fluidised bed requires only one-tenth of the raw material compared with an equal sized conventional fluidised bed and this helps considerably as far as stock holding of colours is concerned.

Other spraying techniques

Flame spraying. The success of this process depends to a very large extent on the skill of the operator. A coating powder is sprayed by means of compressed air, through a suitable flame which both melts the powder and raises the surface temperature of the metal object. Molten particles

Fig. 8. PVC powder applied electrostatically to a mild steel sign surround. Economical powder usage and good 'wrap round' are outstanding features of this process. (Courtesy of Plastic Coatings Ltd.)

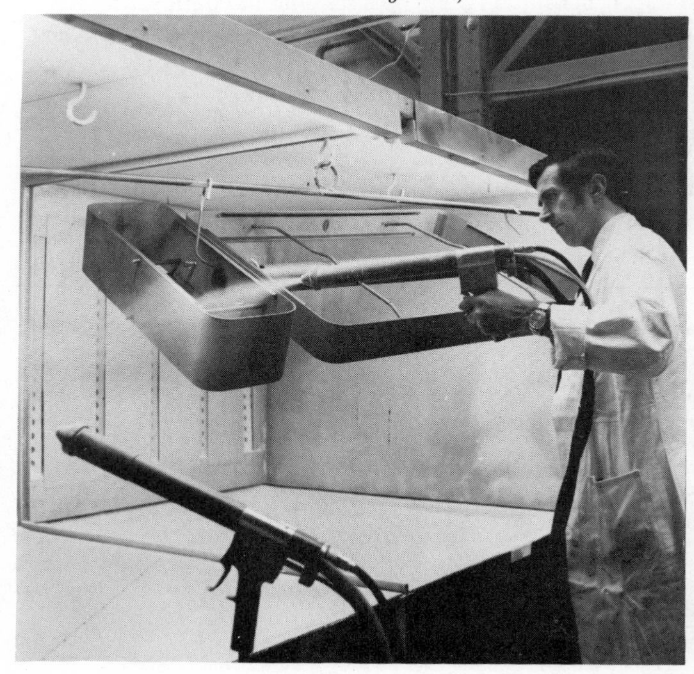

Fig. 9. A battery of rotational lining machines enable one man to line 300 cylinders per 8 hr shift. A specially formulated self-adhesive polyethylene powder is the coating material. (Courtesy of Plastic Coatings Ltd.)

impinge on the workpiece where they fuse together and cool. The problems with this process are that at the high temperatures required to melt thermoplastics, *ie* between 100°C and 300°C, degradation is likely to occur and it is difficult for the operator to obtain an homogeneous and uniform coating. No subsequent sintering is necessary.

Flock spraying. Cold powder is sprayed, under air pressure, onto the pre-heated metalwork. Depending on the type of powder used, so post-sintering may be required. The process is essentially a manual one relying on operator skill and speed and is, as a consequence, rather slow. Powder recovery is not always easy.

Dispersion spraying. The plastics to be applied is dispersed in an aqueous or solvent base and is sprayed in a manner similar to the application of paint. After sintering one or two further coatings are usually applied with intermediate sintering in order to build up a reasonable coating thickness. The significance of this method can be appreciated when fluorocarbon (PTFE, FEP, *etc*) coatings are required since they can only be applied from dispersions.

Vacuum coating

Especially suited to the internal lining of chemical plant (pipework and vessels), this process relies on a high vacuum (about 30 in Hg) to draw powder from a fluidised bed into the pre-heated article so dispersing it over the entire surface. Two openings in the article are desirable, one for the vacuum connection and the other for admitting the coating material. When the powder has entered the article and has been dispersed the residual vacuum is maintained whilst the powder fuses onto the hot metal.

Atmospheric air, when admitted through a filter, has the effect of raising the pressure, thus forcing the hot coating material into firm contact with the surface thereby eliminating air inclusions. This is a patented process.

Rotational lining

Internal surfaces of cylinders (*eg* fire extinguishers) are coated by this technique which is akin to cast lining. A predetermined quantity of coating powder is introduced into a primed, heated container. The container is rotated on one or more axis ensuring a uniform coverage of all internal surfaces. Fig. 9 illustrates a machine developed for this purpose, allowing the cylinders to be pre-heated and the coating to be sintered on a time cycle basis.

RAW MATERIAL

The materials currently applied by the techniques described above are PVC, polyethylene, nylon, chlorinated polyether, epoxy and the fluorocarbons (see Table 1).

Polyethylene

This is the cheapest of the thermoplastic coatings. The low density grade is normally used, since species with better melt flow indices are available in this grade. It is not as durable as the other coatings listed, as it is susceptible to environmental stress cracking, particularly when used over large flat areas. It finds many uses for coating wirework and expanded metal where complete envelopment occurs, such as domestic wire articles and cable tray.

Special self-adhesive grades of polyethylene have been formulated for use in the rotational lining process and are used for preventing corrosion of the cylinder walls internally.

High density polyethylene offers superior resistance to acids and is used on chemical equipment or where a smooth semi-matt finish is required.

Table 1. Properties of plastics used for coating.

Material	Normal thickness range/in	Continuous working temperature range °C		Chemical resistance			Toxicity	Abrasion resistance
				Acids	Alkalis	Solvents		
PVC plastisol	0.015–0.50	−30	+65	Good	Good	Poor	Very slight	Very good
PVC powder	0.01–0.03	−10	+60	Good	Good	Poor	Very slight	Very good
Nylons 11 and 12	0.01–0.03	−50	+100	Poor	Good	Good	Nontoxic	Excellent
Polyethylene LD	0.015–0.075	−70	+70	Fair	Fair	Poor	Nontoxic	Fair
Polyethylene HD	0.015–0.065	−70	+70	V.Good	Good	Good	Nontoxic	Good
Chlorinated polyether (Penton)	0.015–0.035	−60	+120	Excellent	Excellent	Excellent	Nontoxic	Good
Epoxide	0.002–0.008	−50	+40	Good	Good	Fair	Nontoxic	Good
PTFE	0.0005–0.005	−80	+50	Microporous	Microporous	Microporous	Nontoxic	Poor
PTFCE	0.005–0.015	−70	+200	Complete	Complete	Excellent	Nontoxic	Fair

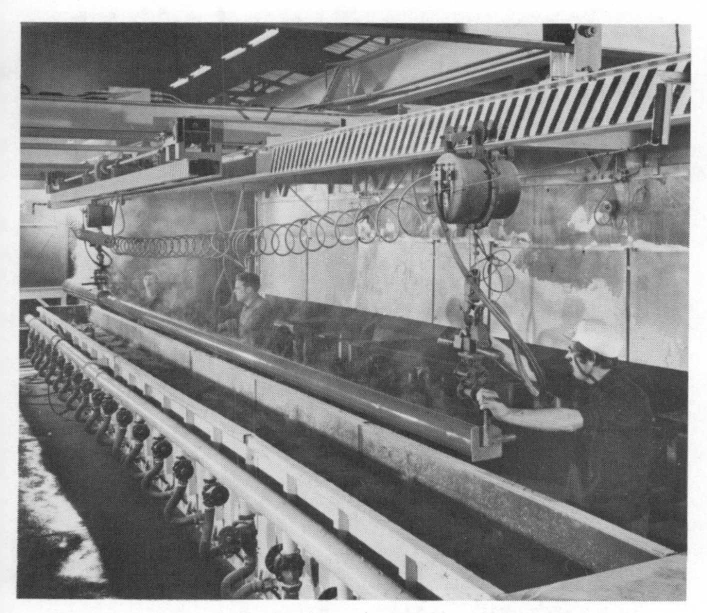

Fig. 10. A 12 m long fluidising tank used for the application of PVC powder to street lighting columns. Used outdoors, PVC powder coatings provide exceptional resistance to weathering and corrosive atmospheres. (Courtesy of Plastic Coatings Ltd.)

PVC

PVC is available in a wide range of hardnesses, from soft and rubbery to extremely hard. This range is achieved by the addition of varying amounts of plasticiser to the host PVC resin. In addition, by suitable choice of plasticiser type, it is possible to achieve oil resistant and exceptional low temperature flexibility grades. PVC resin is extremely resistant to chemical attack and once again the careful choice of plasticisers results in a flexible material with similar resistance to the host resin. Plastisol coating of chemical plant has now largely replaced the alternative, more costly, rubber lining. The resilience and oil resistance has led to the coating of stillages for the motor industry. The outdoor durability of PVC has been developed to a very high degree by the correct choice of stabilisers, with the result that numerous large items, such as lamp posts, have been coated for exterior use (Fig. 10).

Nylons

Nylon 11 and nylon 12 are the only commercially available grades of nylon at present used as coating materials. Nylon has excellent natural adhesion to a variety of metal substrates and, in consequence, a coated article has remarkable impact resistance. The combination of high abrasion resistance and impact resistance has led to its use for coating furniture designed for stacking; in addition to ensuring long life for the product, coating has cut down transport rejects and eliminated the need for wrapping.

Nylon has a very low coefficient of friction which has led to its use for the coating of aircraft pulley wheels, thus reducing cable wear. The good chemical resistance of nylon has resulted in its use in corrosive atmospheres. The furniture at swimming pools being an excellent example – conventional paint finishes deteriorate rapidly. Hospital furniture and equipment coated in nylon is maintenance-free and hard wearing and the heat resistance of the material means that it is unaffected by sterilisation temperatures. The exterior durability of correctly pigmented nylon is of a very high order and this factor, combined with other properties, has led to its use for balustrading and other such exposed articles.

Chlorinated polyether

Chlorinated polyether offers outstanding resistance to acids, alkalis and solvents up to 120°C. It is an expensive material and this tends to limit its application to the field in which it is pre-eminent, namely corrosion resistance. It is not used for decorative coatings. It is frequently used for coating valves and pipework for chemical plant (Fig. 11).

Epoxy

Fully cured epoxy coatings are thermoset materials and the initial coatings applied by fluidised bed or electrostatic techniques are generally partially cured epoxides. They are sufficiently thermoplastic at this initial stage to form continuous coatings. The curing schedules for epoxy coatings have tended to be long compared with the sintering times for thermoplastic coatings, and this means costly capital expenditure for fully atuomatic plant. Coatings applied by electrostatic methods are thinner than conventional fluidised bed coatings and, in the case of epoxy, this is necessary since at high coating thicknesses conventional epoxides are quite brittle and chip easily.

PTFE

When first discovered, it was visualised that the major outlet for polytetrafluoroethylene (PTFE) would be related to its outstanding resistance to chemical attack. However, since it has not been possible to obtain porous-free coatings of sprayed PTFE dispersions, its major growth as a finish has been due to its other remarkable properties, those of non-

Fig. 11. A Penton (chlorinated polyether) coated sectional tank for storing chemicals is spark tested to 10 kV d.c. to ensure the absence of porosity. Penton is non-toxic and obviates the need for using costly stainless steel. (Courtesy of Plastic Coatings Ltd.)

Fig. 12. Domestic holloware being sprayed with non-stick PTFE on a rotary spindle automatic machine. This dispersion coating is subsequently stoved at 400°C for ½ hr on a conveyor type oven. (*Courtesy of Plastic Coatings Ltd.*)

stick and release. It is non-toxic which has led to the release coating of items used in the bakery and confectionery trades. The coating thickness necessary to afford the release properties is extremely low, of the order of 0.0005 in (0.013 mm). The release properties are used on domestic non-stick frying pans and other cooking vessels (Fig. 12).

PTFCE

Polytrifluorochloroethylene is the most recent member of the fluorocarbon family to achieve commercial usage. The replacement of one of the fluorine atoms in PTFE by a chlorine atom has slightly reduced the chemical and heat resistance. However, much more significantly, PTFCE can be applied from dispersion to obtain porous-free coatings. In view of this it is used for coating such items as bursting discs, thermometer pockets and articles to be both chemically inert and electrically insulating.

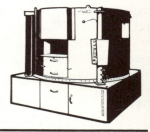

Chapter 14

Cellulose Film

D O Richards *BSc* and **W Scott** *BSc ARIC*
British Sidac Ltd

Cellulose film is produced from high purity cellulose in the form of 'dissolving' pulp. This is obtained from natural wood which consists mainly of cellulose and lignin, roughly in the ratio 70:30. To remove the lignin the wood, having been debarked and chipped, is 'digested' with various chemicals. Two of the most common processes for producing dissolving pulps are the 'sulphite' process and the 'sulphate' process.

1. In the sulphite process the wood is digested with an acid solution of sulphur dioxide and an alkali or alkaline earth bisulphite.

2. In the sulphate process a solution of sodium hydroxide and sodium sulphide is used as digestor. The raw material for the sodium sulphide is sodium sulphate from which the process derives its name.

These chemicals, although dissolving mainly lignin, do remove some of the cellulose, this means that only 40–50% of the wood is obtained as cellulose. After various further treatments to bleach and remove impurities and adjust the final degree of polymerisation of the cellulose, the wet 'pulp' is formed into a sheet and dried. In this form it is received by the manufacturer for conversion into transparent cellulose film.

FILM MANUFACTURE

Bales of the pulp containing sheets of a suitable size are fed, usually from a conveyor belt, sheet by sheet into a 'steeping' tank into which sodium hydroxide solution is also pumped (see Fig. 1). A large stirrer ensures good mixing and the sheets are broken-up forming a thick slurry. This slurry is pumped away continuously and the rate of feed of pulp and sodium hydroxide are adjusted to maintain desired conditions in the tank.

The slurry is pumped to a press fitted with perforated rollers; as the slurry passes between the rollers excess

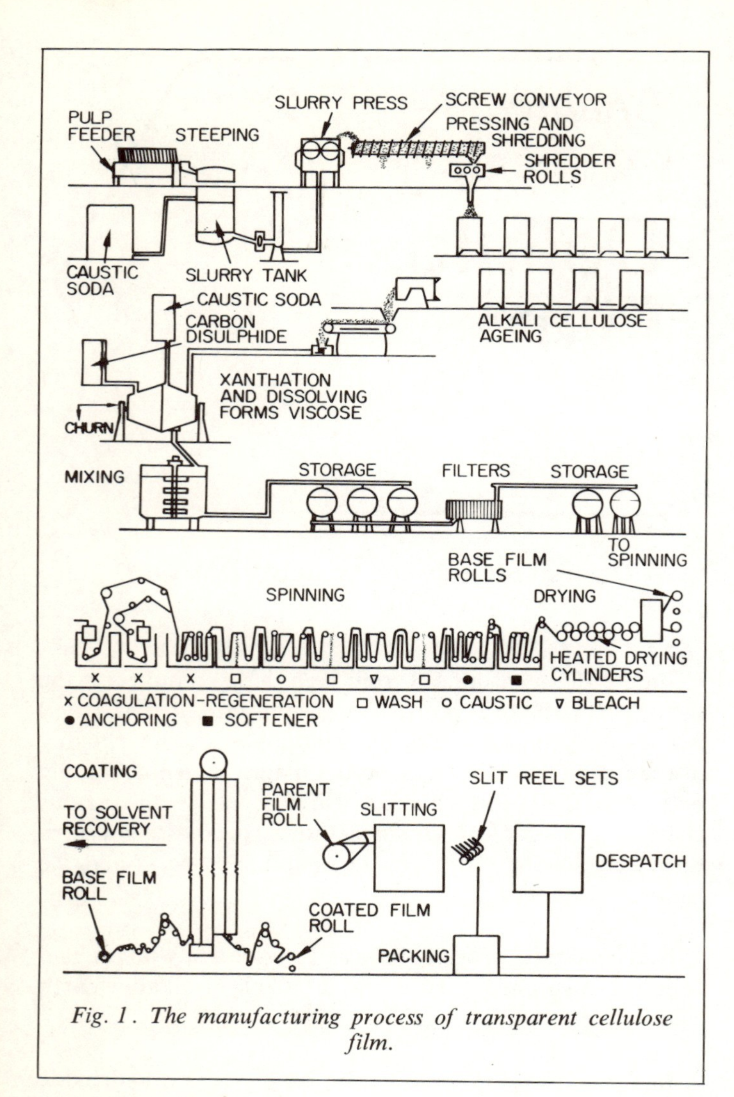

Fig. 1. The manufacturing process of transparent cellulose film.

sodium hydroxide liquor is squeezed out. After being brought back up to strength with strong sodium hydroxide solution the liquor is used to steep fresh pulp. The steeped pulp, after removal of excess caustic (sodium hydroxide) liquor, is in the form of a 'mat'; this is broken up mechnically and the 'crumbs' so formed are called 'alkali-cellulose'.

The formation of 'alkali-cellulose' is very important for the rest of the process. Wood pulp (purified cellulose) as received from the pulp manufacturer has a degree of polymerisation (DP) (*ie* number of glucose units per cellulose molecule) of some 800–1000. This is too high for subsequent processing and needs to be reduced. 'Alkali-cellulose' is attacked by atmospheric oxygen and such attack results in a lowering of the degree of polymerisation. The longer the 'alkali-cellulose' is kept in contact with air the lower is the final DP. Thus, the 'alkali-cellulose' is aged by keeping for various times until the DP has been reduced as required.

The next step in the process is to turn the cellulose into a form in which it can subsequently be dissolved in water. Alkali cellulose is stirred with carbon disulphide in closed churns. At the end of churning it has been converted from a white to an orange colour. At this stage it is called 'sodium cellulose xanthate' or simply 'xanthate'. The xanthate is a soluble form of cellulose and is dissolved in weak sodium hydroxide solution to give a solution containing about 10% cellulose. This solution is orange-brown and 'syrupy' and is known as 'viscose'.

The cellulose xanthate in viscose is unstable and slowly decomposes back into cellulose, carbon disulphide and sodium hydroxide. It thus becomes more and more insoluble, a fact which is helpful to the film manufacturer. His ultimate aim is to get the cellulose out of solution so that by allowing the viscose to 'ripen' optimum conditions can be chosen. The 'ripening' is carried out by storing the viscose in large tanks.

Film formation is carried out by pumping viscose through a long narrow slit into a bath containing sulphuric acid. Sulphuric acid neutralises the caustic soda in the viscose and destroys the cellulose xanthate causing the cellulose to be precipitated in a continuous sheet. This process is known as 'casting'.

The newly cast film is milky white in appearance due to sulphur particles and entrapped gas bubbles. The casting process removes these and produces the familiar transparent cellulose sheet. From the casting bath the film is passed through other acid baths to ensure that precipitation of cellulose is complete, then through a series of wash baths to remove entrained acid. Final traces of sulphur and sulphurous compounds are removed by a two stage treatment. A sodium hydroxide bath converts them into sodium sulphides and a subsequent bleach bath oxidises the sodium sulphides to sodium sulphate which is odourless.

The film, as it now stands, if dried would be extremely stiff and impossible to use, therefore, softening agents have to be added to it. This is achieved by passing the film through a bath of softener solution, typically glycerol and glycols are used, prior to drying, where the excess water is driven off leaving the softener in the film. The dryer consists of a large number of rotating heated cylinders through which the film is threaded.

Coating

The final product can either be used as it is or subjected to further treatment such as the addition of a coating. In this case it is usual to apply a 'sizing' agent to the film prior to drying to improve the coating adhesion to the 'base' film. Dye baths can also be incorporated if a coloured film is required.

In order to confer such necessary properties as heat sealability and moistureproofness, the base cellulose film is topcoated with a lacquer of either modified nitrocellulose or vinylidene chloride copolymer. The former system was developed in the late 1920's and is based on nitrocellulose plasticised with organic esters. In addition, paraffin wax and resins are added to confer moistureproofness. This type of coating can be modified to give only partial moistureproofness or, if necessary, can be made non-heat sealing.

The coating is applied by dissolving the components in suitable organic solvents, and then applying the solution to the base film with doctor rollers. The separation of these rollers controls the weight of coating on the film. The lacquer wet film is then passed through a heated chamber which allows the solvents to evaporate. The solvent vapours are drawn off for recovery and subsequent re-use. In order to restore the original moisture content of the film, and maintain flexibility, it is now necessary to recondition the film by passing it through a second chamber containing high humidity air. After conditioning the film is cooled and reeled in readiness for slitting.

The nitrocellulose coated films were the only type available until the early 1950's, when vinylidene chloride copolymer coatings were introduced. The method of coating is essentially the same as that used for the nitrocellulose type. This later type of coating has the advantage that subsequent conversion, *eg* printing and bag making, does not have as marked an effect on barrier properties as on the nitrocellulose coatings where the water vapour permeability can be increased quite appreciably.

FILMS AVAILABLE

Although the last decade has shown a large increase in the number of plastics films available to the consumer, cellulose film has maintained its position as one of the principal flexible transparent packaging films due primarily to its great versatility of formulation.

Both coated and uncoated films are available in a wide range of thicknesses, and may be coloured, plain or opaque. In order to classify the properties of individual film types the UK cellulose film manufacturers, in conjunction with many of the European producers, have developed a system of nomenclature, the principles of which are outlined below:

M	Nitrocellulose coated on both sides. The coating is heat sealable only when the letter 'S' is included.
QM	Nitrocellulose coated on both sides but less moistureproof than M.
PS	Nitrocellulose coated on both sides but less moistureproof than QM.
DM	Nitrocellulose coated on one side only.
P	Non-moistureproof.
MXXT	Vinylidene chloride copolymer coated on both sides.
MXDT	Vinylidene chloride copolymer coated on one side only.
B	Opaque.
C	Coloured (the name of the colour follows the code, *eg* 325 PFC Red).
F	Suitable for twist wrapping.
S	Heat sealable.
U	For adhesive tape manufacture.
V	High surface slip.

In addition to the above letters, films are also defined by

	280 MXXT	350 MS
Yield (m^2/kg)	35.71	28.57
Substance (g/m^2)	28.0	35.0
Tensile strength (kgf/mm^2)	12 MD 6 CMD	12 MD 6 CMD
Elongation at break	20% MD 50% CMD	20% MD 50% CMD
Youngs modulus (dynamic) (Dynes/cm^2)	6.5×10^{10} MD 2.5×10^{10} CMD	6.5×10^{10} MD 2.5×10^{10} CMD
Heat seal (g/38 mm)	200	150
Water vapour permeability (g/m^2/24 hr)	12	8

Table 1. Typical physical properties of cellulose films.

their yield, and by a suffix denoting any special features. Typical suffixes are:

12 For tobacco.
30 Standard film for frozen foods.
36 Standard film for conversion.

350 MS/36 is then defined as a film of 35 g/m^2, nitrocellulose coated on both sides, heat sealable and impermeable to water vapour, having greater flexibility than MS films thus making it suitable for general conversion.

APPLICATIONS

Cellulose film's greatest advantage over its competitors is probably its machine performance. Due to its stiffness, rigidity and lack of electrostatic charge it is possible to have troublefree running on high speed packaging machines. In addition, there is no necessity for rigid temperature control of the machine sealers since, unlike most plastics films, cellulose film will seal over a very wide range of temperatures.

The facility to vary the moistureproofness of cellulose film makes it suitable for wrapping a wide range of products, *eg* where the requirement is for dirt protection only and to confer sparkle and gloss to the end product, as is the case in shirt wraps, then the normal material to use would be P grade. At the other extreme, the over wrapping of potato crisps requires a product with excellent oxygen, moisture and grease barriers and suitable material here is MXXT. An important additional feature of cellulose film's barrier properties is its ability to retain volatile flavours in such products as sweets and spices without imparting to these products any off flavours.

More recent developments have seen cellulose film extend

Fig. 2. A modern film composite of cellulose, coated on one side by vinylidene chloride copolymer and the other, printed, side coated with polyethylene.

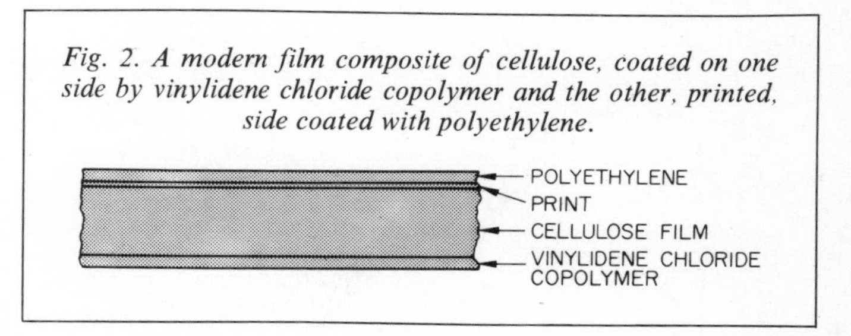

its function from that of a single ply wrapping material to its use in laminates. The most common composite currently in use is copolymer coated cellulose laminated to polyethylene. This type of laminate can consist of polyethylene adhesive or wax laminated to MXXT or, on the other hand, polyethylene extrusion coated onto the uncoated, printed side of MXDT (Fig. 2). This latter type has all the advantages of glossy print without any of the disadvantages of printing ink solvents in contact with the product.

In addition to the well-known uses to which cellulose film is put, there are some other more diverse applications, *eg* cellulose film flakes are used in oil well drilling for blocking cracks in porous strata; in the shredded form, cellulose is widely used as a cushioning material in the packing of fragile goods; and in the uncoated form as the membrane in artificial kidney machine dialysers. It is as a consequence of the vast range of end uses to which it may be put that a healthy future for cellulose film is ensured.

Chapter 15

Machining Plastics

T Lawrence and **R Godwin**
Tufnol Ltd

Although the ease with which plastics can be moulded and formed frequently offers a major advantage in manufacturing processes, it is often more desirable, or even necessary to machine. Thc decision to machine can be based on a variety of factors, for example:

i. For economy, where the quantity involved does not warrant large expenditure on moulds or forming tools. When a particular component could be easily machined from available stock sections of sheet, rod or tube, careful costing of quantities and potential may be necessary before embarking on moulding tool costs, always provided of course that suitable moulding machinery is available.

ii. Where the geometry of the component involves difficult undercuts or other features requiring complex inserts or collapsible moulds which would make moulding impracticable.

iii. Where the required degree of accuracy or quality can only be achieved by machining.

iv. Where the nature of the plastics material being used does not permit moulding, *eg* in the case where laminated plastics are chosen.

Most plastics can be machined with excellent results, using conventional wood or metal working machinery, with only minor alterations to tooling. The main consideration when machining plastics is their much greater sensitivity to heat compared to metals, the effect of which is exaggerated by their much lower thermal conductivity and, to some extent, by their higher rate of thermal expansion. To minimise the build up of heat, coolants can be used and tools should be designed to provide adequate clearances. Light cuts should be taken with high surface cutting speeds and slow feed rates.

Most of the common machining operations can be carried out on plastics, but when proceeding to study these in detail, it should be noted that the common types of plastics can, for convenience, be grouped for each machining operation, according to their machining properties. For example, for many operations, nylon behaves in a similar manner to acetal, polypropylene and PTFE since all of these have a similar type of surface, with a waxy 'feel', and produce swarf which is in the form of long, stringy shavings. Polyethylene could be included in the same group, but when machining this material, greater care should be taken, owing to its greater heat sensitivity. In contrast to this, for example, acrylics and styrenes are relatively hard and brittle and come away in the form of chips when cut.

Fig. 1. Facing on a spindle moulder.

SAWING

For production work, plastics are normally cut using circular, band, jig or fret saws of the type commonly used for cutting metals but, in some cases, conventional wood working blades can be used. As with all plastics cutting tools, it is important to keep saws sharp so that they will run cool, and where long runs are involved, carbide tipped teeth can prove more economical, or even essential in the case of glass or asbestos filled material. Opaque acrylic materials are also more abrasive than transparent or translucent versions and it is preferable to use tipped saws when these are being cut in any quantity.

Circular saws

When using circular saws, finer toothed blades with little or no set are used when the harder and more brittle plastics are being cut and for thinner sections of all plastics material. Acrylics and styrenes up to $\frac{3}{8}$ in (10 mm) thick show best results with hollow ground blades with 12 to 15 teeth per inch ($4\frac{1}{2}$ to 6 teeth per cm), while on heavier sections, blades with 10 teeth per inch (4 teeth per cm) or less are more satisfactory. Thinner sections of nylon, acetal or phenolic laminates for instance, would also be cut with fine toothed blades of 8 to 12 teeth per inch (3 to $4\frac{1}{2}$ teeth per cm) but much coarser blades with 4 teeth per inch ($1\frac{1}{2}$ teeth per cm) or less give good results with thicker sections. Thick sections of nylon or polypropylene, for example, can be successfully cut with blades having 1 tooth per inch ($\frac{1}{2}$ tooth per cm) or less. The hand saw shape of tooth, in which the cutting edge lies along the radius of the saw, is commonly used, but other shapes, with negative cutting angles, as in the peg tooth saw, also give good results. As a general rule, when difficulty is experienced with chipping, a finer toothed saw with less set, will improve the finish obtained.

Table 1. A general guide to suitable speeds and feeds for machining plastics.

	SAWING		TURNING				DRILLING			ROUTING	MILLING
	speed ft/min	blade teeth/in	cutting speed ft/min	feed in/rev	front clearance	top rake	speed $\frac{1}{8}$in dia rev/min	speed $\frac{1}{2}$in dia rev/min	speed 1 in dia rev/min	speed rev/min	speed ft/min
Phenolic laminates	1500 to 7500	2 to 11 (skip tooth)	600	0.005 to 0.010	12° to 15°	0° to 30°	8000	2000	700	18000 to 24000	4000 to 6000
Moulded thermosets	2000 to 6000	8 to 20	250 to 800	0.005 to 0.010	15°	0° to 15°	8000	1500	600	18000 to 24000	4000 to 6000
Glass filled laminates	300 to 400	3 to 10	300 to 500	0.004 to 0.008	0° to 15°	15° to —5°	1000	500	300	18000 to 26000	150 to 400
Polyethylene, PTFE	2000 to 6000	4 to 9 (skip tooth)	300 to 500	0.004 to 0.010	18° to 30°	0° to —5°	1000	500	300	15000 to 24000	800 to 1500
Acetal, Polypropylene, Nylon	3000 to 6000	4 to 9 (skip tooth)	500 to 1000	0.004 to 0.010	18° to 30°	0° to —10°	1200	800	500	15000 to 24000	800 to 1500
Rigid PVC	4000 to 11000	4 to 11	300 to 1000	0.008 to 0.024	15° to 22°	0° to —5°	5000	1000	500	15000 to 24000	600 to 900
Acrylics	5000 to 10000	4 to 12	100 to 500	0.005 to 0.010	15° to 20°	0° to —5°	6000	1000	600	15000 to 24000	700 to 1200
Polystyrene	2500 to 4000	10 to 24	300 to 1000	0.002 to 0.008	15°	0°	1500	700	300	12000 to 24000	400 to 1000

NB. 1 ft = 0.305 m, 1 in = 25.4 mm

Although hollow ground blades are necessary for the harder plastics, especially in thin sections, for production runs it is often prudent to use blades with the widest set consistent with good finish, since a large set will prevent clogging and assist in reducing heat, thus allowing a faster feed rate. For example, phenolic laminates over $\frac{3}{8}$ in (10 mm) thick can be cut with blades having approximately 0.025 in (0.6 mm) set on each side. Incidentally, it should be noted that phenolic laminates are somewhat exceptional, since, although they are fairly hard plastics materials, the paper or fabric base tends to bind the resin together and inhibit chipping. This allows the use of saws with relatively coarse teeth and wide set, which would otherwise normally be reserved for softer plastics. It is very difficult to present hard and fast rules regarding the feed rate for plastics materials and it is common practice to hand feed, which enables the operator to 'judge' the correct speed. The peripheral speed of the blade is not of prime importance, but high speeds produce best results and speeds from 4000 ft/min (1200 m/min) to 12 000 ft/min (3 650 m/min) are common.

Band saws

Band saws are generally used for rough cutting of discs or gear blanks and curved pieces. It is essential to have the saw sharp and lined up accurately with the guides set close to the saw and the blade just touching the thrust block or wheel when the machine is running light. The width of the saw band may range from $\frac{3}{8}$ in (10 mm) to more than 1 in (25 mm) and is determined mainly by the thickness and radius of curve of the material being cut. Most of the above comments regarding tooth pitch, set, *etc*, also apply here, but cutting speeds are generally lower, 6000 ft per min (1800 m per min) being used for thin sections reducing to 1600 ft per min (490 m per min) for thicker work. Special saw blades, called skip or buttress tooth band saws, have been developed to produce particularly good results with plastics and similar soft materials. These have hardened teeth with very deep gullets to help remove the chips formed during cutting and overheating is much less likely to be a problem with these blades.

Jig and fret saws

Jig and fret saws are used mainly for cutting intricate shapes in material up to, say, $\frac{1}{2}$ in (13 mm) thick, where no other machining method is practicable. Cutting speeds of between 500 and 800 strokes per minute are recommended but the feed must be very light to prevent excessive heat build up and coolants are desirable where thermoplastics are being cut.

Abrasive wheels

Abrasive wheels are normally only used when asbestos or glass fibre thermosetting plastics are being cut, since these give much longer life under the highly abrasive conditions. They are usually supplied up to 12 in (300 mm) in diameter and should be run at approximately 3800 rev/min. These wheels are extremely brittle and must be adequately guarded against the possibility of flying fragments. It is essential that these wheels are properly supported on both sides to prevent fracture.

DRILLING

Plastics may be drilled on any drilling machine but, if small holes are required, it is preferable to use high speed drillers. Standard high speed steel twist drills are commonly used but these must be ground to remove the positive rake on the cutting edges and, for high speed production work, the land behind the drill point should be removed, to improve the flow of swarf. If ground in this way, drills with a normal 118° point angle will produce good results in most cases but, when drilling through holes in softer plastics materials, such as nylon or polyethylene, a better finish can be obtained on the reverse side, by reducing this angle to approximately 80°. However, as a general guide, when drilling thin material the included angle of the point should be such that the full diameter of the drill has entered the hole before the point breaks through and, for this reason, on larger diameter drills, an angle of as much as 150° may sometimes be required. Special drills with extra wide, highly polished flutes are available for thermoplastics and can be advantageous where difficulty is experienced, but for general practice, the benefits gained from using these are minimal. For long production runs or where glass or asbestos filled materials are being drilled, carbide tipped drills are more economical.

Drills must be cleared frequently to remove swarf, using the woodpecking technique, and the feed should be light, to minimise the heat generated. To prevent breaking away at the under edge of the hole, the material should be held firmly against a hardwood or similar backing and the feed

Fig. 2. Putting an external thread on a plastics tube, using a ground thread chaser.

Fig. 3. Milling a laminated plastics block.

of the drill reduced before it breaks through. When very accurate holes are required they should be drilled very slightly undersize and reamed, but where the hole is to be tapped, a better thread will be obtained if the hole is drilled slightly oversize on the core diameter. A feed rate of 0.006–0.008 in (0.15–0.20 mm) per revolution would be quite normal but, where an especially fine finish is required, air drills, which carry a stream of air directly to the cutting edges, may be used with a feed rate of 0.003 in (0.07 mm) per revolution.

For large holes in sheet materials a wing or fly cutter can be used, but to obtain clean edges it is advisable to cut the hole from both sides. A small pilot hole should be drilled to guide the wing cutter and speed should be rather lower than for drilling.

Fig. 4. Cutting a bank of sprockets on a universal milling machine.

TURNING

Standard metal-working lathes are commonly used for most turning operations, but wood-working machinery can also be used. Tool design is important and for most plastics materials tools should have a zero to slightly negative rake and be slightly radiused to prevent scoring of the work. To minimise the build up of heat, tools must be kept sharp, and carbide tipped tools are generally most convenient to use for production runs. Side and front clearances of 12°–15° should be used but the front clearance may need to be increased for the softer materials.

Fairly high speeds, *ie* in the region of 600 ft/min (180 m/min), will be found satisfactory for most work but where efficient water or soluble oil coolants are applied, speeds up to 1000 ft/min (300 m/min) can be used. Feed rates between 0.004 in (0.10 mm) and 0.010 in (0.25 mm) per revolution will be satisfactory in most cases but, if light cuts are taken for final operations, it is rarely necessary to use feeds of below 0.006 in (0.15 mm) per revolution to achieve the desired results.

Screwing

Screw threads with a rounded form, such as Whitworth, are generally most satisfactory in plastics materials, since this assists in reducing the effects of notch sensitivity. Self opening die-heads with ground thread chasers are recommended. Dies should have a negative rake and the top of the thread chaser should be flat to negative and set in line with the centre of the work. Care should be taken, particularly with thermosetting plastics, to ensure that the thread is properly cut and not merely pushed aside, otherwise a poor and weak thread form will result. To obtain a good finish, starting and finishing cuts should be light and it is suggested that, where possible, dies should be reserved for machining plastics only.

TAPPING

Holes drilled or moulded in plastics materials can be easily tapped by hand or using any of the usual arrangements for

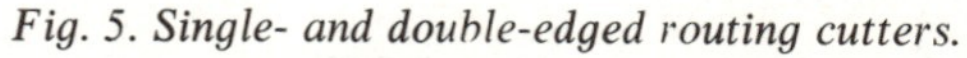

Fig. 5. Single- and double-edged routing cutters.

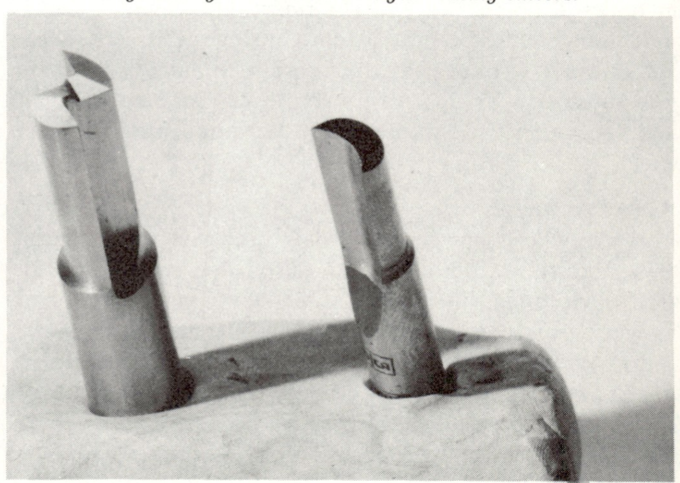

Fig. 6. Turning a bank of gear blanks.

machine tapping. Ground thread taps should be used but, where the softer thermoplastics are being machined, the tap should be chrome plated to 0.002–0.005 in (0.05–0.13 mm) over nominal size to overcome the tendency of the material to close in after the tap has passed. It is advisable for pilot holes to be drilled slightly oversize and, to prevent tearing of the first few threads, the pilot hole should be slightly chamfered. Before tapping blind holes, care should be taken to remove all swarf from previous drilling.

Slower spindle speeds should be used than for drilling and the use of a lubricant will produce a better finish and help to minimise overheating with thermoplastics. Typical speeds for tapping holes from 4BA up to $\frac{1}{2}$ in Whitworth on automatic tapping machines would be 200 rev/min for tapping and 750 rev/min for withdrawing and clearing the tap.

MILLING

No special equipment is required for milling plastics and they can be machined on plain or universal millers or on woodworking spindle moulders. Where a considerable amount of material has to be a removed in one cut, it is advisable to use a plain or universal miller, but when removing small amounts, for example, when facing strip to the required limits or when bevelling edges, it is preferable to use a spindle moulder. Milling cutters with straight or spiral teeth should be used (spiral to obtain a smoother finish) and ideally the pitch of the teeth should not exceed $\frac{5}{8}$ in (16 mm) for a 4 in (100 mm) diameter cutter and proportionately less if the diameter is smaller. Cutters should be of high speed steel but, for extensive production runs, tungsten carbide cutters are recommended. If cutters are kept sharp and in good condition, the use of a coolant is rarely necessary.

Climb milling is always preferable to rip milling and a cutting speed of 4 000 to 6 000 ft/min (1 220–1 830 m/min) should be satisfactory for most work. The material should be fed past the cutter at an even rate and the depth of cut should be fairly small, say, $\frac{1}{16}$ in (1.6 mm) when feeding by hand or up to $\frac{1}{8}$ in (3.2 mm) when machine fed. However, to prevent overheating, some thermoplastics may require shallower cuts. When machining the less rigid types of plastics, it is important to ensure that the whole of the workpiece is fully supported, to prevent distortion caused by the cutting pressure.

ROUTING

Routers are commonly used to form curved sections and profiles and excellent results can be obtained with plastics materials. Although single-edged cutters can be used, double-edged cutters are recommended and these should be made from 18% tungsten high speed steel. However, tungsten carbide tipped cutters will give longer life, particularly when machining mineral filled materials. Spindle speeds should be high, between 18 000 and 24 000 rev/min and feed should be slow to moderate. In order to keep the wear and tear on the cutters to a minimum, it is advisable to roughly profile the work by band or fret sawing before routing and, where a considerable depth of cut is to be taken, this should be done in stages, gradually lowering the cutting head.

When double-edged cutters are being used, it is advisable to grind away towards the centre of the cutter to facilitate removal of swarf and, although the use of a coolant is rarely necessary, a jet of air directed onto the work will also assist in clearing swarf. Certain plastics, such as fabric laminates, nylon, *etc*, make ideal material for producing jigs, since they are lightweight, easy to handle and withstand considerable usage. Also, the cutter will not be damaged by accidental contact with jigs made from plastics materials.

Fig. 7. Centreless grinding a plastics tube.

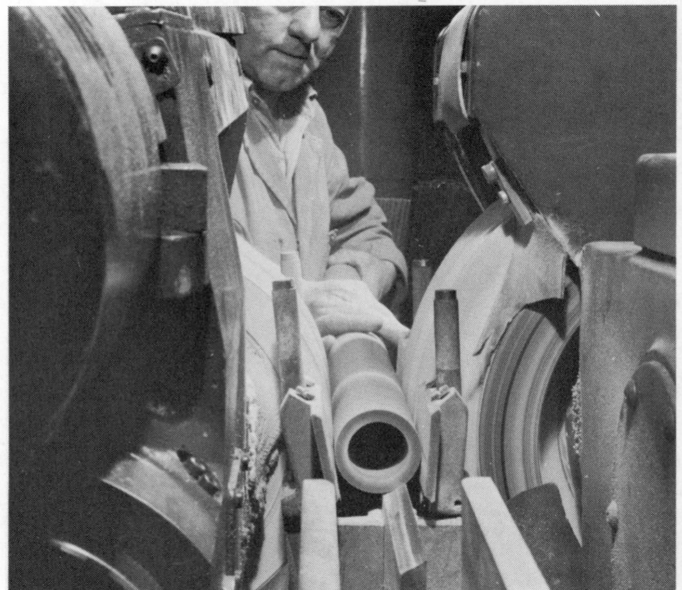

GEAR CUTTING

Machining of plastics gears is normally limited to nylon, acetal or laminates, and gear teeth can be cut on a milling machine or a gear shaper with the usual cutters and tools. Feeds and speeds vary with the shape and size of teeth to be cut, but 140 ft/min (43 m/min) is a good average peripheral speed when a high speed milling cutter is being used. When plastics gears are to be stressed in service to nearly their maximum strength, it is advisable to cut them on a gear planer or shaper. When small teeth are being cut, say up to 10 DP (2.5 module) the teeth can be finished at one cut, but when larger teeth are being machined, greater accuracy can be obtained by taking two or three cuts, removing not more than 0.005 in (0.12 mm) on the last cut. However, when machining, it is important to ensure that plastics gears are well supported to the full diameter of the gear by a hardwood or soft metal backplate to prevent the cutting pressure from causing distortion or damage.

GRINDING

Most types of plastics rods and tubes can be production-sized by centreless grinding and tolerances of plus or minus 0.005 in ($\pm$ 0.13 mm) or less can usually be readily achieved. An open grain silicone carbide wheel is recommended. Care should be taken to prevent the wheel from clogging and its surface should be periodically trued. Although thermosets, such as laminated phenolic rods and tubes, should be ground dry, the use of a water jet will be advantageous when grinding thermoplastic rods. A diameter reduction of 0.010 in (0.25 mm) per run would be normal, with finer cuts for finishing.

SHEARING (PUNCHING AND GUILLOTINING)

Most plastics can be punched or guillotined in relatively thin sheets with reasonable success using hardened steel tools, but some materials, such as acrylics and rigid vinyl, may require to be heated to avoid cracking or crazing. It is most difficult to lay down hard and fast rules for tool design since this varies considerably with the grade of material and the profile being punched and, in some cases, trial and error is the only sure way of determining the most correct tool design.

Normally, the punch and die should be made to give a clearance of 0.0005 in (0.013 mm) or less and the punch should be perfectly smooth and parallel. When punching laminates or fairly hard thermoplastics, flat bottomed punches should be used and these should be set so that the punch enters the die by approximately 0.0005 in (0.013 mm) at the end of its stroke, or rather more if such materials as thin nylon or polypropylene sheets are being punched. In some cases, particularly when softer materials, such as polyethylene or flexible PVC are involved, it will be necessary to use knife-edged tools and to allow the punch to enter the die for some distance, to separate the piece being punched from the strip. However, with all materials, it is important to ensure that the material surrounding the area to be punched is held flat during the punching operation and, although a standard stripper plate can sometimes be used when punching simple shapes from thin, rigid materials, it is often necessary to use rubber pressure pads acting in conjunction with the movement of the punch. Where punches pass through a rubber pad, it should be remembered that ample clearance must be allowed for distortion of the rubber under pressure. Punches should not project from the body of the tool before pressure is applied and faces should be flush. When producing tools, it should be noted that plastics materials tend to close in after the punch has passed and a hole left by a punch will be slightly smaller than the actual size of the punch. Allowance should also be made for thermal expansion of tools and plastics materials when heat is to be applied. Location pegs should be parallel for a distance equal to the thickness of the material being punched.

Fig. 8. A 3-stage punching tool for thin sheet material, showing rubber pressure pad.

Sharp corners should be avoided in the design of components to be produced from rigid materials since these greatly promote cracking. When highly abrasive materials, such as glass fabric laminates, are to be punched, it is advisable to produce the die in the form of a removable insert in the tool, thus allowing worn punches and dies to be replaced without scrapping the whole tool.

Rigid polystyrene cannot be successfully punched owing to its tendency to craze rather badly and for the same reason cast acrylic sheets, high impact polystyrene and rigid vinyl sheets will require heating, unless very thin. These should be heated to between 80° and 90°C and in some instances it is also preferable to heat the tools. Heating should be thermostatically controlled and in many cases a standard hot plate will be satisfactory. However, if difficulty is experienced in cold punching nylon sheet, this may be overcome by heating it in water, but nylon and acetal can normally be punched cold in thicknesses up to $\frac{1}{8}$ in (3 mm) without great difficulty. PTFE and polyethylene should punch satisfactorily in thicknesses up to $\frac{1}{4}$ in (6 mm) without heating. Thin plastics sheets can also be cut on guillotines designed for dealing with thin sheet steel. As for punching, the material must be firmly clamped in position during cutting and it is essential that the cutting edges of the blades be kept keen.

Chapter 16

Welding Plastics

H R Stilton
General Industrial Plastics (Research & Development) Limited

The fabrication of many plastics products often involves welding, sealing or adhesion of plastics sheets or preformed components. Several methods of achieving satisfactory bonds are currently in use, these include: high frequency welding; ultrasonic welding; heat sealing; hot gas welding; solvent welding; and spin welding.

The choice of method is determined mainly by the type of plastics, speed and quantity of production required, which in turn leads to the capital cost of the equipment and the end use of the fabricated product.

HIGH FREQUENCY WELDING

High frequency (hf) welding is the most widely used form of sealing together two or more sheets of plasticised PVC. The basic principle of the technique is that the materials to be welded are clamped between two metal plates, or electrodes, a high frequency field applied to the electrodes, which in effect become capacitor plates of a dielectric heating circuit. The high frequency field applied between the electrodes passes through the material, causing heat to be generated at the interfaces of the materials by molecular friction. A typical circuit using a rectified power supply is illustrated in Fig. 1.

In practice, however, the complete hf generator is more sophisticated, allowing for refinements such as automatic matching between the output and oscillator circuits, the voltage across the electrodes and thus power into the work load. Refinements such as low voltage control circuits, safety switching and the provision of sensing devices to instantly prevent oscillation if the resistance between electrodes is reduced to a level that would otherwise cause damage to the work and electrodes, are additional features of modern generator design.

Currently a major change in generator design has become necessary because of legislation, which is already in force in some countries. This is the limiting of radiation within

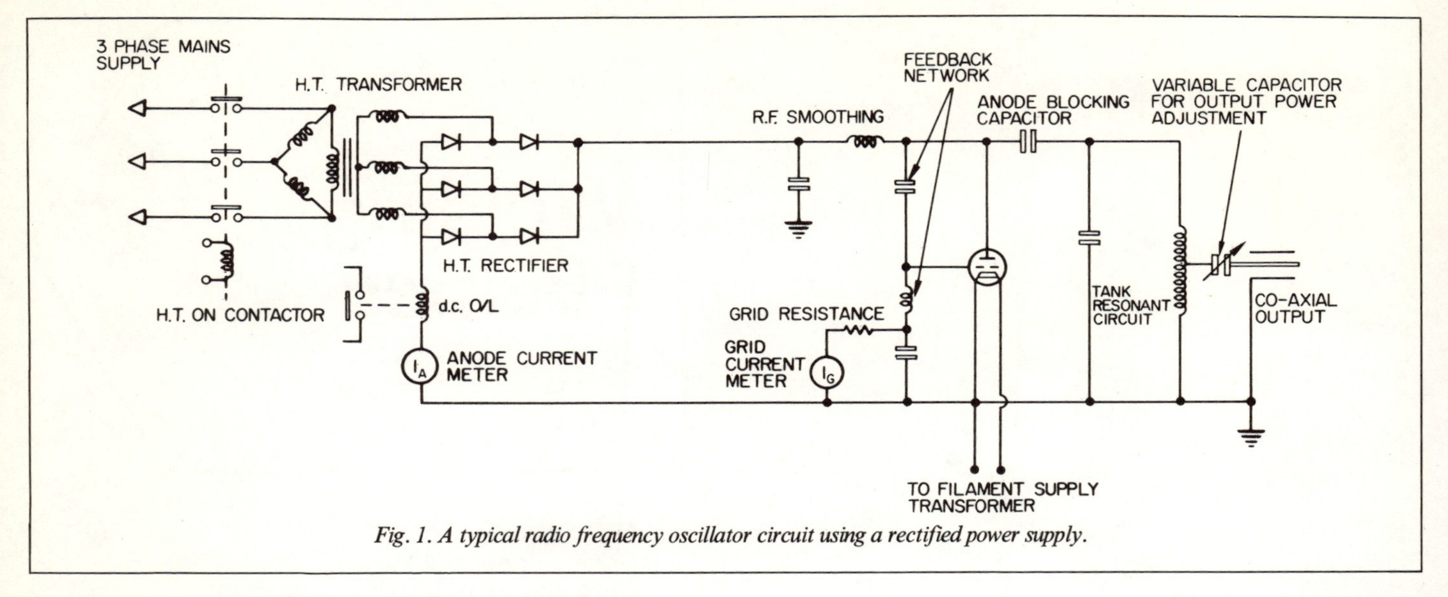

Fig. 1. A typical radio frequency oscillator circuit using a rectified power supply.

wave bands specified by international agreement. There are several bands available but, in effect, the limits imposed are such that the only practical band is 27.12 MHz, $\pm 0.6\%$.

Basically, the differences between generators with stabilised and non-stabilised output frequency are the inherent losses of power into the resonant or output circuit. In the case of non-stabilised generators, about 85% of the power available from the oscillator valve is transmitted to the work load. With frequency stabilised generators, this is reduced to about 66%, because of increased losses within the resonant circuit. Consequently, a larger and more expensive oscillator valve and resonant circuit is required in a stabilised generator, to provide equivalent power to that given by a non-stabilised generator. Furthermore, to prevent radiation leakage, the frame and panelling of frequency stabilised and harmonic suppressed generators must be securely bonded mechanically, all supply and control circuit leads associated with the generator must be adequately filtered. Harmonic filters and 'stop' filters covering a band of frequencies up to 2000 MHz are also necessary between the resonant circuit and the work load.

Mechanically, the work press to which the electrodes are attached is conventional in as much as the requirements are basic, *ie* application of pressure by pneumatic, hydraulic or motorised means, or a simple pedal action. In practice, high frequency (hf) plastics welding presses are somewhat more sophisticated in as much as most presses are used for a variety of applications requiring varying controls, such as pressure adjustment, accurate stroke limiting stops and timed cycles. It is imperative that platens, which currently can be as large as 2.5 m × 1.8 m, are accurately made and aligned parallel to each other to limits of less than 0.075 mm. As pressures of up to 80 tons are sometimes required, it can be seen that substantial press frame and platen structures are necessary.

Electrically, hf welding presses need to have one electrode assembly, and this normally includes one of the press platens, isolated from the press frame. Post insulators cast from ceramic, polyester or epoxide, which have the necessary tensile and compression strength and, at the same time, possess the necessary electrical and hf insulating qualities are therefore used.

An important aspect of high frequency welding is the tooling and handling gear. The latter is an adjunct to the press and generator and is a means of loading the tools or dies and materials quickly into the welding position, to maximise production. Several types of orthodox handling mechanisms are in use, these include: work trays which can be manually operated or powered; rotary tables with multi work trays; and material indexing units which accurately progress materials from rolls to the welding station. Frequently, specialised machinery is purpose built, but the basic requirements of pressure, hf power and time are always inevitable.

Tooling or dies are often referred to as electrodes. This is technically correct as they are the means by which the high frequency voltage is applied to the work. Electrodes can be made from any metal, but 'half-hard' brass is the most acceptable since it is fairly easily machined, has a reasonably high electrical conductivity and is sufficiently hard to stand up to the mechanical pressures exerted.

ULTRASONIC WELDING

It is known that if two surfaces are rubbed together vigorously, frictional heat will be produced. This is the basis of ultrasonic welding. As far as plastics welding is concerned, ultrasonic energy results from the conversion of mechanical vibrations into heat.

Whilst it varies from material to material, all solids have an internal resistance to vibration. This is, in effect, similar to an electrical resistance, thus as a higher impedance is introduced, as with the joint between two material faces, intense ultrasonic energy is transmitted and the temperature across the interfaces will rise.

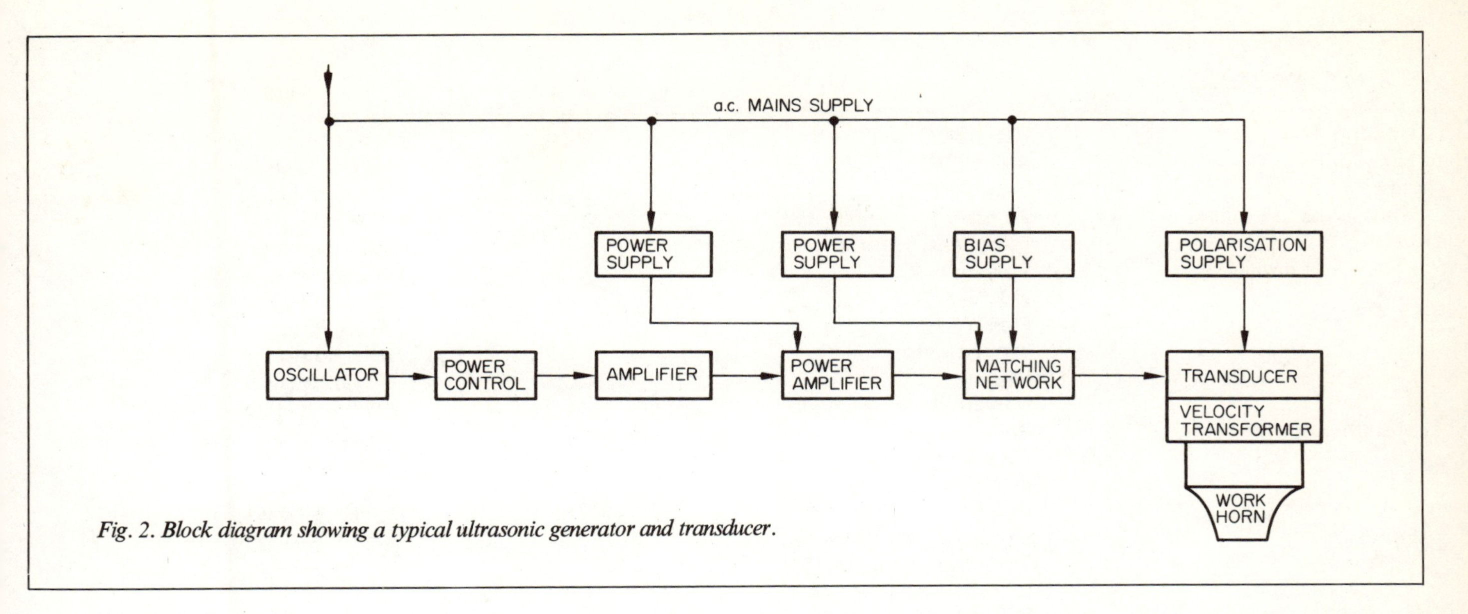

Fig. 2. Block diagram showing a typical ultrasonic generator and transducer.

An ultrasonic welder basically consists of a generator which converts electrical mains frequency to the required ultrasonic frequency, normally in the 20 kHz range and a transducer which converts electrical input to the ultrasonic output. A typical layout for an ultrasonic welder is illustrated in Fig. 2. As with high frequency generators, most ultrasonic generators in use employ refinements such as overload circuit breakers and other safety devices, as well as power, timing and tuning controls.

There are two types of transducer, the piezoelectric or crystal type and the magnetostrictive type. Crystal transducers use natural or synthetic materials that will change in physical dimension if an electrical charge is applied across opposite faces. The disadvantage of transducers using quartz crystals or synthetic crystals is that if the temperature rises above the point known as the Curie point, it can have an effect on the life span of the crystal. Magnetostrictive transducers use the property of certain metals, such as nickel, where a changing magnetic field produces the change in the physical dimensions of the laminate stack which, whilst losing some efficiency if heated above the Curie point, will return to normal on cooling.

In construction, a transducer is similar to a small transformer. Two coils of wire are wound onto a nickel lamination stack. The work horn is screwed to one end of a stub, the other end of which has been bonded to the nickel stack. The assembly, together with a cooling fan, is mounted into a small press head which allows vertical pressure, normally under 10 kgf/cm^2, to be applied by the applicator or work horn to the work. The applicator can be changed for undertaking different applications such as spot, ring or bar welds or more complex shapes.

Two types of ultrasonic seal are possible, *ie* contact or near field such as with high frequency welding, where the electrode or applicator is applied directly to the material where the weld is to be made, and remote or far field which is suitable for joining rigid mouldings. With the latter, the tip of the work horn is applied to a point of the work which may be some inches away from the faces to be welded, and the vibrations are transmitted through the material.

If the dimensions from the top of the work horn to all parts of the faces to be welded are equal, a single application will suffice. If there are gross differences from the tip to the weld face, the work horn may have to be applied more than once, in different places.

Table 1. The relative merits of the processes for welding specific plastics materials. The classifications listed are general and depend on the precise material-formulation and thickness and the shape and rigidity of the component.

MATERIAL	HIGH FREQUENCY	ULTRASONIC	HEAT SEALING	SPIN	SOLVENTS	HOT GAS
Rigid PVC	Poor	Good	—	Good	—	Excellent
Flexible PVC	Excellent	Poor (c)	Poor	—	—	—
Polyethylene sheet	—	—	Excellent	—	—	Good
HD Polyethylene	—	Poor	Poor	Good	—	Good
Polystyrene	—	Excellent	—	Good	Excellent	—
Acrylic	—	Poor	—	Good	Excellent	—
Nylon	Good	Good	Good	Good	—	—
Polypropylene	—	Good	—	Poor	—	Good
Polyvinyl acetate	Good	—	—	—	Good	—
ABS	Good	Excellent (c)	—	—	—	—

NOTE: With ultrasonic welding (c) denotes contact or near field welding only.

PRODUCT	MATERIAL	PROCESS
Ducting	Rigid PVC	Hot gas welding
Tank liners	Rigid PVC	Hot gas welding
	Flexible PVC	HF welding
'Tarpaulins' and large sheets	Reinforced flexible PVC	HF welding
	Polythene sheet	Roller welding
Loose-leaf covers	Polystyrene	Ultrasonic
	Flexible PVC	HF welding
Upholstery	Flexible PVC	HF welding
	Woven nylon	HF welding
Liquid filled sachets	Flexible PVC	HF welding
	Polyethylene laminates	Hot bar sealing
Moulded components	Polystyrene	Ultrasonic or
	ABS	Solvent or
		Spin or
		Hot knife welding
Extrusions	Extruded polyethylene	
	Extruded polypropylene	Hot knife welding
	Extruded polystyrene	
	Extruded PVC	
Packaging	Polyethylene	
	Acetate	Impulse sealing

Table 2. A cross section of plastics welded applications and recommended processes.

SOLVENT WELDING

Many plastics can be softened and fused with the aid of solvents, although its use is mainly restricted to rigid forms of plastics. Solvents are probably used mostly for the fabrication of polystyrene. Many toys are made in this manner and the majority of model kits employ the principle.

Solvents used for bonding polystyrene include: ethylene dichloride; methyl ethyl ketone; methylene chloride; and trichlorethylene. These are fairly quick drying solvents, but if a longer drying time is required, perchlorethylene or toluene can be used. The latter can take about ten minutes to dry. It is possible to mix different solvents to obtain a required drying time and water or airtight joints can be achieved by dissolving polystyrene in the solvent before applying it to the faces to be bonded. Bonds made with adhesives that contain polystyrene solvents, should not be stressed until they are completely hardened.

Acrylics are often fabricated into display signs, ornaments and models, using such solvents as: acetone; chloroform; and toluene.

The disadvantages of solvent welding are the need for much manual operation and the risk of disfiguring parts of an object that are not being bonded, as a result of misapplication of solvent. The most simple method of applying a solvent is to use a brush but, where single faces are to be coated, a solvent impregnated sponge can be employed. Longer production runs may allow the use of a spray gun with a stencil.

The majority of other plastics can be bonded to each other or to other materials by the use of solvent based bodied adhesives, or adhesives based, for instance, on water. Numerous proprietary adhesives are available and manufacturers can be consulted to advise on the most suitable. It should be noted that many solvents are inflammable and give off toxic vapours.

HOT GAS WELDING

The hot gas welding process is a means of welding together certain rigid plastics, preferably not less than 1 mm thick, using a filler rod of a compatible material. The surfaces of the materials to be welded should be prepared to accept the filler rod, *ie* single or double V butt or an angled fillet, should be provided. The surfaces and the filler rod are heated and softened by a stream of hot gas, directed from a welding torch. The hot gas, which should be dry and oil free, can be air or nitrogen, the latter being preferred for polyethylene.

Gas pressure, rate of flow and temperature may vary slightly, depending upon the plastics being welded, but 0.2 to 0.75 kgf/cm^2, 15–30 litres/min and 300°C respectively can be taken as a general guide. The gas temperature should be recorded by a thermocouple 3 mm from the nozzle of the torch.

The torch, as well as directing the gas stream, is also the means of heating it. This can be done electrically, or by gas. With electrical heating, the gas passes through a tube made of an insulating material capable of withstanding high temperatures (usually ceramic) on which is wound the heating element of up to about 500 watts power rating. With gas heating the compressed air or inert gas is passed through a coil which in turn is heated by the flame of a gas, such as butane, acetylene, *etc*. There is obviously no direct connection between the two gasses, or between the flame and the plastics being welded. The temperature of the gas being directed to the work can be controlled by altering its rate of flow at the inlet to the torch.

Filler rods are generally circular or triangular in section, but in any case should, if possible, match the profile of the weld preparation. For rods of circular section, about 3 mm diameter is the size most generally used. Preferably, welding rods and the faces to be welded should be roughened before welding, and must be clean and grease free.

During welding, hand pressure should be applied to the filler rod, and if this is of a semi-flexible nature, it may be necessary to have some device for pushing the rod into the grooves or corners. A small roller attached to the welding torch may suffice.

The advisable number of weld runs, using a 3 mm diameter rod, depends upon the sheet thickness; and whether or not a single or double V butt (*ie* welded from both sides) is necessary depends upon the required strength. Using rigid PVC as an example, a single V butt will give up to 40% joint efficiency and a double V butt up to 80%. In the case of angled fillets, it is always advisable to make a weld run on both sides of the material.

Material thicker than 3 mm will benefit in joint strength if multiple weld runs are made on each side, *ie* 6 mm—3 runs, on each side; 12 mm—6 runs on each side. The same applies to double V butts, although to obtain optimum strength from single V butts three weld runs should be made on 3 mm thick material and five on 10 mm thick. In all cases the prepared welding groove angle should be about 70° with a root gap of about 0.6 mm. Fig. 3 illustrates the various weld types and runs.

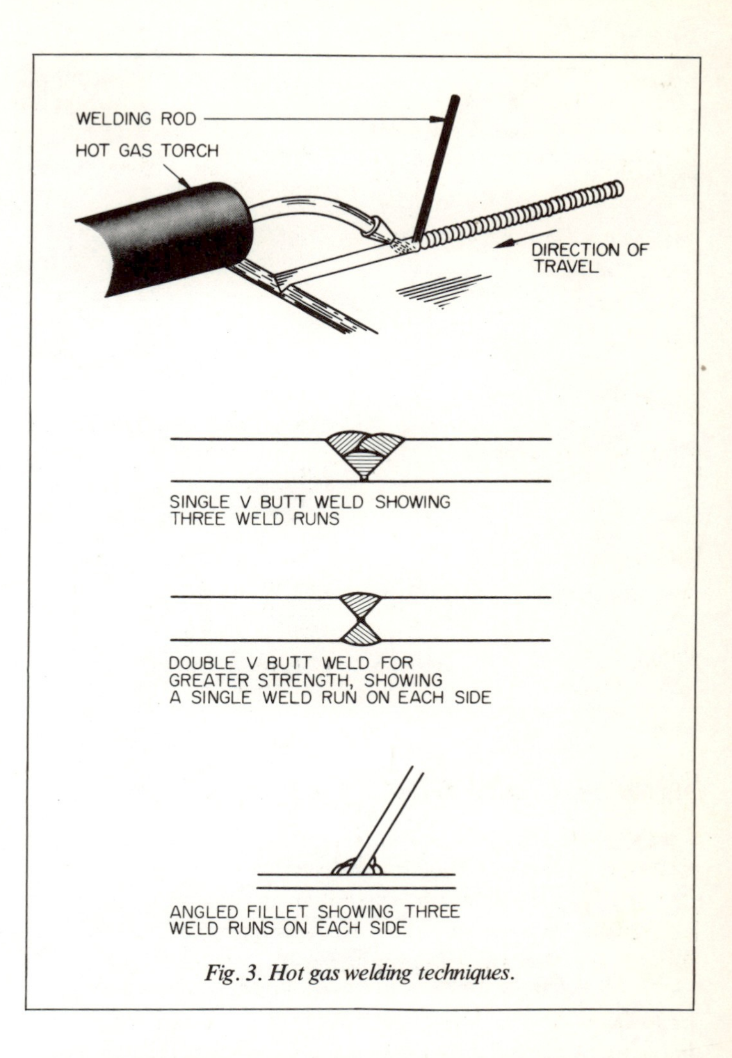

Fig. 3. Hot gas welding techniques.

HEAT SEALING

Several categories of plastics welding are discussed under this general heading of heat sealing, some of which are as follows:

(a) *Hot bar sealing,* where the plastics to be sealed is clamped between two bars, one or both of which are electrically heated.

(b) *Impulse sealing* which employs a resistance element that is electrically heated only during the sealing cycle, and which is applied to the work under pressure for the whole of the required sealing cycle.

(c) *Roller welding* in which the interfaces of the plastics materials are passed over a heated 'shoe' immediately before being bonded together by pressure rollers.

(d) *Heated tool welding*, which is in effect an electrically heated platen or knife, against which parts to be welded are held before being placed together under pressure until cool.

There are numerous extensions of these basic techniques, such as hot wire instead of a hot bar, pressure clamps instead of rollers, *etc*. Methods (a), (b) and (c) are applicable to sheet welding, whilst method (d) is more readily adaptable to mouldings and extrusions.

Hot bar welding employs simple heating jaws or bars, one or both of which are continuously heated to a temperature of between 180°C and 230°C. The temperature is thermostatically controlled. To prevent the plastics being sealed from sticking to the heated bars, a barrier material such as PTFE is interposed between the plastics film and the heated bars. The bars are normally attached to a small pedal press or hand tongs, but in the case of high output requirements, the heated bars are an integral part of a purpose made equipment. An example is a polyethylene bag making machine, which operates at well over one seal/second per webb run.

Impulse sealing employs an impulse generator which feeds power to the heating element via a low voltage isolating transformer, for pre-set time cycles. Fig. 4 shows a simple impulse generator circuit. An advantage of impulse heat sealing is that the heating platen or jaws will cool after the power has been switched off, allowing pressure to be maintained on the work, which for some applications is necessary.

The heating element, which within the area of the electrical power available can be made to many configurations, is best made from a metallic material, such as nickel chrome in the region of 0.01 mm thick. The thickness is, however, dependent to some extent upon the materials being sealed.

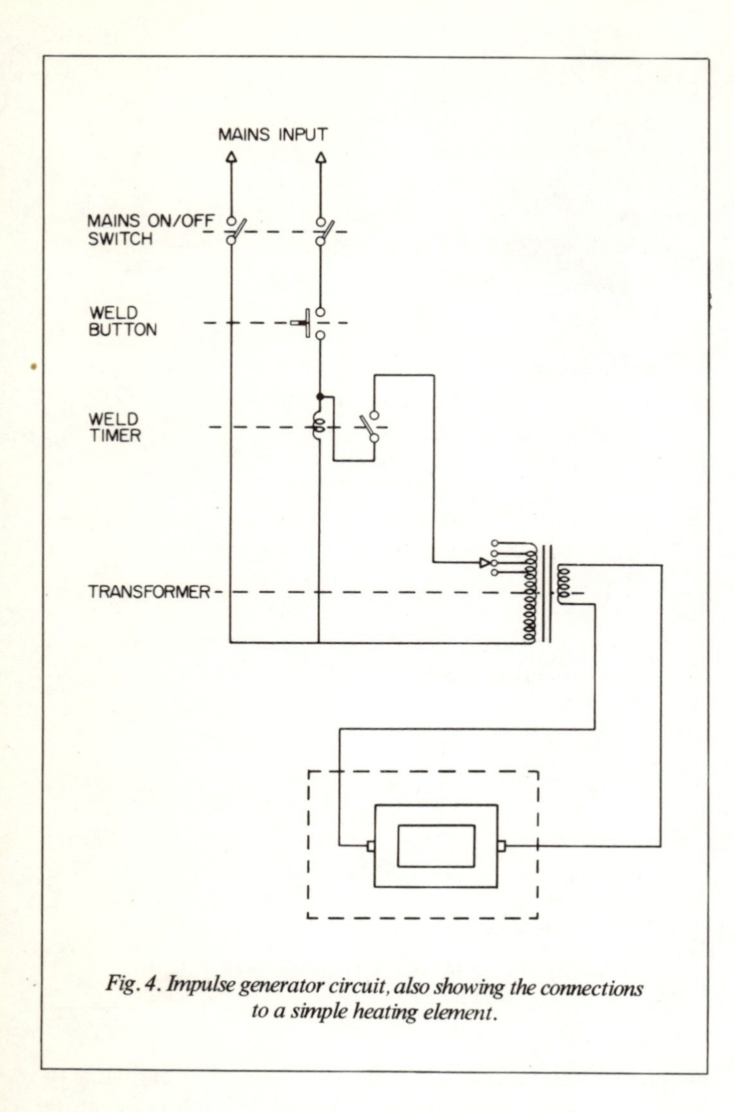

Fig. 4. Impulse generator circuit, also showing the connections to a simple heating element.

An important point in designing impulse sealing jigs is that the electrical path between the connections from the generator must be equal (see Fig. 4).

Pressure of the heating elements is also important and, whilst this again depends upon the material type and thickness, it is imperative in all cases that the pressure be equalised over the heating area. Therefore, the elements which, of course, are mounted onto a non-conductive material, should be rigidly fixed on a flat machined surface, or be completely flexibly mounted on a suitable rubber so that under pressure 'it finds its own level'.

Roller welding is probably more similar to sewing than other forms of plastics welding. Its disadvantage is that it requires similar skills from the operator, although such attachments as hemming and alignment jigs are available to assist the operator. It is used for welding flexible thermoplastic sheeting by heating the interfaces and immediately applying pressure by rollers. The interfaces are heated by an electrically-powered, thermostatically controlled shoe or wedge over which the material is continuously fed. A pair of pressure rollers immediately adjacent to the off take side of the wedge applies the required pressure, ensuring a bond whilst the interfaces of the material are still plasticised.

Heated tool welding can in practice be more complex than the other forms of heat sealing already described. Its most simple form is an electrically heated plate or knife with polished surfaces. The parts to be bonded are held against the polished surfaces, thus conducting the heat to the interfaces of the plastics components, which become plasticised, after which they are held together under pressure until cool. The temperature requirements vary from about 180°C to 230°C.

Most plastics components when being held against the heating plate will leave a residue which may oxidise. This can affect subsequent welding, therefore, clean plates are essential. A material to which the molten plastics will not stick, such as PTFE glass cloth, is therefore advisably used to cover the surfaces of the heated platens. PTFE glass cloth, in the region of 0.25 mm thick, will normally act as a partial thermal insulator reducing the temperature to the components being welded by about 10°C. Compensation must therefore be made for this.

For **pipe welding** or **butt welding** of rigid sheets, semi-automatic machinery is available. Polyethylene or polypropylene pipe lengths of over 300 mm diameter are easily welded on machinery which ensure accurate pre-set heating temperature and time cycles and pressure requirements. Whilst the principles of butt welding thermoplastic sheets are the same as pipe welding, it will be obvious that the mechanical design, especially in relation to the material clamping, and heat distribution will be very different.

Heating, under pressure, the surfaces of thermoplastic materials, will force the plasticised material outwards. This will smooth out any minor discontinuities in the surfaces to be welded, but too much displacement may lead to distortion of the product. Pressure against the heating element should therefore be gradually reduced during the

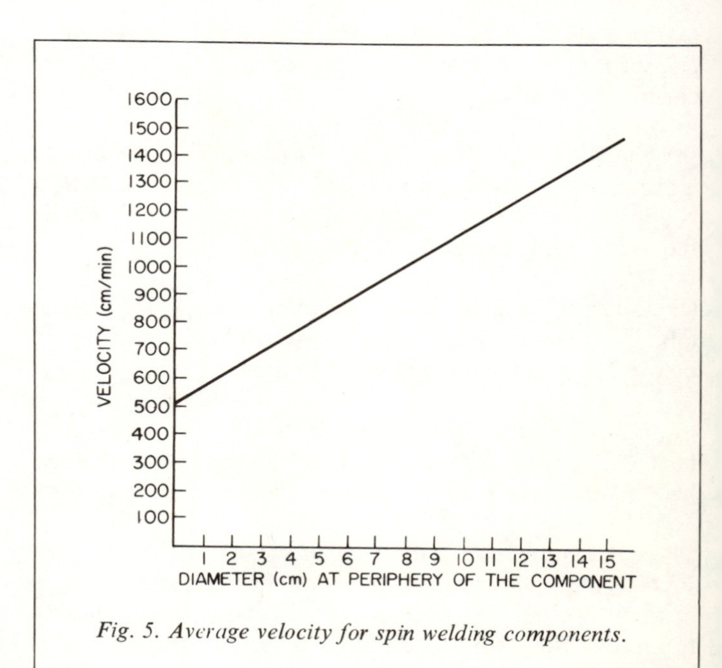

Fig. 5. Average velocity for spin welding components.

heating cycle. Heating periods are seldom less than 10 seconds, and can be as long as 80, particularly when stressed materials are being welded. Poor quality welds can result from heating too quickly.

A good general rule is that, when the material surface being heated has bulged outwards to about 50% of the thickness of the section, the heating element should be removed and the plasticised surfaces held together until the heat has reduced to room temperature. In some cases forced cooling (air or water) may be employed. Average pressure requirements are between 1 kgf/cm^2 for polypropylene and 3.5 kgf/cm^2 for ABS, depending upon material thickness.

FRICTION WELDING (SPIN WELDING)

The heat required for welding is obtained by pressing the surfaces of the two parts together and rotating one of them. Because of the low thermal conductivity of some thermal plastics materials, the welding temperature is rapidly attained. The rotation ceases, of course, when the welding temperature is reached, but the pressure of the two parts against each other must be maintained until they are cool.

The pressure between the components should be in the region of 7 kgf/cm^2 and Fig. 5 indicates the average velocity at the outer edge of the component. Friction welding is best suited to assemblies in which at least one of the components is of a circular cross section.

One of the problems associated with friction welding is the difference in the heat pattern caused by the variation in velocity from zero at the centre to maximum at the periphery. One solution is for the surface of one component to be slightly domed from the centre, so that this will make the first contact, thus starting heating at this point. The heating will spread towards the periphery. This also enables dirt, *etc*, to be expelled from between the components. Hollow sections do not present the same problems. Components being welded together should preferably be of identical material. The joint efficiency of a correctly made friction weld can be 100%.

Chapter 17

Assembly Techniques

F A Dixon *FIWM*
Creators Limited

The choice of the most suitable assembly method for a particular application will depend on a variety of factors, including: the type and number of components to be assembled; the end use of the assembled unit; and the conditions under which the finished assembly has to operate. Only if the type of methods to be used are examined in a critical manner at the design stage will it be practical to achieve efficient and economical methods of assembling thermoplastic components. The modern engineer has available to him an ever increasing array of assembly techniques for joining thermoplastics. Unless he is an expert in this particular field, the advice of manufacturers of adhesives and fasteners should be obtained at an early stage when contemplating manufacturing any new product involving assembly.

ADHESIVE BONDING

It is often considered that the long established method of bonding two surfaces together with an adhesive has been superseded by new, apparently superior, processes. Nothing can be further from the truth. There are many occasions when the use of adhesive can prove to be comparatively cheap and simple. The bonds produced by adhesive have maximum resistance to shearing loads (those in a plane parallel to the adhesive line). To obtain the greatest mechanical strength between two components that are to be joined, it is advisable that the largest possible areas be placed in immediate contact.

Effective bonding will only be obtained if proper preparation of the surfaces to be joined is achieved. Surfaces should preferably be roughened before bonding, but it is essential that the areas to be bonded together are absolutely clean. Two points requiring particular attention are the presence of silicones, which are used widely today as mould release agents and which prevent adhesion, and the possibility of plasticiser migration.

Surfaces	Basic adhesive types recommended
Acetal	Epoxy, Nitrile
Cellulose Acetate,	Nitrile, Urethane, Cellulose Nitrate.
Cellulose Acetate Butyrate	Polyester
Polyethylenes	Epoxy, Rubber Latices
Polystyrene	Epoxy, Polyester, Styrene-Butadiene, Acrylic, Urethane
Polyurethane	Nitrile, Epoxy, Urethane, Polyester
Polyvinyl Chloride	Nitrile, Urethane, Polyvinyl Acetate, Polyester, Resin Emulsions
Nylon	Epoxy, Neoprene, Polyester, Phenol-Resorcinol
Polypropylenes, Polycarbonates	Information available from material manufacturers

Table 1. Guide to the types of adhesives suited to various plastics.

To determine the correct adhesive to give the greatest bond strength, the following points should be taken into consideration:

1. The strain to which the joint will be subjected; the type of strain; its size and possible duration.
2. The overall dimension of the part to be joined; the configuration and geometry of the area to be bonded can be critical.
3. Any time factors involved in making the bond and the amount of heat and pressure that can be used.
4. The types of equipment available to assist in the bonding operation, *ie* jigs, fixtures, adhesive rollers, ovens, other types of heating equipment, presses, *etc.*
5. The cost involved in comparison with other methods.
6. The service conditions under which the assembly will operate. Some adhesives become brittle at low temperatures or tacky at high temperatures and the bond strength is affected.
7. The chemical resistance required from the adhesive if the assembly is to work in a corrosive atmosphere.

Table 1 gives a general idea of the normal recommended methods to be used. However, it is advisable to contact a supplier of adhesives before arriving at any final conclusions. Most suppliers have a wide practical experience in the formulation and use of adhesive for many special requirements and will normally have an intimate knowledge of plastics fabrication.

Methods of application

Adhesives should be applied only in a thin film, and care must be taken to avoid large deposits occurring. Initial tack is usually sufficient to hold the areas together but frequently the maximum bond strength can take up to 24 hours to develop. There are a number of methods of application, and for each component a suitable and appropriate method is available. The main methods that can be recommended are as follows.

Dipping. This is satisfactory for large areas. Automatic equipment for mass-production is comparatively simple to make. For smaller production runs, hand dipping can be efficient.

Spray application. For large or complex shapes, spraying provides a fast method of application and is possibly the most satisfactory and economic method. It gives accessibility to difficult areas. As with dipping, it is when the process is automated that the maximum economic advantages can be obtained.

Brushing. This is the most simple method of applying adhesive, and is satisfactory for comparatively small production runs or where the area to be bonded is small. It is useful where masking is impractical. A thin even film is difficult to obtain by brushing.

Hand operated glue guns. These allow a thin film to be applied with speed and accuracy. The adhesive is supplied under pressure by the use of a hose and by adjusting the air pressure the speed of application can be matched to the desired flow.

Adhesive can play an important part in the process of assembly, and some of the advantages obtained are:

1. Good weight to strength ratios.
2. An equal distribution of stress over the assembled areas.
3. Practical for use on flexible plastics materials.
4. The assembly of small or complex shapes is comparatively simple.
5. Mass production techniques are practical giving low production costs.

No single adhesive is satisfactory for every application, and it cannot be over-emphasised that the advice of one of the many highly competent adhesive manufacturers should be obtained at an early stage.

MECHANICAL FASTENING

The basic theory of assembly, be it by mechanical fastening or any other method, is comparatively simple, as it is purely a process of joining together by one means or another two or more components to form a completed or partially

Fig. 1. Non-collapsable threaded inserts for assembly after moulding. (Courtesy of The Precision Screw & Manufacturing Co Ltd.)

Thread Sizes	Material Tested BXL's X52 (General Purpose Phenolic Wood Filled Moulding Compound)		Material Tested ABS	
	Pull-out load	*Installation Pressure*	*Pull-out load*	*Installation Pressure*
6BA range	100 lb (46 kg)	260 lb (118 kg)	45 lb (20 kg)	110 lb (50 kg)
4BA range	120 lb (55 kg)	280 lb (127 kg)	55 lb (25 kg)	125 lb (57 kg)
3BA range	190 lb (87 kg)	390 lb (178 kg)	62 lb (28 kg)	160 lb (73 kg)
2BA range	215 lb (98 kg)	450 lb (204 kg)	95 lb (43 kg)	180 lb (82 kg)
0BA range	260 lb (118 kg)	540 lb (246 kg)	100 lb (46 kg)	200 lb (91 kg)

Table 2. Pull out loads and installation pressures for press-in inserts for two plastics. (Reproduced from details supplied by the Fastex Division of ITW Ltd).

completed product. The technology used in the process of mechanical fastening of thermoplastic components is continually improving and developing and plastics themselves are being used today to produce new types of fasteners.

To achieve the most suitable method of mechanical fastening, the following list of points should be considered:

a. The types of thermoplastics which are to be joined—rigid, flexible, hard or soft; and whether other materials such as metal and wood are involved.
b. The possibility of vibration occurring in the assembled article or, because of safety factors, is a self-lock necessary?
c. The permanence of the joint. Some articles need to be taken apart and reassembled.
d. The weight of material in the boss or side wall to preclude cracking caused by fatigue.
e. The cost of the fastener and the completed joining operation.
f. The size of production run which is visualised.
g. The capital costs anticipated.

Expanded inserts assembled after moulding

These inserts are normally made from aluminium or brass, and are fitted into the component after moulding, providing self-locking threads in a blind hole (Figs. 1 and 2). A standard type of insert has an internal thread, a slit along two-thirds of its length, and carries grip patterns upon its surface. The insert is pressed into a drilled hole and as the screw is driven home, the insert expands, the grip pattern on its outer surface penetrates the inside wall of the plastics and the bursting strain is applied to the plastics wall. The strain will vary dependent upon the number of points on the grip pattern spread along the length of the insert. When an insert has been correctly fitted into a hole of the right size, it should, when tested to destruction in the moulding, hold in position up to the moment the plastics boss pulls away from the parent moulding.

Automatic machinery is available for high speed application, giving insertions up to the rate of thirty a minute. However, for short production runs, hand insertion equipment is comparatively easy to manufacture.

The possibility of assembling inserts at the side of the press between cycles whilst the material is hot should not be overlooked. This gives cost savings and uses operator idle time.

Non-collapsable inserts assembled after moulding

When a non-collapsable insert is driven or pushed into the recommended size of hole, the press-in insert will cut or broach into the material up to the desired depth, providing a high load bearing, free running thread. Care must be taken to ensure that the screw does not bottom in the hole, as this can cause the insert to jack out. High assembly speeds are obtainable by using either a fly press or specially designed insertion machinery. Table 2 gives details of the pull out load and installation pressure necessary when press-in inserts are used on two standard materials.

Threaded bushes fitted after moulding

Sleeves which are threaded both internally and externally, and are fitted to pre-threaded holes, and although high performance is obtained, particularly in respect of tensile

Fig. 2. Expanding multivine self-locking insert. (Courtesy of The Precision Screw & Manufacturing Co Ltd.)

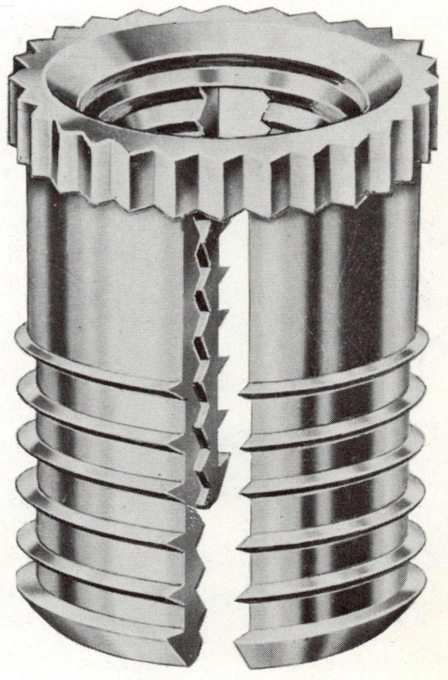

strength, installation costs tend to be high. Pitches that are coarser than 32 turns/in are recommended for use with thermoplastics, as heavy duty buttress threads reduce the possibility of stripping. As most thermoplastics are notch sensitive, a rounded thread is preferred.

Moulded-in inserts

These are inserts assembled into the female half of the mould before the moulding operation and eliminate post assembly operations. Care is necessary to ensure the insert is fitted squarely into the mould to obviate expensive mould damage. The types of insert used are normally of the external knurled type, although splined thread bushes are satisfactory. At the design stage, care must be taken to ensure that high stress concentrations are not allowed to build up around the insert. Tool damage is possible, caused normally by the incorrect fitment of the insert into the mould, and there is the possibility of flashing over the hole which will require post moulding operations. The economics of lost moulding time due to insertion must not be overlooked, although automatic insertion methods have been engineered for long production runs.

Riveting

Riveting is one of the oldest methods of joining components. The technique has been modernised so that almost any type of fastening is possible at fast rates.

The size of the hole to be used is important to the strength of the final joint. Too large a hole will cause a loose joint and too small a hole will result in a fracture. Manufacturers of rivets supply specifications giving the correct size of hole to be used.

Fig. 3. Automatic riveting machine illustrating the assembly of rivets in a relay component on a special fixture manufactured by Bifurcated Engineering Ltd for Monogram Electrical Co Ltd.

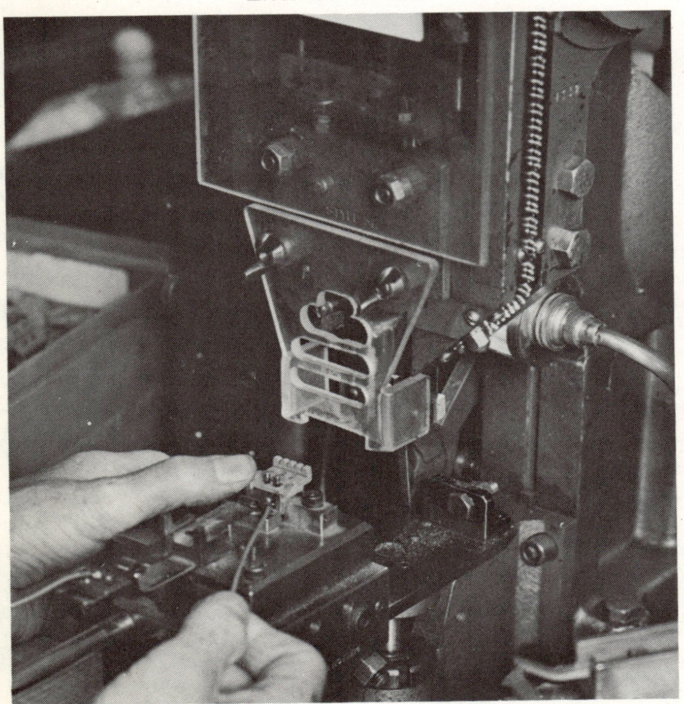

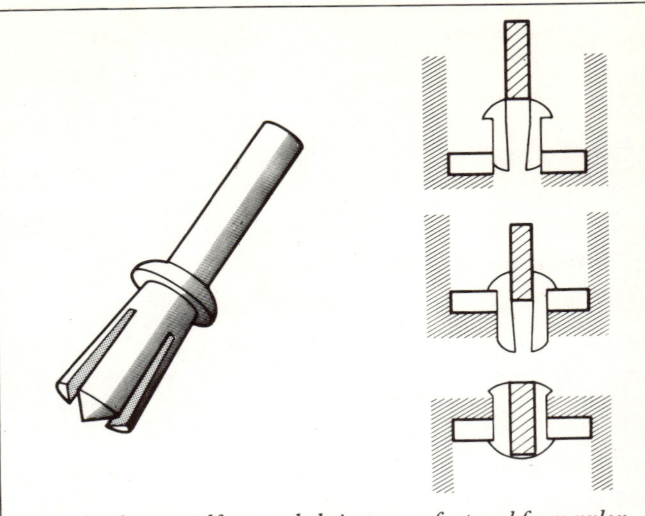

Fig. 4. Plastics self-expanded rivet manufactured from nylon. the method of assembly is also shown. (Courtesy of ITW Ltd, Fastener Division.)

The joint given by the use of rivets is a strong mechanical joint rather than leakproof. Thermoplastic materials have memories, therefore rivets tend to relax when the load is removed. There are many types of automatic and foot operated riveting machines available, which allow fast fixing rates (Fig. 3).

Solid rivets. Solid rivets are the oldest type of rivet, and have been superseded in many respects by the wide variety of metal and plastics rivets that have been developed.

Bifurcated rivets. This is a self-piercing rivet, which is used on comparatively soft materials and is shaped like an old-fashioned clothes peg. After the rivet has been driven through the material, the ends are turned over, giving a strong clinch.

Tubular rivets. This is a standard solid rivet with a hole drilled along the length of its shank to a depth of over 112% of the shank's diameter. The idea of a hole is to allow for a recess that will accept punched out material, as the rivet was originally designed as a self-piercing fastener.

Semi-tubular rivets. This comes half way between the solid and tubular rivets, and has advantages over both. In using either tubular or semi-tubular rivets, care must be taken to ensure a good clinch is obtained. This will be affected if the rivet is either too long or too short for the hole being used.

Pop rivets. This is a hollow rivet pre-assembled onto a steel pin or mandrel. The mandrel is designed to fracture at a pre-determined point during the setting operation, when the parts being fastened have been drawn close together and the joint is tight.

Plastics self-expanding rivets. One piece self-expanding blind plastics rivets are normally made of nylon (Fig. 4), although most thermoplastics including polyethylene, polypropylene, acetal and polystyrene can be used. The rivets are assembled from one side of the workpiece using a manual or air operated assembly tool. This gives a fastening method at a comparatively low cost and, because of the self-expanding principle, gives good resistance to vibration.

Drive fasteners. These are made from nylon and offer a simple method of holding together two or more panels or for fixing identification plates. They are used in the electronics industry to fasten insulation materials and for attaching circuit boards to chasses.

Plastics itself can be used to provide a reasonable riveted joint, and this should not be overlooked at the design stage. If the component is designed so that lugs protrude from one side of the mating surface whilst the other component has matched holes, it is possible to obtain a satisfactory joint by the application of heat and pressure on the lug once it has been inserted through the hole.

Spring steel fasteners

Spring steel fasteners are usually made from close annealed carbon steel strip, and after being formed are hardened and tempered. A conventional sheet metal screw or self-tapping screw is used with the basic type of spring steel fastener, which has an arched base and two arched prongs (Fig. 5).

The method by which the fastener operates is that two distinct forces are exerted on the screw as the fastener is tightened. Firstly, a compensating thread locks as the two arched prongs move inwards to engage and lock the flanks of the screw thread. Secondly, a self-energising spring lock is created by the compression of the arch as both the prongs and base of the screw are tightened.

A wide range of steel fasteners are available in which a screw or bolt is not used. The basic principle of these normally is a push-on fix where the prongs are not pitched to follow the helix of a thread, and the two sheared arms are of equal height. The push-on fix is forced over a plain stud, the fixing locks bite into the surface, finally depressing the arched base. This method of assembly is extremely cheap and easy-to-use.

Self-tapping screws

The use of self-tapping screws is limited by the number of times the component can be dismantled for servicing or maintenance. The number of occasions this can be done should be limited to no more than six, and care must be taken to ensure cross threading or stripping does not occur.

Fig. 5. Spring steel fasteners in operation showing the double lock action. (Courtesy of Firth Cleveland Fasteners Ltd.)

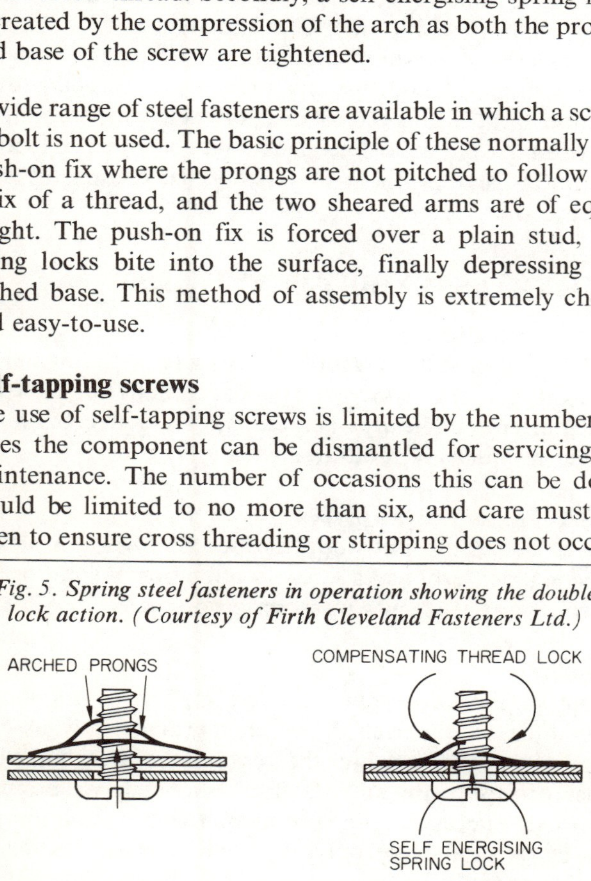

ARCHED PRONGS

COMPENSATING THREAD LOCK

SELF ENERGISING SPRING LOCK

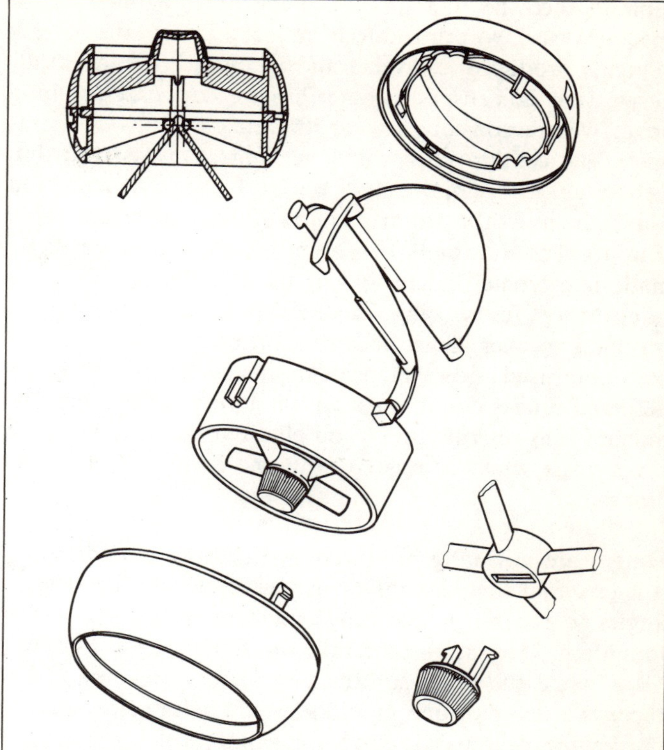

Fig. 6. Air vent eye ball assembly manufactured by Creators Ltd, illustrating the snap fit process.

For the purposes of producing threads in most thermoplastics, a standard metalwork tap and die can be used but it is recommended that the drilled hole be countersunk prior to tapping. Thread cutting screws are most suitable for use with thermoplastics, and screws with thread pitches of standard machine screws are to be recommended.

Snap fit methods

The basic method of snap fitting is a moulded undercut which is situated in one part of an article and is joined to another component which has a lip (Fig. 6). This gives a reasonably strong mechanical joint, which will only become pressure-proof or leak-proof if O-rings or washers are fitted during the assembly operation. The types of material most suitable for this application are nylon, ABS, acetal resin (Fig. 7), polycarbonate and polypropylene, which are primarily materials which have good stretch and recovery characteristics. This process may appear to be cheap, and therefore an attractive method of joining two or more components together. However, consideration must be given to the possibility of high tool costs, as the provision of the undercuts involved necessitate injection moulding tools having moving parts. When attempting to use the snap fit application on different thermoplastics, care must be taken to examine the shrinkage factors that occur after moulding. Even when using the same material this particular problem must not be overlooked. The advice of injection moulding companies, toolmakers and material suppliers should be sought at an early stage of development when considering the snap fit process.

Integral hinges

The use of polypropylene integral hinges has increased

rapidly since the first product using an integral hinge, a spectacle case, was introduced in 1958. Today, the variety of items produced is diverse and includes brief cases, adhesive tape dispensers, cabinet hinges, dart cases, fishing boxes, ventilation louvres, car air vents and powder compacts. Basically, a polypropylene hinge consists of a thin web of material joining two parts of a component, and normally it is the thinnest part of the entire assembly (Fig. 8). The durability of polypropylene hinges correctly made is extremely high, and at the ICI Plastics Division Laboratory tests using flexing apparatus have tested products to over 23 million times without failure. It must be emphasised, however, that design is critical, as badly designed hinges could crack within a very short time. The integral hinge is rust proof and eliminates assembly which saves costs when compared with the fitting of orthodox hinges.

Hinges are normally produced by injection moulding. As it is essential that the polypropylene molecules lie at right angles to the hinge, the hinge must be flexed soon after moulding. It is important that the flexing does not take place while the hinge is hot. Best results are obtained if there is a delay of one moulding cycle before flexing. The production of separate hinges is a practical operation. If comparatively large articles are being produced, such as brief case halves, a separate hinge can be used successfully to join the two halves together.

Expanded polystyrene has been successfully used commercially for integral hinges. Its design differs from the polypropylene hinge inasmuch as it takes the form of a strap with a number of parallel hinges, each of which accommodates part of the flexing.

Ultrasonic assembly of inserts

Ultrasonic vibrations are transmitted through the insert by the use of a standard ultrasonic welding horn. The heat which results from this action allows the insert to be fused into the hole. Care must be taken when deciding upon this process to ensure that wall thicknesses are sufficient to allow for the stresses that are caused when the mating parts are tightened. Hole sizes in relationship to the insert are vital to achieve a torque resistance and pull out characteristics that are possible by this method. If the hole is too small, the plastics residue will spread over the top of the surface and, if the hole is too large, the insert will not be fixed firmly into the wall side. The initial cost of ultrasonic plastics assembly equipment is higher than almost every other joining method, but this process can be competitive because of the high speeds of assembly that can be obtained, the elimination at the design stage of other assembly operations and by the versatility of the equipment once purchased.

Fig. 7. A cam cleat (left) and a bull's-eye fairlead (right) moulded by RWO (Marine Equipment) Ltd. from Kematal. ICI's acetal copolyer, for use on yachts. (Courtesy of ICI Plastics Division.)

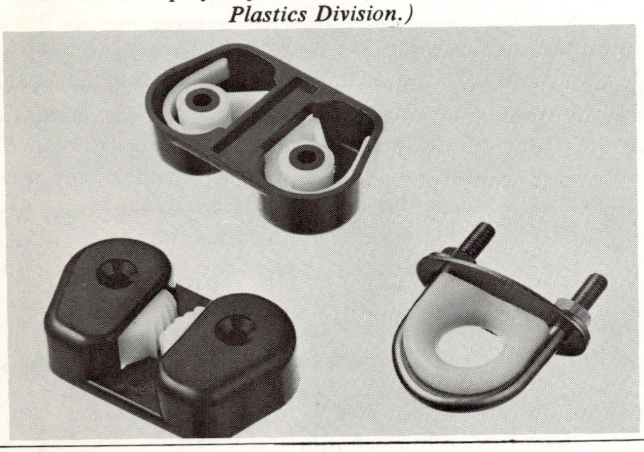

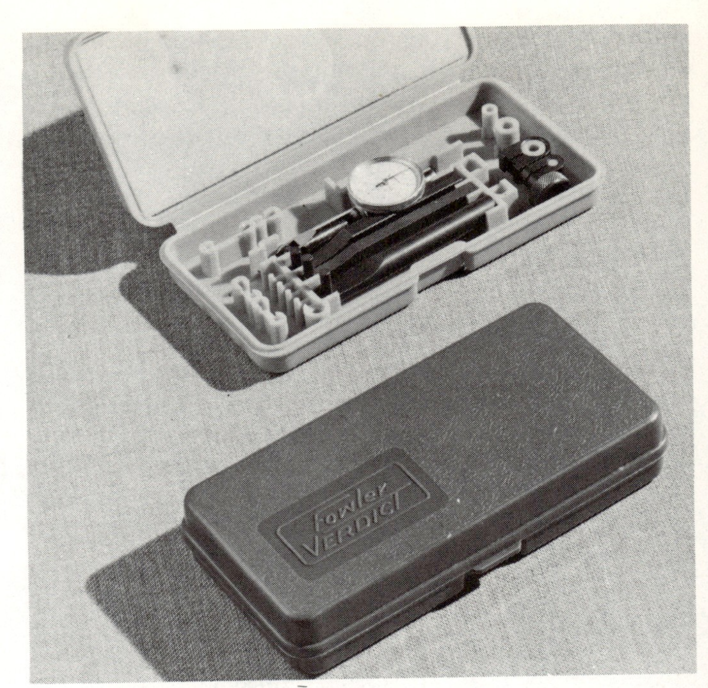

Fig. 8. Instrument box moulded for Verdict Gauge Limited by the London Association for the Blind. (Courtesy of ICI Plastics Division.)

OTHER ASSEMBLY METHODS

Mechanical stitching

The joining of soft thermoplastics to give a stitching effect is possible by the successive applications of two electrodes which are mechanically operated and are connected to the output terminals of radio frequency generators.

Metal caps for plastics bushes

This is a snap-on split stopper which is keyed to a plastics boss on the insertion of a screw. Elimination of splitting and cracking on the screw's insertion occurs.

Stapling

Metal staples have only a limited application for joining thermoplastic materials. They are normally only used on thin walls where a need for the joint is not critical and where other processes are not practical. Staples have, however, been used successfully in conjunction with heat sealing for joining two halves of polypropylene demister nozzles for the motor trade.

Chapter 18

Reinforced Plastics

D Pickthall
Fibreglass Ltd

In a reinforced system there are two principal components, a fibre and a matrix. The function of the fibre is to carry the load applied to the system. The matrix is there primarily to transfer the load from the point of application into the load-bearing fibres which have higher strengths and modulus than the matrix. Of particular commercial interest is glass fibre which is a brittle, as opposed to a ductile, fibre. Besides glass there are other reinforcements, asbestos has been used to some extent and more recently carbon fibre has become very well known.

At the present time the usage of reinforced plastics is still relatively small, with a consumption in the United Kingdom of about 50000 tonnes per annum. However, the growth rate over the last decade has averaged 17% per annum. This growth is greater than the expansion of plastics materials as a whole and the long term prospects for plastics composites incorporating fibrous reinforcements is expected to be considerable.

TYPES AND FORMS OF REINFORCEMENT

The ultimate strength of a composite of fibre and polymer is basically dependent upon the stress-strain relationship of its components. Any fibre whose strength and modulus of elasticity is greater than a polymer is capable of theoretically providing reinforcement to that polymer. Practically, however, the difference in properties has to be considerable to obtain efficient reinforcement and the effectiveness of the reinforcement is dependent upon the degree of adhesion which can be achieved. Other factors which also influence the properties of a composite are relative volumes of reinforcement and matrix, their physical and chemical properties, the temperature resistance and, of very considerable importance, the fibre length which is being used.

Because of the importance of fibre length, glass fibre has dominated the reinforcement scene because of the relative ease with which glass can be drawn into fibrous form.

PROPERTY Composition %	'E' GLASS
SiO_2	52.4
Al_2O_3 Fe_2O_2	14.4
CaO MgO	21.8
B_2O_3	10.6
Na_2O K_2O	0.8
Specific gravity of fibres	2.56
Tensile strength of freshly drawn, undamaged fibre lbf/in² kgf/cm²	530 000 37 200
Young's Modulus of Elasticity lbf/in² kgf/cm²	11×10^6 7.73×10^5
Coefficient of thermal expansion deg F deg C	2.7×10^{-6} 4.9×10^{-6}
Coefficient of thermal conductivity Btu in/ft²h deg F W/m deg C	7.2 1·04
Refractive index $n^{20}D$	1.545–1.549

All figures are approximate and average

Table 1. Properties of 'E' glass fibre.

Although there are many possible glass compositions only one is widely used—this is 'E' glass. Its properties are given in Table 1.

Carbon fibre has only been on the scene for a few years and a great deal of work has now been done on the manufacture and utilisation of this unique fibre. In the UK there are three grades commercially available and typical properties for these are shown in Table 2. Carbon fibre can be made both in continuous and staple fibre forms but is considerably more expensive to make continuous.

Reinforcements are used principally in planar forms. The orientation of the fibres within that plane have a considerable effect on the properties of the resulting composite and products are available to satisfy all possible constructions. For the highest strengths, uni-directional reinforcements are frequently used but although these give extremely high strengths in the direction of the fibres the properties perpendicular to the fibres are little better than those of the resin. By weaving yarns and rovings two-directional fabrics can readily be produced and a very wide range of fabrics have been developed. Of greatest importance, however, is random reinforcement where fibres usually in short lengths (50 mm) are distributed randomly and bonded to form a non-woven mat. This form of material, although developing lower strengths, has the major advantage in giving uniform properties in the plane of the composite.

TYPES OF MATRIX

In Chapters 3 and 4 the basic plastics materials were described. In reinforced plastics both thermosetting and thermoplastic polymers are used although their method of use is very different. Very many thermoset resins can and have been used with reinforcements. Unsaturated polyester resins are normally looked upon as the main material used in conjunction with reinforcements, and the majority of this chapter will be devoted to the use of these materials. Many other thermosetting resins, epoxides, phenolics, silicones, and some of the exotic polymers such as the polyimides have been used extremely successfully.

With most thermosetting resin systems additives such as catalysts and hardeners have to be added to the resins to cause polymerisation to occur. Many other additives are also used with thermosetting systems and these include inorganic fillers, internal release agents, pigments, *etc.* These were dealt with in detail in Chapter 2.

Although the amount of thermoplastic polymer used with reinforcement is still small there is an increasing interest in the use of reinforcements with thermoplastic as the matrix to provide improved properties, particularly strength, stiffness, dimensional stability, shrinkage and heat resistance. Virtually all thermoplastics have been reinforced and a great many are now available commercially. The best known is nylon but it is expected that great advances will be made in the use of styrenes, olefins and some of the more specialised polymers, such as polycarbonate and polysulphone.

PROPERTIES

A great deal has been published on the properties of reinforced plastics. Tables 3 and 4 summarise the range of mechanical properties that can be obtained with glass and carbon reinforced plastics. The long term properties and the performance of composites under specialist conditions such as corrosion, electrical stress and fatigue have been well documented.

PROCESSING

Over the last fifteen years there has been a great deal of research work on the mechanical, physical and chemical performance of composites. Although considerable effort

Table 2. Typical mechanical properties of reinforcing carbon fibre. (Courtesy of Fothergill & Harvey Ltd.)

Fibre	Specific gravity	Ultimate tensile strength		Youngs modulus		Specific Youngs modulus	
		GN/m²	lbf/in²	GN/m²	lbf/in²	GN/m²	lbf/in²
Type 1 High Modulus	1.95	1.89	275×10^3	379	55×10^6	194	28.2×10^6
Type 2 High Strength	1.75	2.58	375×10^3	262	38×10^6	150	21.7×10^6
Type A	1.85	1.89	275×10^3	207	30×10^6	112	16.2×10^6

GLASS FIBRE REINFORCED	Glass Content % by Weight	Glass Content % by Volume	Density kg/dm³	Thermal Coefficient of Expansion degC⁻¹ × 10⁻⁶	Tensile Strength in MN/m²	Tensile Modulus in GN/m²	Compressive Strength inMN/m²	Flexural Strength inMN/m²	Elongation %	Rockwell Hardness	Dielectric Strength MV/m	Heat Distortion °C	Continuous Heat Resistance up to . . . °C
Uni-directional Roving													
wound epoxide	60–90	40–80	1.7–2.2	4–11	550–1700	28–62	310–480	690–1860	1.6–2.8	M98–120	12–15	175–200	260
extrusion polyester	50–75	32–59	1.6–2.0	5–14	410–1200	21–41	206–480	690–1240	1.6–2.5	H80–112	8–15	165–190	260
Bi-directional Fabric													
satin weave polyester	50–70	32–52	1.6–2.9	9–11	240–400	14–24	206–275	340–520	0.5–2.0	M80–120	14–20	N.Av.	175
woven roving polyester	45–60	28–41	1.5–1.8	11-16	220–340	8.6–16	96–137	200–296	0.5–2.0	M80–120	14–20	N.Av.	175
Random Distribution													
preform polyester	25–50	14–32	1.4–1.6	18–33	68–165	5.5–12.4	124–206	137–310	1.0–1.5	H40–105	8–15	175–200	200
hand lay-up polyester, spray-up polyester	25–40	14–24	1.4–1.5	22–36	62–137	5.5–11	124–170	137–275	1.0–1.2	H40–105	8–15	175–200	175
continuous strand mat polyester	20–50	5–32	1.3–1.6	22–36	68–210	4–12.4	69–206	124–303	1.0–1.5	H40–105	8–15	175–200	175
Compounds													
DMC polyester	6–26	10–24	1.8–2.0	24–34	34–68	11–14	137–179	41–179	0.3–0.5	H80–112	8–15	200	230
prepreg (polyester)	20–35	13–25	1.8–1.85	18–33	48–96	5–8.6	206	117–172	N.Av.	N.Av.	10–12	N.Av.	110–150
glass nylon	20–40	10–23	1.3–1.5	13–33	117–190	5.5–14	103–165	145–275	1.5–7.5	E64–75	15–20	220–275	200
THERMOPLASTICS													
Nylon			1.14	100–114	79	1.4–2.8	34–90	55–103	25–30	R108–118	12–18	95–180	150
Polyethylene (high density)			0.96	110–130	30	0.55–1.0	16	13.7–21	50–600	R15	15–40	50–65	120
Polypropylene			0.90	100–200	39	1.0–1.7	58–69	28–34	10–700	R85–110	29–30	100–110	160
Polystyrene (high impact)			1.08	40–100	45	3–4	110	21–34	1–25	M65–80	20–28	65–90	80

Table 3. Comparison of properties of glass reinforced plastics with some thermoplastics. (Courtesy of Fibreglass Ltd.)

Carbon		Type 1 High Modulus				Type 2 High Strength			
Resin Systems		Code 61 LY558 HT973	Code 63 ERLA 4617/DDM	Code 67 ERLA 4617/MPD	Code 73 P13N	Code 61 LY558 HT973	Code 63 ERLA 4617/DDM	Code 67 ERLA 4617/MPD	Code 73 P13N
Flexural Str.	GN/m²	0.827	0.896	0.841	0.83	1.38	1.58	1.52	1.45
Flexural Mod.	GN/m²	200.0	206.8	206.8	206	124.1	124.1	124.1	110
Tensile Str.	GN/m²	0.724	0.703	0.758	0.72	1.38	1.45	1.31	1.37
Tensile Mod.	GN/m²	193.0	206.8	200.0	193	117.2	131.0	124.1	117
Compressive Strength	GN/m²	0.655	0.655	0.648	—	1.16	1.31	0.965	0.75
Compressive Modulus	GN/m²	206.8	206.8	206.8	—	137.9	137.9	124.1	124
Interlaminar shear strength s/d 5:1	GN/m²	0.045	0.055	0.048	0.045	0.083	0.103	0.090	0.08
Poissons Ratio		0.32	—	—	—	0.34	—	—	—
Density	g/m³	1.701	1.708	1.719	1.740	1.558	1.565	1.576	1.597

Table 4. Typical mechanical properties of carboform composites (65 % fibre by volume). (Courtesy of Fothergill & Harvey Ltd.)

has been devoted to improving the mechanisms of combining fibre and matrix together this has not been as comprehensive as for end properties. There is today a much greater awareness of the need to production engineer the processes of combining the raw materials and the importance of total design and value analysis is now better appreciated. Traditionally, in the United Kingdom, contact moulding is the process which is synonymous with reinforced plastics. This is not the case in some other countries. For example, in the United States mass production techniques of press moulding are the most widely used.

Because of the very wide range of techniques available for converting fibre and matrix into structural parts, it is necessary to differentiate between the basic groups of processes which are involved. To this end the processes must be considered under two main groups:

a. Batch or low output systems.
b. Mass production and continuous systems.

Reinforced plastics can cope perfectly satisfactorily with component needs ranging from one-off to millions. However, this cannot be done by one single process and unfortunately the reinforced plastics industry has acquired the image of low output batch production whereas the truth is that the materials are capable of almost any output requirement. The characteristics of the major processes are shown in Table 5.

Batch or low output systems

The most simple process of producing reinforced plastics parts is the hand lay-up contact moulding process. This is the traditional technique sometimes affectionately known as 'bucket and brush'. For hand lay-up moulding (Fig. 1) only low capital investment is necessary and this on many occasions results in production engineers looking down on it as a technique. The only major capital investment is in the tool or mould and as with most moulding the quality of the product is nearly always dependent on the quality of the tool. Only from goods tools can high quality com-

	Contact Moulding		Matched-Die Moulding		
	Hand lay-up Spray-up	Filament winding	Dough moulding	Mat, Preform Pre-preg.	Cold press
Maximum part size determined by	Mould size, Transport of part	Machine dimensions	Press rating and size	Press rating and size	Press rating and size
Shape and styling limitations	None	Usually surface of revolution	Mouldability	Mouldability	Mouldability
Translucency	Yes	Yes	No	Yes*	Yes
Volume of production category	Low to medium	Medium to high	High	High	Medium
Number of finished surfaces and quality	One excellent	One excellent	All excellent	Two very good	Two good
Typical glass content by weight	20-35%	65-90%	10-35%	25-45%	20-40%
Strength category	Medium	Very high	Low-medium	Medium-high	Medium
Strength orientation	Random except fabric types	Follows winding pattern	Random throughout moulding	Random	Random
Resin rich, corrosion resistant surfaces	Yes, by Gel-coat &/or surface mat	Yes, by Gel-coat &/or surface mat	No	Yes, by Gel-coat &/or overlay mat**	Yes, by Gel-coat &/or surface/overlay mat
Practical thickness range (inches)	0.030 to 1.000	0.010 to 2.000	0.060 to 1.000	0.030 to 0.250	0.060 to 0.500
Common moulding tolerance on thickness (inches)***	±0.020	±0.010	±0.002	±0.008***	±0.020
Local thickness increase	As desired	As desired	As desired	Usually 2 ⩽:1	Usually 2 ⩽:1
Metal inserts and/or edge stiffeners	Possible	Possible	Possible	Possible	Possible
Built-in cores	Possible	Possible	Possible	Possible	Possible
Minimum radii for ease of moulding (inches)	0.500	0.125	0.030	0.125	0.250
Undercuts	Yes	No	Yes	No	No
Minimum recommended draw angle (degrees)	2	3	1	1	2
Holes moulded in to avoid material waste	Yes large	Yes (to suit wind pattern)	Yes	Yes	No
Trim in mould	Yes (rough trim)	Yes	Yes (except fine flash)	Yes (except fine flash)	No
Moulded in signs and labels	Easy	Easy	Difficult	Difficult	Possible
Combination with thermoplastic liners for corrosion resistance	Yes	Yes	No	No	No

*Except Pre-preg. **Except Pre-preg. ***Dependent upon laminate thickness.

Table 5. Characteristics of major thermoset moulding processes. (Courtesy of Fibreglass Ltd.)

ponents be produced with low re-work levels. Using reinforced plastics tooling very large and very complex shapes can be produced and only by this technique can these components be moulded. Good workshop control with very good supervision of materials and labour is essential for efficient production. Care in preparation of specifications and in the handling of all raw materials and in the training of operators is essential. Workshop conditions must be controlled both for consistent quality and safety. Polyester resins are combustible, catalyst and accelerator additives are potential fire risks and adequate precautions against fire and for fume control are necessary.

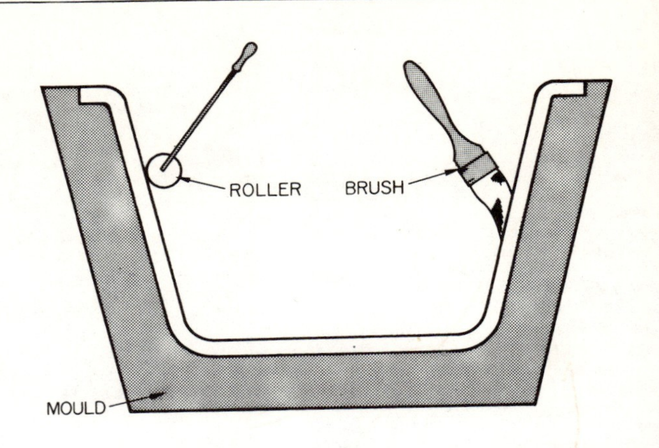

Fig. 1. Hand lay-up method of producing reinforced plastics products.

In contact moulding only one smooth finished surface is produced, this is the mould face. Gel coats (resin rich layers) are used against the mould surface to achieve a high surface finish and the layers of reinforcement are built up successively behind the gel coat until sufficient strength or rigidity has been obtained. In contact moulding the control of the amount of reinforcement and the proportion of reinforcement to resin has to be controlled not only for weight and structural purposes, but also to ensure that cost targets are achieved as material costs are a high proportion of finished part cost.

Various techniques have been developed for speeding up the contact moulding process. Of most importance is the spray-up technique (Fig. 2). The cutting of random mats

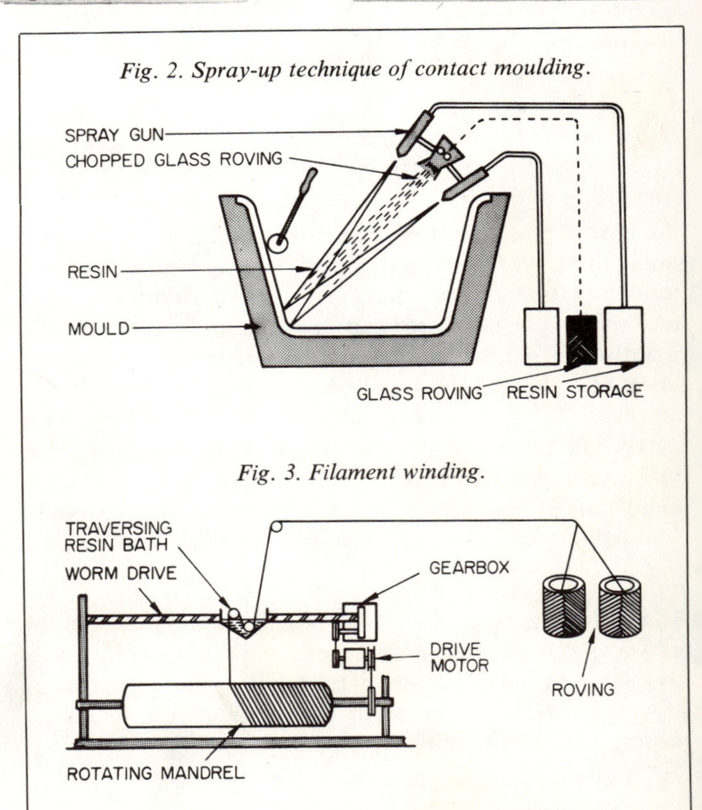

Fig. 2. Spray-up technique of contact moulding.

Fig. 3. Filament winding.

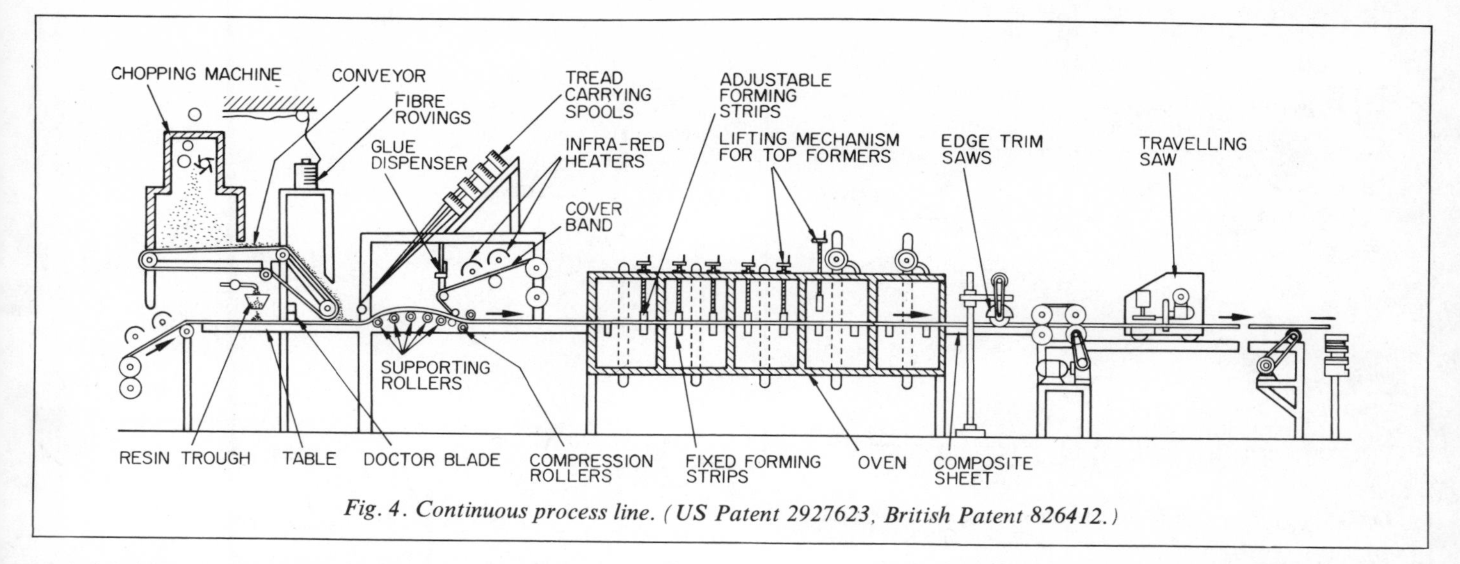

Fig. 4. Continuous process line. (US Patent 2927623, British Patent 826412.)

and fabrics and their placing onto the mould and their impregnation with resin takes a considerable time. By chopping roving and spraying this together with the activated resin onto the mould a considerable amount of the time in getting the materials on to the mould surface can be eliminated. If correctly adjusted and operated the wetting-out, rolling-out and de-airing of the laminate can be achieved rapidly. However, the major disadvantage of this technique is that the operator of the spray gun has the total responsibility for controlling the amount of material which he is depositing and as this can be as high as 5 kg per minute, the operator has to be extremely experienced and well trained in the operation of the gun to achieve uniform, predictable and planned results. Spray-up has been mechanised on some occasions to produce simple panels where the spray gun has been mechanically reciprocated over moving moulds but this has only limited application.

Other methods have been developed for speeding up the contact moulding process and these include heat assisted cure, either by the use of low temperature ovens or even by incorporating heating elements into the mould surface. Special systems have been developed to reduce or eliminate the need for regular use of release agents on moulds.

Speeding up the mould turn-round can become extremely important as moulds can take considerable space in the moulding shop and the curing stage is dead time as far as production is concerned. However, care has to be exercised in trying to speed up moulding processes as mould turn-round is often controlled by the working day and other practical considerations rather than by the moulding cycle.

There are several other batch processes. Vacuum bag and resin injection techniques have been used for specialist application but these are not used extensively in commercial practice. Centrifugal casting has been used for the manufacture of pipe sections but this is again a specialised area. Filament winding (Fig. 3) is essentially a batch process but by good engineering and well planned layout, pipes and tanks made by the filament winding process can be produced at very considerable rates. There are also some continuous processes for producing pipe but these have not yet achieved significant use. Filament winding is an incorrect term for the process that is generally used, the term comes from the aerospace industry where very precise continuous glass strands have been laid down in specific positions for pressure bottles and very high performance applications. Commercially, rovings, roving fabrics, glass fabrics and random mats are used as reinforcement and are wound on to mandrels using less sophisticated machinery.

Mass production

The basic change from batch low output systems to mass production techniques is when the process changes from a low capital investment/high labour cost system to one employing high capital cost and relatively low labour cost. Mass production techniques can be divided into three major groups, two for thermosetting systems and one for thermoplastics. These are continuous processes and matched die moulding for thermosetting and injection moulding for thermoplastic systems.

Fig. 5. Preform manufacture.

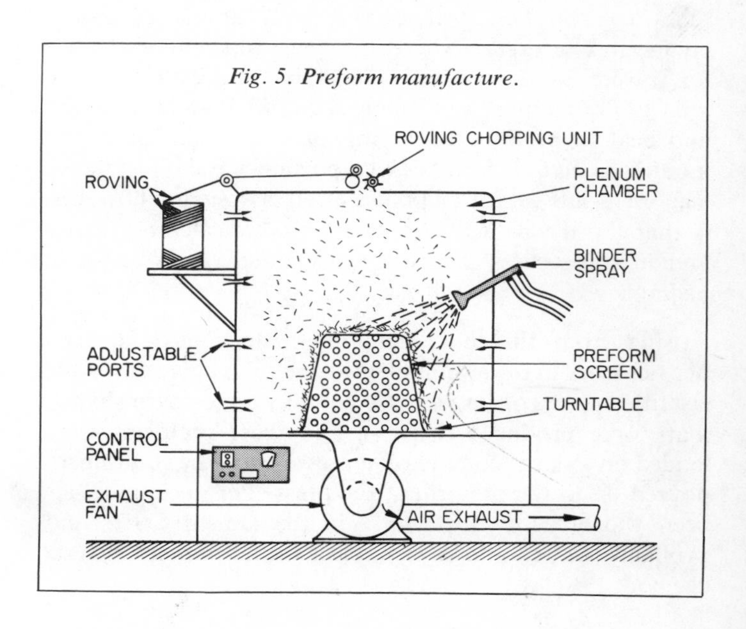

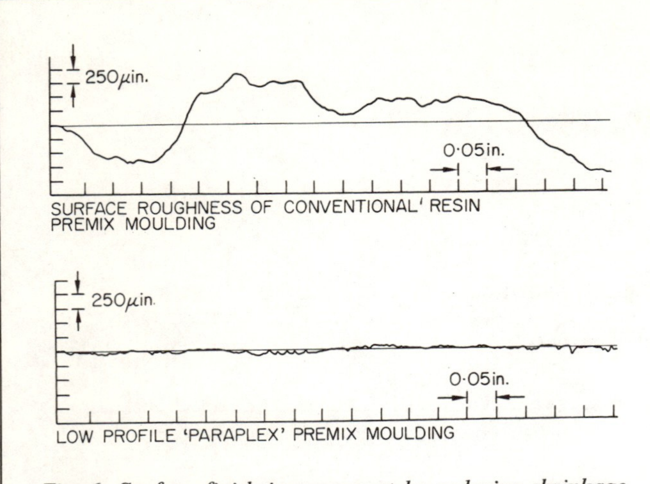

Fig. 6. Surface finish improvement by reducing shrinkage (Courtesy of Rohm and Haas Co.)

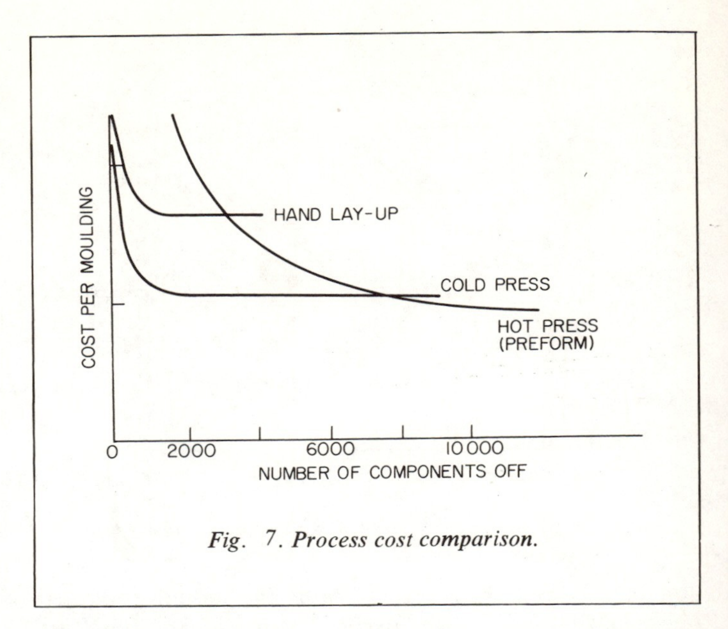

Fig. 7. Process cost comparison.

Continuous processes have been fairly widely used particularly for the manufacture of flat and corrugated sheeting suitable for cladding and roofing applications, and also for the manufacture of rod-blanks, for insulators and fishing rods and special sections for a wide range of uses. All continuous processes are essentially heat cure assisted systems as otherwise the plant would have to be excessively long and it would be difficult to maintain stability of the resin systems before actual use. A typical continuous sheeting line layout is shown in Fig. 4.

Matched die moulding is potentially the greatest opportunity for wider use of reinforced plastics components. Compression moulding was discussed in Chapter 6. Reinforced plastics match die moulding is basically very similar to phenolic or other thermosetting plastics compression moulding except that with the use of polyester resins no by-products are created during cure and therefore lower pressures can be used. This results in larger presses, larger parts and therefore greater opportunities for integrated designs. The earliest compression moulding with reinforced plastics systems was with dough moulding compounds or premix where short glass fibre strands (6 mm long) were mixed with resin and filler slurries to produce a tacky dough mix. This material can flow under pressure and heat to produce very complex shapes similar to the mouldings that can be made with phenolics and UF systems. The great advantage of polyester/glass systems, however, is that the mechanical properties are considerably better and in some cases the electrical characteristics can also be improved.

Arising from this hot moulding system of reinforced plastics a series of developments have taken place over the last fifteen years on larger parts. Initially preforms or shaped mats were produced (Fig. 5), and these preforms were loaded on to large compression presses with the resin slurry poured on to the preform at the press. This technique has been widely used particularly in the United States and probably the best known example is the Chevrolet Corvette car body.

In the early sixties a new material was developed in Germany which is now known as pre-impregnated mat or sheet-moulding compound. Here a random mat of long fibres is pre-impregnated with a polyester/filler slurry system which after impregnation can be matured to give a sheet material. This material can be cut to shape and then loaded into a press which, through heat and pressure, would cause the material to flow to fill the cavity. Mechanical properties equivalent to preform moulding can be achieved by this system with the added advantages of greater versatility of shape and improved processing as only one material is used on the press and the pressing operation can become a much cleaner and more efficient system.

Parallel to these material improvements, the processing equipment has been improved and much stiffer, more stable presses are now available and these have been made to operate very much faster. Automatic loading and un-

Table 6. Production rates for one compartment bus seat under varying resin cure and mechanical assist conditions. (Courtesy of Dow Chemical Co.)

Resin Type	Load and Unload Conditions	Press Open Time, Sec	Press Closed Time, Sec	Parts/hr	Rate Factor	Approx. (c) Parts/yr
Normal Cure	Manual	90	90	20	1·0	48,000
Normal Cure	Manual	60	90	24	1·2	58,000
Fast Cure (a)	Manual	90	30	30	1·5	72,000
Fast Cure (a)	Manual	60	30	40	2·0	96,000
Fast Cure (a)	Automatic	25	30	65	3·3	160,000
Fast Cure (a, b)	Automatic (b)	20	10	120	6·0	290,000

(a) Isophthalic resins based on chlorostyrene/styrene mixtures.
(b) These rates of 20 and 10 seconds have been obtained on smaller (1-2 ft²) parts than the bus seat, and are shown here to indicate the speeds to be expected on larger parts as techniques are refined.
(c) Based on 14 hr/day, 200 days/year.

Polymer	Tensile Yield Strength kgf/cm²	Flexural Modulus kgf/cm²	Izod Impact Notched kgf/cm/cm	Unnotched kgf/cm/cm	Elongation %	Specific Gravity	Water Absorption 24 hr. %	Mould Shrinkage cm/cm	Deflection Temperature 18.6 kgf/cm² @ °C	Co-eff of Linear Expansion cm/km/°C
ABS	1340 (440)	98,000 (21,000)	6.5 (21.8)	32.6	3 (up to 60)	1.28 (1.03)	0.11 (0.2)	0.001 (0.003)	112 (93)	2.9 (9.5)
SAN	1220 (770)	113,000 (25,000)	5.4 (2.2)	19.0	3 (up to 4)	1.3 (1.08)	0.10 (0.25)	0.001 (0.004)	102 (93)	3.2 (3.7)
Polystyrene	880 (560)	84,000 (35,000)	5.4 (1.6)	13.6	1 (up to 3)	1.28 (1.06)	0.05 (0.07)	0.001 (0.004)	102 (102)	3.4 (7)
Polycarbonate	1300 (630)	84,000 (24,000)	20.1 (70.4)	95.2	up to 5 (up to 130)	1.43 (1.2)	0.07 (0.15)	0.002 (0.006)	149 (135)	2.3 (6.6)
Acetal	900 (650)	91,000 (28,000)	4.4 (7.6)	19.0	2 (up to 40)	1.63 (1.41)	0.60 (0.25)	0.005 (0.02)	163 (124)	4.3 (8.1)
Polypropylene	550 (320)	56,000 (15,000)	7.1 (10.9)	24.5	3.2 (up to 700)	1.13 (0.90)	0.03 (0.02)	0.004 (0.017)	146 (57)	3.6 (8.0)
Polysulphone	1270 (720)	84,000 (25,000)	9.8 (7)	75.8	1.5 (up to 100)	1.45 (1.24)	0.20 (0.22)	0.002 (0.007)	185 (173)	2.5 (5.4)
PPO	1480 (770)	77,000 (27,000)	10.9 (9.2)	65.3	5.0 (up to 80)	1.27 (1.06)	0.06 (0.06)	0.001 (0.007)	182 (188)	2.7 (2.9)
Nylon 6	1620 (700)	84,000 (21,000)	12.5 (18.0)	108.8	up to 10 (up to 320	1.37 (1.13)	1.2 (1.6)	0.004 (0.010)	216 (77)	4.0 (8.3)
Nylon 6/10	1480 (600)	77,000 (20,000)	31.1 (6.5)	119.7	up to 10 (up to 300)	1.30 (1.09)	0.2 (0.4)	0.004 (0.010)	216 (77)	4.5 (9.0)
Nylon 66	1830 (740)	91,000 (29,000)	10.9 (8.2)	92.5	up to 10 (up to 300)	1.37 (1.14)	0.8 (1.5)	0.005 (0.015)	255 (85)	4.1 (8)
PVC	1130 (490)	88,000 (28,000)	8.7 (109)	51.7	2 (up to 40)	1.53 (1.40)	0.02 (0.15)	0.001 (0.003)	71 (65)	2.7 (11)

Properties of unreinforced polymers in brackets.

Table 7. Properties of 30% glass fibre reinforced thermoplastics. (Courtesy LNP Corporation)

loading devices have been developed and although these are still in their early stages they should, within the next few years, become standard practice. The development of better presses and a new form of material have also required two other major developments, these are faster curing and low shrink or low profile resin systems. When a resin system cures from the liquid to the solid phase, shrinkage inevitably occurs. This can be partially controlled and minimised by the use of filler systems, but this is not sufficient to prevent sink marks and distortion occurring in the mouldings. Systems have now been developed, particularly by Rohm and Haas in the United States, to chemically reduce shrinkage and this has had a major effect on improving surface finish so that less preparation is needed before post moulding finishing. Fig. 6 illustrates the improvement. The combination of high speed presses, high cure rates, sheet moulding compounds and low profile resin systems has opened up tremendous possibilities for the design of structural plastics components to replace large diecastings and sheet metal pressings. To illustrate the production capacity of today's presses Table 6 has been taken from a recent technical paper.

Before leaving match die moulding, mention must be made of an intermediate process between batch and mass production. This is cold press moulding where reinforced plastics matching tools are used, mounted in low tonnage presses. The polyester resin system utilises a high exotherm so that heat is self-generated within the moulding to speed cure. The tools themselves heat up and quite rapid mould turn-round can be achieved on large mouldings. Finish is not as good as with hot press moulding and tool life is limited.

It is impossible to generalise on cost of moulding reinforced plastics components or the relative effect of the different processes as there are too many factors, but a general indication of the effect of numbers of mouldings on cost is shown in Fig. 7.

Injection moulding has been briefly described in Chapter 8. The use of filled materials does not significantly effect the moulding process although higher barrel temperatures and greater injection pressures may be required. Table 7 compares the properties of some reinforced and unreinforced thermoplastics and illustrates how the presence of reinforcing fibre improves most of the mechanical and physical characteristics of polymers.

BIBLIOGRAPHY

Marine Design Manual. Gibbs & Cox.
Glass Reinforced Plastics. Edited B. Parkyn.
Fiberglass Reinforced Plastics. R. H. Sonnebonn.
Handbook of Reinforced Plastics of the S.P.I. S. Oleesky & G. Mohr.
Polyester Handbook. *Scott Bader & Co. Ltd.*
F.R.P. Design Data. *Fibreglass Ltd.*

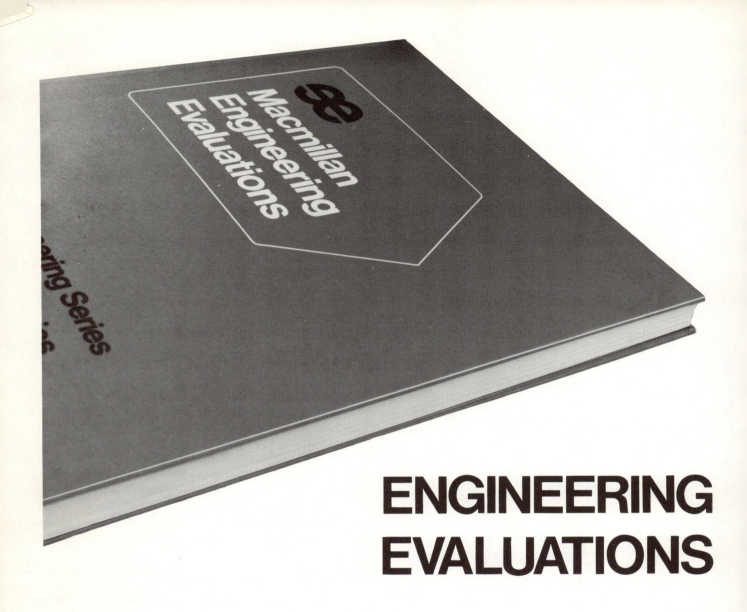

Chapter 19

Expanded Plastics

H Gibson *ARIC*
Shell International Chemical Co Ltd

Expanded (or cellular) plastics are continually finding new uses in a wide variety of industries. Furniture and bedding, domestic appliances, automotive and building industries, for example, are all using large quantities of cellular plastics and as the various technologies advance, further applications will be found for these plastics, both in new outlets and as replacements for existing materials which for technical or economic reasons will be superseded. Polyurethane and polystyrene foams are currently the most widely used, but phenolic, urea-formaldehyde, silicone, *etc*, foams all have a place in plastics foam technology.

FLEXIBLE POLYURETHANE FOAM

The two major types of flexible polyurethane foams are those based on polyesters and polyethers. These terms refer to the chemical nature of the polyol (or resin) component. Although polyether-polyols are now much more important commercially than polyesters in the manufacture of urethane foams, the differences in physical properties of the finished foams allows the continued use of polyesters in some applications.

Chemistry

The chemistry of polyurethane foam formation is complex and has been reported elsewhere[1,2]. A more simple treatment is given below. Polyurethane foam formation is dependent upon the chemical reactions of the isocyanate group — N C O.

A urethane is produced as a result of the reaction between an isocyanate and an alcohol.

$$\begin{matrix} R{-}N{=}C{=}O \\ + \\ H{-}O{-}R^1 \end{matrix} \longrightarrow \begin{matrix} R{-}N{-}C{=}O \\ \;\;\;| \quad\;\; | \\ \;\;\;H \quad O{-}R^1 \end{matrix}$$

a urethane

This reaction can be extended if the simple isocyanate is

replaced by a diisocyanate and the alcohol by a polyol.

$$\underset{\text{diisocyanate}}{OCN{-}R^1{-}NCO} + \underset{\text{polyol}}{HO{-}R^2{-}OH} + OCN{-}R^1{-}NCO$$

$$\rightarrow OCN{-}R^1{-}NH\overset{\overset{\displaystyle O}{\|}}{C}O{-}R^2{-}O\overset{\overset{\displaystyle O}{\|}}{C}NH{-}R^1{-}NCO$$

Or more generally, for a polyurethane.

$$(m + 1)R^1(NCO)_2 + mHO{-}R^2{-}OH \rightarrow$$

$$OCN{-}\left[R^1{-}NH\underset{\underset{\displaystyle O}{\|}}{C}O{-}R^2{-}O\underset{\underset{\displaystyle O}{\|}}{C}NH\right]_m{-}R^1{-}NCO$$

Isocyanates react with materials having active hydrogen atoms and in the production of a polyurethane foam a whole series of reactions of this type takes place. For simplicity, only one functional isocyanate group, and the functional parts of the reaction products are illustrated.

$$—NCO + H_2O \rightarrow \underset{\text{unstable carbamic acid}}{\left[—NHC\begin{matrix}{/\!\!/}O\\ \backslash OH\end{matrix}\right]} \rightarrow —NH_2 + CO_2 \qquad \text{I}$$

$$—NCO + \underset{\text{amine}}{NH_2—} \rightarrow \underset{\text{a urea}}{—NH\underset{\underset{\displaystyle O}{\|}}{C}NH—} \qquad \text{II}$$

$$—NCO + \underset{\text{a urea}}{—NH\underset{\underset{\displaystyle O}{\|}}{C}NH—} \rightarrow \underset{\text{a biuret}}{—NH\underset{\underset{\displaystyle O}{\|}}{C}—\overset{|}{N}—\underset{\underset{\displaystyle O}{\|}}{C}NH—} \qquad \text{III}$$

$$—NCO + \underset{\text{polyol}}{HO—} \rightarrow \underset{\text{a urethane}}{—NH\underset{\underset{\displaystyle O}{\|}}{C}O—} \qquad \text{IV}$$

$$—NCO + \underset{\text{a urethane}}{—NH\underset{\underset{\displaystyle O}{\|}}{C}O—} \rightarrow \underset{\text{an allophanate}}{—NH\underset{\underset{\displaystyle O}{\|}}{C}—\overset{|}{N}—\underset{\underset{\displaystyle O}{\|}}{C}O—} \qquad \text{V}$$

Reactions II–V inclusive result in an increase in viscosity of the reaction mixture, formation of cross linking and, finally, setting of the mixture. Reaction I, besides being a prerequisite to reactions II and III, causes the mixture to foam due to evolution of carbon dioxide. Reactions III and V increase the number of cross links and therefore have a direct bearing on the ultimate properties of the foam. Reaction IV, leading to chain extension, is catalysed by both acids and bases while reaction V causing cross-linking, is catalysed by bases, but is unaffected by acids. Because of this, the presence of bases which could produce premature gelation should be avoided in prepolymer manufacture.

Manufacturing processes

The original 'pre-polymer' process necessitated two stages involving firstly the manufacture of the prepolymer from the polyol and the polyisocyanate and secondly the reaction of the prepolymer with water, catalysts, and stabilisers to produce the foam. More usually, the one-shot process is now used, which as the name implies is a single stage process enabling high throughputs to be achieved.

Raw materials

The raw materials required in the 'one-shot' process are as follows:

Polyol. A wide range of polyols is available for the production of polyurethane foams. Polyether foams normally employ triols of molecular weight 3000–5000.

Di-isocyanate. The di-isocyanate normally used for producing flexible polyurethane foam is an 80 : 20 (parts by weight) mixture of 2 : 4 and 2 : 6 tolylene di-isocyanates (TDI). For some uses a 65 : 35 mixture of TDI is used.

[Structures: 2:4 isomer — benzene ring with CH_3, NCO (position 2) and NCO (position 4); 2:6 isomer — benzene ring with CH_3, OCN (position 2) and NCO (position 6)]

2:4 isomers of TDI 2:6

TDI is a hazardous chemical and manufacturers handling instructions must be followed.

Catalysts. These are usually of two types: (1) Amines, *eg* triethylenene diamine ('Dabco', Houdry Process & Chemical Co.), dimethylaminoethanol, *etc.* (2) Organotin catalysts—normally stannous octoate.

Water. For reaction to produce carbon dioxide gas and thus control the density, simultaneously producing a more cross-linked structure.

Surfactant. Usually silicone oil (L520, L540, Union Carbide Chemical Co) which controls cell structure.

Fig. 1. General view of a slabstock plant (Courtesy of Viking Engineering Co Ltd.)

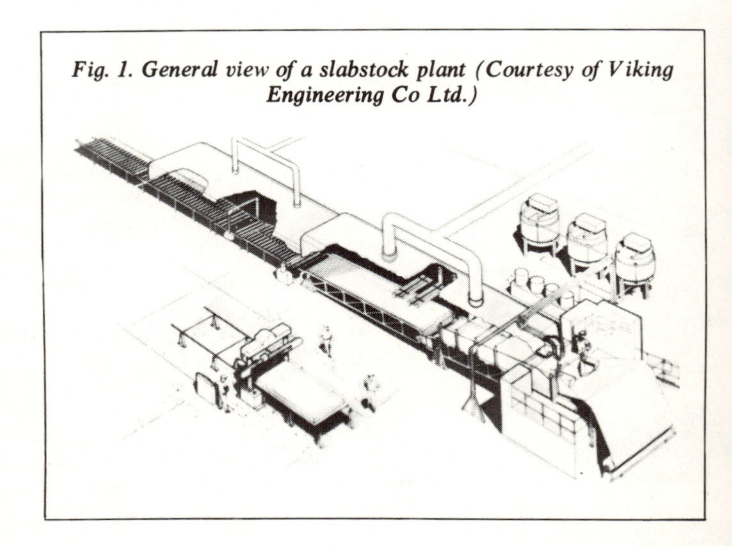

Fig. 2. Liquid foam mix is dispensed on to moving belt. (A Shell photograph.)

Auxiliary blowing agent. Optional—used only if lower density foam, or softer foam is required. Usually trichlorofluoromethane (Refrigerant 11) is used.

Processing techniques

Essentially the process comprises the mixing together of the raw materials in an enclosed mixing head in such a way that thorough mixing is achieved. The resultant mixture is discharged in the liquid state onto a moving belt or into a mould where foaming occurs. The chemical reactions occurring during the foaming process are complex, the reactions (I–V) shown above taking place simultaneously and competing with each other for isocyanate. The balance between foaming reaction and polymerisation reaction is critical, and is achieved by adjustment of the formulation.

Basic equipment for factory production of one-shot polyurethane foam consists of a machine having component storage tanks, metering equipment for each component, and an efficient mixing head. A conveyor belt is required to carry away the foam as it is forming, whether in the form of slab-stock or in moulds. Facilities for ventilation of the space above the mixing head and the conveyor are necessary for the removal of toxic fumes. A heating arrangement to produce a tack-free foam surface, thus facilitating earlier handling, is desirable.

Slabstock production

Most large slabstock production machines meter each component separately but plants can be simplified by restricting the number of components. Certain raw materials are then premixed, the most common arrangement being:

a. Di-isocyanate
b. Polyol
c. Tin catalyst
d. Water, amine catalyst, silicone oil.

Care must be taken to avoid pre-mixing raw materials which would react with one another. Colourants and refrigerant are usually metered separately.

The components are maintained at ambient temperatures and are fed continuously through individual metering pumps to the mixing head. It is preferable that components may be recycled. The pumps should meter to an accuracy of $\pm$ 1% over 90% of their range and they should maintain the predetermined ratio of the components over long periods at variable throughputs. The types of pumps employed vary according to the machine, but gear pumps, annular piston pumps and diesel injection pumps are widely used. Total outputs vary between 50 kg and 500 kg per minute.

Mixing is almost always achieved by some type of rotating stirrer, although the design of the latter may vary from a simple peg-type to a complex arrangement with varying degrees of shearing action. The degree of mixing, which is dependent on the stirrer speed, the design of the mixing head, the residence time of the liquid therein, and frequently the amount of nucleating air added, can have a considerable effect on the structure of the final foam. Undermixing is characterised by coarse cell structure, whilst overmixing leads to very heterocellular foam and occasionally foam collapse.

On slabstock machines, the complete mixing head usually traverses the width of the conveyor belt. This results in the liquid being laid down in a series of lines at small alternate angles to one another. The moving conveyor belt slopes away from the mixing head at an angle of 2–6 degrees to the horizontal thus ensuring that liquid mix being dispensed onto the belt does not fall onto material which has already started to foam. Paper is fed continuously to line the conveyor belt and the side walls.

The conveyor speed is related to the width of the block, the throughput of raw materials and the reactivity of the foam system. Belt speeds of 1–6 m/min are common. The toxic fumes produced during the foaming process are removed in a ventilation tunnel which should completely engulf the conveyor system from the mixing head. The tunnel should be sufficient to cover the conveyor during 5–10 minutes running. At this stage, blocks are cut (using horizontal or vertical blades) to convenient lengths and conveyed to a curing area. Typical block sizes are: width 0.8–2.1 m; height 0.5–1.2 m; length 1–60 m depending upon application.

Polyurethane foam formation is a highly exothermic reaction and blocks are self-curing. Stacking must allow air to pass between blocks during this time, otherwise a fire hazard is created. Normal curing time is 8–24 hours after which blocks are transferred to a converting section. Fig. 1 shows a typical view of a flexible polyurethane foam slabstock plant. Fig. 2 shows liquid foam ingredients being dispensed on a slabstock plant.

Some of the flexible polyurethane foam slabstock equipment manufacturers are as follows: Admiral Equipment Corporation, Akron, Ohio, USA; Karl Henneke Maschinenfabrik, Birlinghoven, West Germany; Laader Berg, Aalesund, Norway; Martin Sweets Co Inc, Louisville 12,

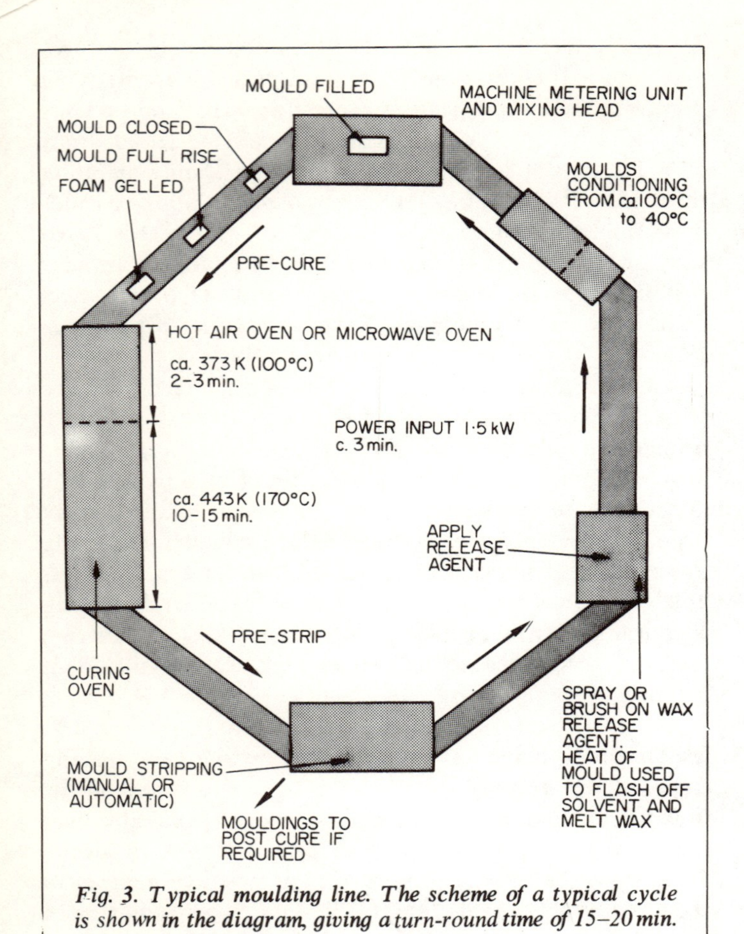

Fig. 3. Typical moulding line. The scheme of a typical cycle is shown in the diagram, giving a turn-round time of 15–20 min.

Kentucky, USA; Plama Ltd, Spjelkavic, Norway; SECMER, La Tronche, Isère, France; Viking Engineering Co Ltd, Stockport, Cheshire, England.

Production of mouldings

Since, essentially, the same raw materials are used for moulding and slabstock production, it is convenient when discussing the production of mouldings to consider the modifications to materials and equipment necessary to change a high output continuous process to one which produces a large number of small discrete pieces. Fig. 3 shows diagrammatically a typical moulding line, and a dispensing machine is shown in Fig. 4.

Fig. 4. Moulding line showing dispensing machine and mould carriers. (Courtesy of Viking Engineering Co Ltd.)

Of the raw materials, the polyols used for moulding are more reactive types, the di-isocyanate is usually the 80:20 blend of TDI, the surfactants and catalysts are usually the same types as are used for slabstock but the concentrations are increased, sometimes by as much as 40–50%. The increase in catalysis compensates for the reduced temperature rise during foam moulding owing to loss of exothermic heat. This reduced temperature rise during mouldings is the basic cause of differences in foam properties between slabstock and moulded foams[2]. Since most mouldings are relatively small, *eg* 0.5–5.0 kg, components are almost always premixed to give 4 or even 2 streams to the mixing head[3].

Surge at the beginning or end of the shot must be eliminated or else mouldings will have areas of unsatisfactory foam. The general principles of overcoming surge[4] are as follows:

(1) All component streams are recirculated to a point as close as possible to the mixer, and the valves are linked so as to operate simultaneously.
(2) The entry velocities of the component streams into the mixing head are equalised by the use of jets.
(3) The streams are pressure balanced so that, at points in the outlet close to the mixer, the pressure on recirculation is equal to the pressure on dispensing to the mixing chamber.

The temperature of the components entering the mixing head should be controlled to ± 1K (± 1°C).

Mould design. Good mould design is of paramount importance if ease of production is to be obtained. Undercuts must be avoided, and thin sections should be kept to a minimum. Uniform wall thickness is necessary if optimum curing is to be obtained.

Curing. Because of the loss of exothermic heat, mouldings are passed through curing ovens before demoulding is possible. The mould heating systems which find application are hot air circulation, steam heating, infra-red heating and occasionally microwave radiation[5]. The circulatory hot air oven is the most commonly used; curing temperatures range from 393K (120°C) to 473K (200°C).

Mould release agent. This is usually a wax which is dissolved in a solvent base and sprayed into the mould. There is a wide choice of release agents and applicators available.

Table 1. Typical properties of polyester based foams.

Density	(kg/m^3)	45	40	35	30	25
Compressive strength (40%)	(kN/m^2)	6.5	6.5	6.5	6.0	5.5
Tensile strength	(kN/m^2)	175	175	180	175	190
Elongation	(%)	310	270	280	220	250
Compression set (90%)	(%)	4	5	8	10	12

Density (kg/m³)	65	50	40	40	35	35	30	30	25	25	20	15
RMA values 25% (lb/50 in²)	50	37	50	45	41	26	40	18	16	9.7	40	9.5
50% (lb/50 in²)	70	56	67	57	54	33	48	22	20	14.0	59	12.5
65% (lb/50 in²)	95	83	88	77	71	47	61	31	28	21.0	73	17.2
Sag factor	1.9	2.2	1.75	1.7	1.75	1.8	1.55	1.7	1.75	2.15	1.8	1.8
Compression set 75% (%)	4.0	2.3	2.1	2.5	3.8	7.5	2.9	8.5	5.2	8.0	6.5	8.9
90% (%)	—	—	2.9	3.8	4.4	—	3.7	—	6.9	—	8.0	—
Resilience (%)	55	55	65	54	53	54	49	58	62	61	38	50
Tensile strength (kN/m²)	125	170	88	163	106	146	123	91	108	62	140	66
Elongation (%)	300	400	200	400	240	400	330	300	440	300	300	380
Tear strength (kN/m)	0.44	0.63	0.20	0.57	0.35	0.55	0.5	0.4	0.47	0.3	0.65	0.40

Table 2. Typical properties of polyether based foams.

Pre-conditioning of moulds. The optimum mould temperature varies with the formulation being employed but is usually between 310K (37°C) and 333K (60°C). It is important that the mould surface temperature is uniform at the time of filling and hence a cooling/conditioning unit must be employed. Water cooling followed by warm air conditioning is frequently carried out. Fig. 3 shows a typical flexible urethane foam moulding line.

Properties of flexible polyurethane foams

Polyester foams are invariably manufactured on slabstock equipment. The obvious difference between polyether and polyester based foams is the lower resilience of the latter. Compared with polyether foams of similar density, polyester based foams exhibit higher hardness, higher tensile strength and higher elongation at break; they also show somewhat better resistance to dry cleaning solvents.

Polyether foams show higher resilience, better indentation load characteristics, lower compression set figures and a wider choice of properties. Typical properties of foams are given in Tables 1–3.

Applications of flexible foams

Polyether foams show higher resilience, better indentation where high tensile strength and solvent resistance for dry cleaning is essential. Laminates are produced using adhesive or flame laminating techniques. Other applications where stiffness and high tensile strength are required include padding in clothes and luggage linings. A further use is in the production of, for example, sun-visors for motor cars.

Polyether foams are widely used in the furniture and upholstery industries. The ease with which block foams can be fabricated and sculptured, and the favouring in much modern design of simple rectilinear shapes has afforded a ready acceptance of foam in these outlets. The bedding industry is another major outlet for slabstock foam, as is the automotive industry where cushioning, crash padding, roof lining and filters are all fabricated from slabstock foam. Other uses include carpet underlays and backing, artificial sponges, packaging and padding, draught excluders, *etc.*

Conventionally, almost all flexible foam mouldings are used in the automotive industry[6], although difficult furniture shapes are also moulded.

Recent developments

The most important developments recently have been the production of 'high-resilience' or 'latex-like' foams. These foams, which are based on mixed or modified isocyanates, are a much closer match for latex rubber foams than has hitherto been possible. The use of these novel formulations gives rise to foams which are also more readily made flame retardant by the use of additives. Although the large scale production of these foams on slabstock plants has not yet been optimised, one major benefit has accrued in that mouldings based on these formulations can be removed from the mould without the application of heat ('cold-cure' mouldings). Such mouldings are finding application in the furniture industry and can be expected to penetrate the automotive industry when future safety requirements dictate better flame retardant properties. Typical properties for such foams are given in Table 4 and the improved hysteresis of such foams is shown in Fig. 5.

SEMI-FLEXIBLE FOAMS

This section can reasonably be divided into: (a) conventional semi-flexible foam moulding, where PVC or ABS skins are first formed and then filled with the semi-flexible

Table 3. Typical properties of flexible mouldings.

Core density (kg/m³)	33	30	28
RMA value 25% (lb/50 in²)	45	35	30
50% (lb/50 in²)	62	50	41
65% (lb/50 in²)	81	73	57
Sag factor	1.8	2.1	1.9
Compression set (90%) (%)	6.5	6.0	4.1
Tensile strength (kN/m²)	170	154	123
Elongation (%)	210	230	215
Tear strength (kN/m)	0.38	0.53	0.36

Table 4. Typical properties of 'high resilience' foams.

Core density (kg/m³)	55	47	40	34
RMA value 25% (lb/50 in²)	30	22	17	10
50% (lb/50 in²)	46	34	26	17
65% (lb/50 in²)	72	52	39	27
Sag factor	2·4	2·4	2·3	2·7
Compression set (75%) (%)	< 10	< 10	< 10	< 10
Resilience (%)	75	74	74	71
Tensile strength (kN/m²)	90	76	63	51
Elongation (%)	100	105	110	115

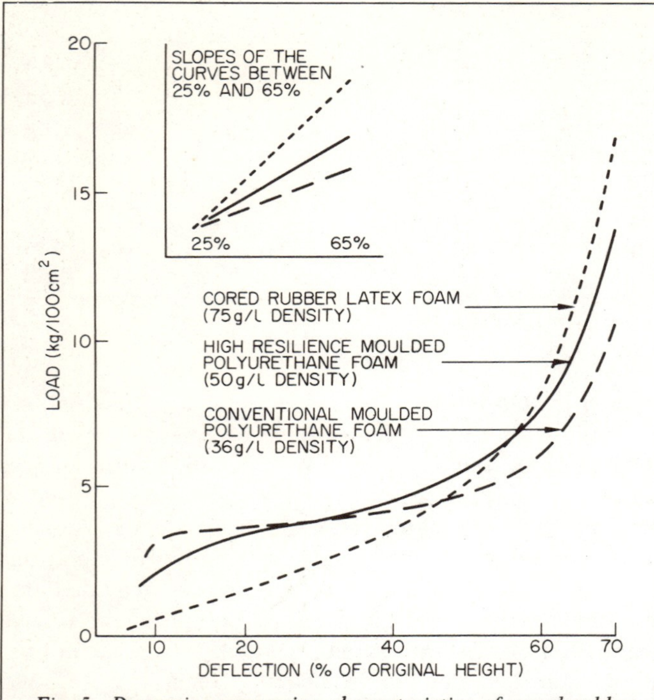

Fig. 5. Dynamic compression characteristics of cored rubber latex, conventional moulded polyurethane and high resilience moulded polyurethane foam.

foam liquid mix which expands to fill the skin; and (b) integral-skin semi-flexible foam moulding where the foaming polyurethane mix produces its own skin.

Conventional semi-flexible foam

The basic chemistry of these foams is similar to those of flexible foams but the raw materials used differ somewhat as described below.

1. The polyol is usually an active triol of molecular weight 4000–5000.
2. The di-isocyanate is that more commonly used in rigid foams, *ie* MDI (see later).
3. Formulated water content lies between 1.0 and 2.5 parts per hundred polyol.
4. A cross linking agent is used in the formulation to give harder foams, *eg* triethanolamine, glycerol, trimethylol propane.
5. Silicone surfactants and tin catalysts are usually omitted since their presence causes high closed cell contents, which leads to foam shrinkage.

In production[7], vacuum formed ABS or rotationally cast PVC skins are placed in a jig and the liquid foam mix is dispensed into it. It is quite common to use inserts in such mouldings. A small shot 2 or 4 component machine is used. Typical properties of such foams are given in Table 5. The major application for such mouldings has been in the production of automotive trim, *eg* arm-rests, crash pads and facia panels.

Integral-skin semi-flexible foams (ISSF)

This newer process[8] allows the foam expansion to be controlled so that a cellular core and a non-cellular outer skin are obtained in one moulding process. The raw materials used are similar to those used in conventional semi-flexible foam but water is ommitted. The foam reaction is provided solely by the exothermic heat causing an auxiliary blowing agent (*eg* Refrigerant 11 or methylene chloride) to vaporise. The heat of reaction will be removed at the mould surfaces, retarding foam formation and producing a coherent urethane skin.

Overall densities of such foams can vary widely from 200 kg/m^3 to 800 kg/m^3 depending upon the quantity of blowing agent used and the degree of mould overpacking employed. Typical properties are given in Table 6.

Applications where ISSF mouldings are being used include arm-rests, facias, steering wheels and inserts, crash pads and bumpers in the automotive industry. Inner soles for shoes and decorative furniture components are other examples.

RIGID POLYURETHANE FOAMS

The evolution of rigid foam technology has seen a change from TDI based foams, frequently made by a prepolymer method, to MDI based foams which can be manufactured in slabstock or moulded form, in continuous laminates (sandwich structure), or can be foamed *'in-situ'* without the hazards of toxicity associated with TDI. Urethane foams can also be sprayed, but positive air pressure hoods should be worn by operatives.

Chemistry and raw materials

The basic chemical reactions are similar to those described at the beginning of the chapter and are dealt with in great detail elsewhere[1,2,9]. The polyols employed are generally of higher functionality (*eg* 3–8) than those used for flexible foam and have much lower average molecular weights (*eg* 300–1200).

The isocyanate used is of the less toxic type and may be

Table 5. Typical properties of conventional semi-flexible foam.

Core density (kg/m³)	100	80	65
Compressive strength (40% compression) (kN/m²)	38	25	16
Compression set (50%) (%)	10	15	15
Tensile strength (kN/m²)	175	115	100
Elongation (%)	44	40	50

Table 6. Typical properties of integral skin semi-flexible foam.

Core density (kg/m³)	200	140	110
Compressive strength (40% compression) (kN/m²)	59	17	10
Compression set (50%) (%)	11	8	8
Tensile strength (kN/m²)	300	105	90
Elongation (%)	85	90	100

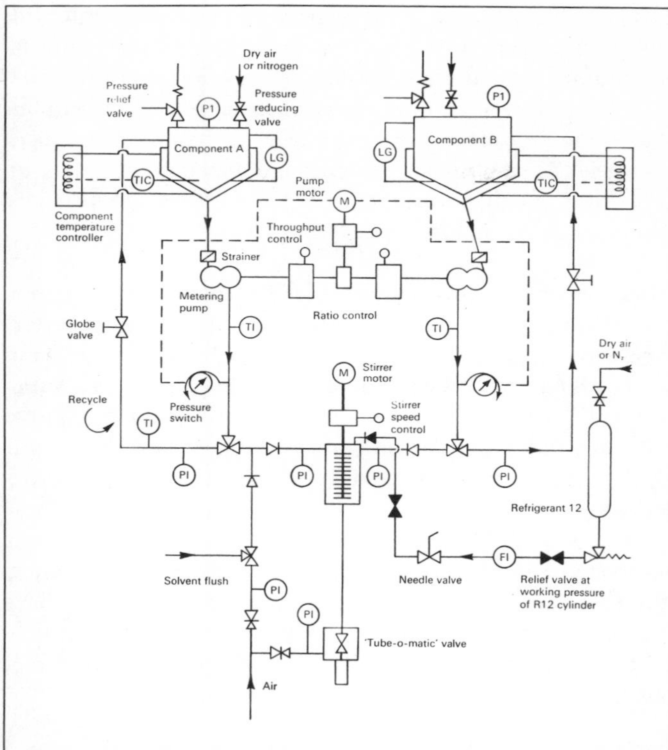

Fig. 6. Flow diagram of two component dispensing machine (equipped with frothing facilities) used for manufacture of rigid urethane foam.

Fig. 7. Typical rigid foam dispensing machine. (Courtesy of Viking Engineering Co Ltd.)

represented by 44′ di-isocyanate-diphenylmethane (MDI).

$$OCN-C_6H_4-CH_2-C_6H_4-NCO$$

MDI.

The MDI used is usually a somewhat crude product and will contain other aryl and polyaryl isocyanates.

Foam manufacture and dispensing equipment

The principles of manufacture are the same as those discussed for flexible foams. It is much more common, however, to use a two component dispensing machine for rigid foam and a flow diagram for such a machine is shown in Fig. 6, and a typical machine is shown in Fig. 7.

Apart from the usual liquid dispensing technique, two other techniques are used in the application of rigid foam:

(*a*) *Frothing.* A partially pre-expanded mixture is dispensed. This is achieved by metering Refrigerant 12, boiling point 245K (− 28°C), through a Rotameter into the mixing chamber under sufficient pressure to keep it in the liquid state. This pressure is regulated by using a control valve. After leaving the nozzle, the mixture froths to the consistence of a shaving lather.

(*b*) *Spraying.* In this case the metering unit delivers the liquids to a mixing head which comprises the spray gun. The gun may be equipped with a mechanical mixer, in which case a solvent flush system is required to prevent gun blockage. Air is used to spray the components on to the substrate.

Mixing may also be achieved by means of high component pressures or air. However, in the latter case the losses are larger as a result of the high degree of atomisation. The different dispensing techniques find uses in the applications discussed below.

Properties of rigid polyurethane foam

Rigid urethane foams are used principally for their remarkable thermal insulation properties. The thermal conductivity is the lowest of any insulation material in common use, average values being of the order $\lambda(u) = 0.022\,W/m.K$ ($k = 0.16\,Btu.in/ft^2h°F$). This property is due to the enclosed cellular structure of the material which entraps gases such as CO_2 and Refrigerant 11 occurring during the manufacturing process which themselves have very low thermal conductivities. In addition, however, density is of prime importance since mechanical properties vary according to the density.

Density. Polyurethane foam used for insulation is normally of low density 28–40 kg/m^3, but for structural uses much higher densities are required.

Thermal conductivity. This value varies slightly depending upon whether the foam is CO_2 blown (from water in the formulation) or Refrigerant 11 blown. The average value after ageing is 0.022 W/m.K.

Closed cell content. Rigid polyurethane foam normally contains about 90 % closed cells.

Mechanical properties. Polyurethane foam has a high

Density	(kg/m³)	30	40	55	64	96	160
Compressive strength	(kN/m²)	185	250	440	560	1000	2300
Tensile strength	(kN/m²)	330	350	600	850	1200	2300
Shear strength	(kN/m²)	200	220	350	500	700	1100
Flexural strength	(kN/m²)	300	400	700	1000	1800	3200
Thermal conductivity	(W/m.K)	0.022	0.023	0.024	0.024	0.029	0.035

Table 7. Typical properties of rigid polyurethane foam.

strength-to-weight ratio due to the rigidity imparted by the cell structure. Typical values at a density of 40 kg/m^3 are:

Compressive strength (10% compression), 207–276 kN/m^2 (30–40 lbf/in^2).
Flexural strength (12.7 mm deflection, 25.14 mm section), 241–310 kN/m^2 (35–45 lbf/in^2).
Tensile strength (approximately), 345 kN/m^2 (50 lbf/in^2).

Strength increases with density and decreases with increase in temperature. Since polyurethane foam is anisotropic the strength properties across the rise will be 60–80% of those found in the direction of foam rise.

Water vapour transmission. Rigid polyurethane foam is inherently resistant to the ingress of water and the transmission of water vapour. Water vapour transmission is of the order of 70 g/m^2.24h (311K (38°C) 88%RH) for a foam of 40 kg/m^3. A vapour barrier should be used on the hot side when cold plant is being insulated.

Adhesion. Polyurethane foam forms a strong adhesive bond with a wide range of substrates including metal, glass and wood.

Flammability. Rigid polyurethane foams, along with other organic plastics, have been the subject of much work in the field of flame retardance. A variety of phosphorus and/or halogen containing additives (reactive and non-reactive) are now available which improve the flame resistance of polyurethane foams. The choice of additives and of test methods is however best decided upon in the light of the application envisaged. Table 7 shows some typical properties.

Applications of rigid polyurethane foam

Slabstock and blockfoam. Slabstock production is analogous to the flexible foam production method described earlier. Blockfoam is the production of discrete blocks by dispensing liquid mix into a large box, *eg* 2 × 1 × 0.5 m. In the latter technique, the foam should not be removed from the box for a few hours. In both cases, the foam should be allowed to cure for 48 hours before cutting. Foams made in this way are usually cut into sheets for general insulation or into specific shapes, *eg* pipe sections.

Laminating. The foam is sprayed onto a continuously moving belt of, for example, paper or metal foil. A top paper or foil is then allowed to come into contact with the top of the foam and the total sandwich passes through an adjustable press. The laminate is automatically trimmed and cut to the desired board size. These laminates find use in cold stores, building panels, roofing panels, *etc*.

Cavity filling. Liquid or frothed mix is injected into the cavity and rises to fill the available space. This type of operation has many possibilities (*eg* refrigerated fish wagons, rail cars, *etc*) but the most common application is the filling of refrigerator or freezer cabinets. The space between the metal outer casing and the inner plastics or foil lining is filled with foam (Fig. 8). This operation is discussed in detail elsewhere[10,11].

Moulding. This is analagous to flexible foam moulding, but it should be appreciated that this is a cavity filling operation where the foam is later removed. Moulds in this process should have quick fastening clamps and be able to withstand 200–300 kN/m^2 pressure. For insulating purposes, pipe sections are frequently made in this way. Furniture shells are also manufactured by this technique[12]. A typical mould construction for a chair shell is shown in Fig. 9.

Spraying. This is perhaps the fastest method of insulating any large area, *eg* storage tanks, farm buildings, *etc*. The thickness of the insulation is dependent upon the number of layers sprayed. An automatic process for spraying tanks has been patented[13]. Such a tank is shown in Fig. 10.

Building panels. This is a cavity filling application but is usually considered separately. The two facing sheets, one of which may have been coated with a vapour barrier on its inner face, are located in a simple mould and held apart by spacers. The mould is then moved into a press (*eg* a multi-daylight press). The mixed liquid is injected from a dispensing machine into the moulds where foam fills the moulds completely and bonds itself to the facing sheets. The moulds are held under pressure for a time and then

Fig. 8. The cavity between the outer metal casing and the inner lining is injected with rigid urethane foam in this refrigerator, giving good insulation and a strong wall. (A Shell photograph.)

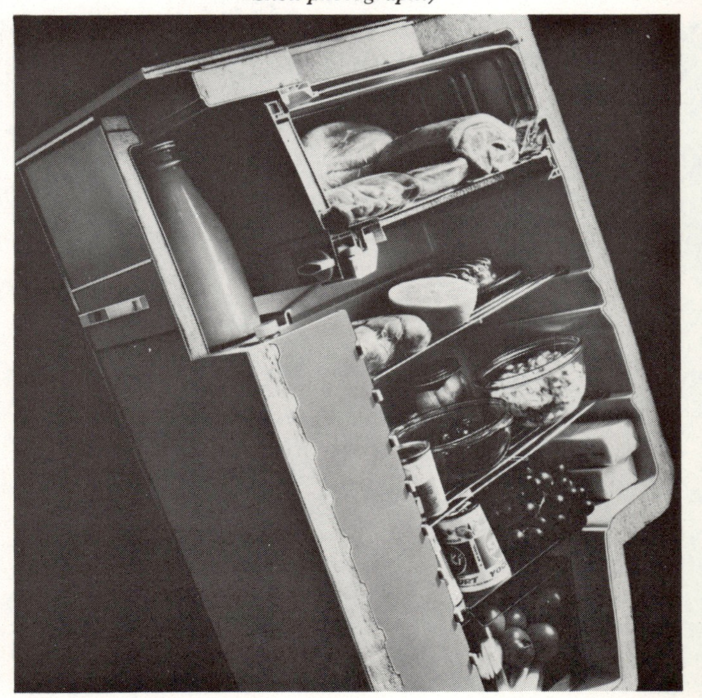

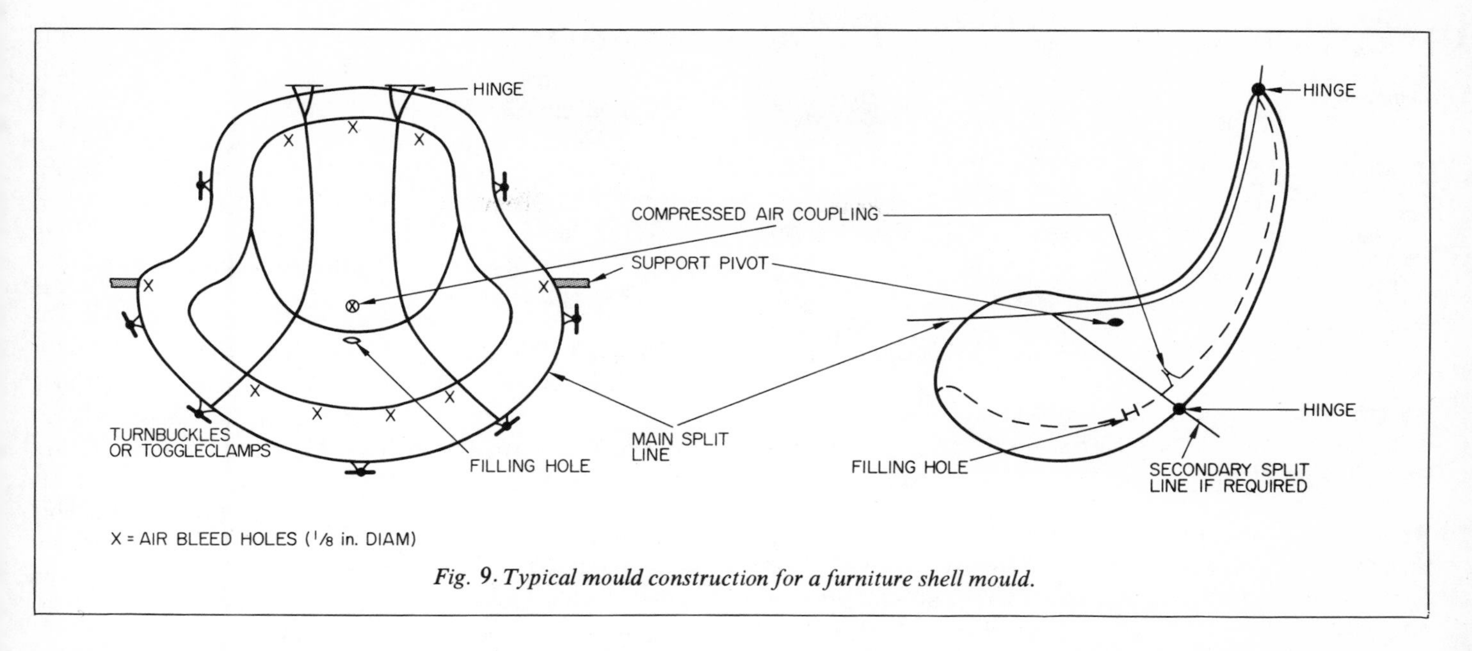

Fig. 9. Typical mould construction for a furniture shell mould.

unloaded where the panels are removed for finishing operations.

Developments

The most interesting development in rigid polyurethane foams recently has been the advent of integral-skin rigid foams[14]. These foams are formed in a similar way to integral-skin semi-flexible foams (discussed earlier) and are finding application in decorative mouldings, furniture pieces, domestic appliance cabinets and the like. In future they may even find application in the automotive industry.

POLYSTYRENE FOAM

Manufacturing processes

Expandable polystyrene for steam moulding is usually supplied as free flowing, translucent polymer beads containing an expanding agent. The various grades are characterised by bead size, *eg* packaging bead diameter 0.4–0.7 mm, general moulding 0.7–1.2 mm, block moulding 1.0–2.4 mm.

To obtain good quality mouldings, the expansion and moulding is carried out in three stages[9] :

(*a*) *Pre-expansion.* Expandable beads are introduced continuously into the base of a steam chest, generally a vertical cylinder with a mechanical stirrer. On meeting the steam, the beads expand and are carried upwards as their density decreases. The density of the expanded beads depends on their residence time in the chest, and this is adjusted by using different take-off levels.

(*b*) *Maturing.* After pre-expansion, the volatile expanding agent condenses and pressure inside the bead decreases. It is essential to allow air to diffuse into the bead and raise the internal pressure to atmospheric level. Maturing may take place in open hoppers or open top drums generally during 24 hours.

(*c*) *Final expansion and moulding.* The mould is firmly packed with pre-expanded, matured beads. Steam is injected into the mould and the pressure allowed to rise to 70–100 kN/m^2. The time taken for this may vary considerably, depending on the density and the size of the moulding. The steam line supplying the mould should be fitted with a steam gauge. A rise in steam pressure indicates that fusion is taking place. When fusion is complete, the steam is shut off and the mould cooled to about 308K (35°C).

Expandable polystyrene can also be extruded using conventional equipment to form a continuous sheet. This sheet may later be formed (*eg* vacuum forming, pressure forming) to produce articles such as fruit trays, *etc.*

Properties of expanded polystyrene

The main physical properties of expanded polystyrene foam are dependent on the density as is seen in Table 8. Unlike the thermoset rigid polyurethanes, thermoplastic poly-

Fig. 10. Demonstration of the automatic tank spraying process[13]. (A Shell photograph.)

Density	(kg/m³)	16	32	48	80	144
Compressive strength	(kN/m²)	70	190	420	800	1100
Tensile strength	(kN/m²)	210	350	600	850	1200
Shear strength	(kN/m²)	120	240	400	700	—
Flexural strength	(kN/m²)	210	420	700	1400	2000
Thermal conductivity	(W/m.K)	0.034	0.031	0.034	0.035	0.038

Table 8. Typical properties of expanded polystyrene foam.

styrene foam is quite deformable and resilient at low densities and non-abrasive. Extruded sheet has a very smooth appearance.

Resistance of polystyrene foam to solvents is more critical than that of polyurethanes. Halogenated hydrocarbons, esters, ketones and aromatic solvents are all solvents for polystyrene.

Applications of expanded polystyrene

Insulation. Polystyrene foam at densities below 110 kg/m³ exhibits a thermal conductivity λ (u) of 0.037 W/m.K (k = 0.26 Btu.in/ft²h°F). As a general rule, therefore, polystyrene foam insulation must be 70% thicker than a urethane foam to provide equivalent heat loss characteristics, but since material costs are substantially lower for polystyrene both foams are used depending upon requirements.

Where minimum insulation thickness is required or *in situ* filling is preferred, *eg* domestic refrigerator manufacture, polyurethane is almost always used, but for lining the walls of refrigerated stores and other large constructions, insulating air conditioning equipment, *etc*, expanded polystyrene board is used in large quantities. Expanded polystyrene ceiling tiles find markets both as condensation preventors and as a decorative finish.

Perhaps the fastest growing market for expandable polystyrene is the packaging field where trays for fruit, fish and other perishable goods are required in quantity. Delicate machinery parts, medical instruments, *etc*, are also packaged in preformed foamed polystyrene.

As with urethanes, furniture shells are mouldable in foamed polystyrene. The choice of material depends upon the design of the chair to some extent but more importantly upon the numbers to be produced.

PHENOLIC FOAMS

Phenolic foams are thermoset, partially open cell foams with good dimensional stability and heat resistance[9]. Like polyurethanes they are foamable *in situ*, at densities from 16–320 kg/m³. Phenolic foam properties are relatively unaffected by heat and moisture, although this is dependent upon the density. Chemical and solvent resistance is very good. Phenolic foams have been used as insulants and also in packaging applications.

UREA-FORMALDEHYDE FOAMS

Urea-formaldehyde (UF) foams are available in densities ranging from 80–160 kg/m³ both as *in situ* foams or preshaped. They have a narrow working range of temperature 243K (– 30°C) to 320K (47°C). In addition, the foams readily absorb a variety of liquids including water. This leads to one important application—cut flower arrangements: UF foam of low density is easily pierced by the stems of flowers and the foam will absorb sufficient water to keep the flowers fresh for several days.

SILICONE FOAMS

Thermoset silicone foams are available at densities between 50 and 250 kg/m³. Although the mechanical properties of these foams are not impressive, they are exceptionally heat stable, withstanding temperatures up to 618K (345°C). Wherever high temperature thermal or electrical insulation is required, the silicone foams are the best available. Table 9 summarises the properties of a variety of rigid foams.

REFERENCES

1. J H Saunders and K C Frisch, Polyurethanes, Pt 1; *Interscience Div., Wiley & Sons.*
2. J M Buist and H Gudgeon (Ed), Advances in Polyurethane Technology, *Maclaren & Sons.*
3. I Musgrave, *J Cellular Plastics* (1969) 5,4.
4. J M Buist, *Rubber Journal* (1966) 148,3.
5. W J Lanigan, *British Plastics* (1963).
6. H L Ward, *J Cellular Plastics* (1970) 6,5.
7. B Ryall and P G Griffiths, *J Cellular Plastics* (1970) 6,5.
8. H. Wirtz, *Kunstoffe* (1970) 60, 1.
9. T H Ferrigno, Rigid Plastic Foams, *Reinhold.*
10. Kühlmöbeltagung '68, *Farben Fabriken Bayer A.G.*
11. *Shell Chemicals Technical Information.* 1CS(X)/68/15.
12. K D Roome and J W Waters, *Cabinet Maker and Retail Furniture* (1968 January).
13. Shell Patented Automatic Tank Spraying Equipment (*PATSE*).
14. H. Piechota, *Kunstoffe* (1970) 60, 1.

Table 9. Properties of various rigid foams.

	Polyurethanes		Polystyrenes		Phenolics		Ureas	Silicones
Density (kg/m³)	30–50	60–100	30–40	60–100	30–60	100–150	13–20	50–250
Thermal conductivity (W/m.K)	0.023	0.023–0.04	0.035	0.038	0.030	0.037	0.024	0.043
Compressive strength (10%) (kN/m²)	140–350	400–1900	170–210	370–1000	140–600	300–900	30	1200–6000
Tensile strength (kN/m²)	140–500	600–1700	350–400	700–1100	140–380	200–500	—	—
Shear strength (kN/m²)	140–210	400–900	240–300	500–1000	100–200	250–900	—	—
Flexural strength (kN/m²)	400–700	1400–2400	380–520	1000–2000	170–450	1300–2000	120	—
Max. service temp. (°C)	120	140	80	110	130	200	50	345

FLEXIBLE FOAM TROUBLE-SHOOTING GUIDE

Processing

Defect	Description	Recommendations
Boiling	Foam does not rise, large bubbles appear and burst at the surface.	Increase tin catalyst. Check for excess TDI. Increase silicone or check for reduced silicone activity. Lower amine catalyst. Look for errors or fluctuations in metering and proportioning components.
Flashing, sparklers	Excessive effervescence of top surface of the rising foam mass.	Decrease TDI. Increase tin catalyst. Lower amine catalyst. Look for errors or fluctuations in metering and proportioning components. Decrease silicone concentration. Decrease reactant temperatures.
Collapse	Foam rises and then falls.	Increase tin catalyst. Increase silicone or check for reduced silicone activity. Look for errors or fluctuations on metering and proportioning components. Look for possible contaminants in system. Check mechanical factors conveyor speed, through-put, rpm of mixers, conveyor vibrations, guide and support of buns.
Crazy balls	Small liquid bubbles moving rapidly over the surface of the foam.	Increase mixing speed. Minimise splashing on lay down.
Smoking	Excessive TDI vapours can be seen rising from the surface of the foam.	Look for errors or fluctuations in component metering and proportioning TDI polyol.
Under-running	Cream line tends to move toward the pour point followed by flow of the liquid reactants under the already rising foam mass on the conveyor.	Lower amine catalyst. Decrease reactant temperatures. Speed up conveyor. Reduce head throughput. Decrease conveyor angle. Increase total catalyst system level.
Tacky surface	Surface of the bun remains sticky for prolonged period.	Increase total catalyst system level. Heat bun surface. Remove air bleed.
Low catalyst tolerance	Formulation exhibits sensitivity to changes in tin catalyst concentration.	Increase volume of catalyst stream by adding a diluent. Look for errors or fluctuations in component metering and proportioning tin catalyst, amine catalyst.
Gross splits	Large vertical or horizontal separations in the bun.	Increase tin catalyst. Lower amine catalyst. Increase silicone or check for reduced silicone activity. Decrease water. Check mechanical factors conveyor speed, conveyor angle, throughput, rpm of mixers, length of traverse, conveyor vibrations, guide and support of buns.
Small side splits	Small diagonal separations at the corner of the bun.	Increase tin catalyst. Increase TDI. Increase Silicone or check for reduced silicone activity. Check mechanical factors conveyor speed, conveyor angle, throughput, rpm of mixers, length of traverse guide and support of buns.
Settling back	Bun rises to maximum height and then settles slightly.	Increase tin catalyst. Increase TDI.
Shrinkage	Bun rises to maximum height and upon standing, contracts and forms wrinkles and indentations.	Decrease tin catalyst. Decrease silicone concentration. Decrease TDI. Increase amine catalyst.
Windows, mirrors	Cell walls remain intact, as evidenced by the reflected light (noticeable on cut surfaces of finished foam).	Decrease tin catalyst. Decrease TDI. Decrease silicone concentration. Find optimum mixing speed for throughput. Decrease cell size.

Properties

Defect	Description	Recommendations
Poor cell size	Foam is composed of large cells.	High pressure system- Increase silicone or check for reduced silicone activity. Decrease mixing speed. Check mechanical factors. Low pressure system- Increase silicone or check for reduced silicone activity. Increase mixing speed.
Closed cells	Cell walls remain intact and do not rupture at the point of maximum foam rise.	Decrease tin catalyst. Increase amine catalyst. Decrease TDI. Decrease silicone concentration.
Dead foam	Foam has low resiliency.	See closed cells.
Bad fingernail	Foam recovers slowly when indented with a sharp object.	Increase water to blowing agent ratio. Decrease tin catalyst. Increase silicone or check for reduced silicone activity. Closed cells (see above).
Friable	Foam is crumbly and does not build polymer strength.	Look for errors or fluctuations in component metering and proportioning TDI polyol. Increase TDI. Decrease polyol.
Scorching	Discolouration of interior of bun.	Decrease TDI. Decrease water to blowing agent ratio. Decrease total catalyst system level. Look for errors or fluctuations in metering and proportioning components.
Poor tensile strength	Tensile values lower than normal.	Increase mixing speed. Decrease TDI.
Poor elongation	Elongation values lower than normal.	Decrease water to blowing agent ratio. Decrease TDI.
High compression set	Compression set values higher than 10%.	Decrease TDI. Increase water to blowing agent ratio.
Low load bearing values	Formulation produces lower load bearing values than desired or expected.	Increase TDI. Increase water to blowing agent ratio. Increase tin catalyst. Look for errors or fluctuations in metering and proportioning components.
High load bearing values	Formulation produces foam with higher load bearing values than expected.	Look for errors or fluctuations in metering and proportioning components. Decrease water to blowing agent ratio. Decrease tin catalyst.

Chapter 20

Reprocessing Plastics

G Cheater
Rubber and Plastics Research Association of Great Britain

The practical need to recover waste from plastics production is self-evident to all who have witnessed the operation of any extruder, and since extrusion is, in various guises, involved in a number of production techniques it can logically, and correctly, be assumed that the ability to reprocess is an important factor in profitable plastics conversion. Injection moulding, blow moulding and film production are all extensions of the extrusion process and the only known method of checking their correct operation is by processing material, usually to waste.

Setting up or changing polymer or colour in a 4 oz (113 g) injection moulding machine can easily consume 2 or 3 lb ($\sim$ 1 kg) of material. Similarly, waste is produced during production. Sprues and runners can often represent as much as 80% of the shot weight and in blow moulding the trimming will commonly represent 40/50% of the product weight. Additionally, the extrusion process is subject to many variables which are difficult to anticipate and can result in the production of large quantities of 'out of tolerance' products whilst corrective adjustments are being made to the tool.

The different properties of thermoset and thermoplastic materials inevitably affect the reprocessability of the materials. Thermoset materials have been 'cured' during manufacture, they often contain high proportions of 'fillers', and their only possible second use is as a filler when ground to a powder and blended with new material for unimportant applications. This chapter will therefore be mainly concerned with the re-use of thermoplastic materials.

Thermoplastics comprise simple long chain molecules without the crosslinked structure associated with thermosets. They can therefore be reformed to powder or granule under suitable conditions without incurring exceptional chemical/physical property changes. Research work has, in recent years, been aimed at reverting to the original monomer or the ingredients used to make the monomer,

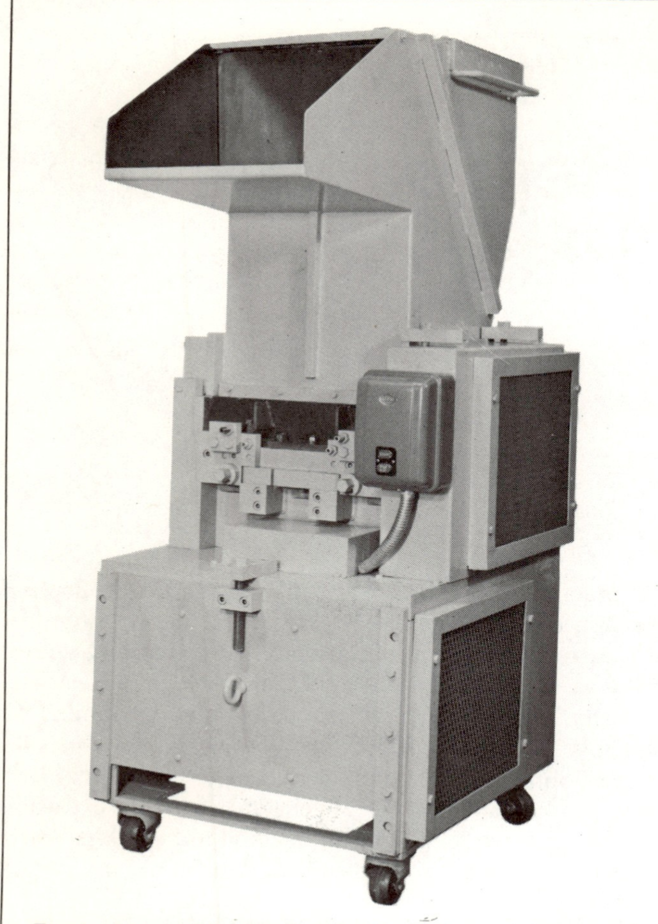

Fig. 1. A typical 'beside the press' granulator. (Courtesy of USI Engineering Ltd.)

however these processes are very specialised and are beyond the scope of this book. Plastics reprocessing can therefore be considered simply as a production processing technique where the 'raw material' may be in an infinite variety of shapes and sizes and the 'product' is in powder, chip or granule form.

To achieve this form of reprocessing, large sections must first be reduced to a size which can conveniently be fed into granulating machinery. After granulating, any of three options can be taken: (a) the chips can be fed as raw material without further treatment; (b) they can be further ground to a fine powder; or (c) they can be fed through a granulating extruder. Options (b) and (c) can be further extended so that corrective additions are blended with the reprocessed material. Variations of this system are adopted in the vast majority of companies who handle thermoplastic materials, either by machinery within their factory or by contract with a specialist reprocessor.

Certain polymers are not readily amenable to this simple sequence and chemical processes are being increasingly adopted which reprocess the material to an earlier stage of manufacture. Transparent sheet of polymethymethacrylate (PMMA) is one instance where, to retain the transparency and colour, reclaimers must resort to distillation to recover the monomer. However, these situations are exceptional and usually arise in companies which operate on a large scale.

REPROCESSING MACHINERY

Reduction of a variety of components to a common size is inevitably a difficult task to mechanise. Where scrap components are of heavy cross section and bulky, the best choice is often a handsaw or a machine band saw. If a machine is to be used, care is necessary if excessive heat generation is to be avoided. Alternatively pre-crushing equipment is available which has a sufficiently large throat to receive and deal with a comparatively large size range.

The granulator

The rotary granulator is the most commonly used type of reprocessing machine. Models are available in a wide range of sizes such that they can either be mobile and located beside an injection moulding machine (Fig. 1) or installed permanently in a central reclamation department (Fig. 2), and there are a number of companies who specialise in their manufacture. Table 1 shows the wide range of sizes and the considerable power which is needed to drive them. Fig. 3 shows the massive proportions of the rotor and its bearings and also the screen that determines the maximum size of particle. Various configurations of fixed and rotating blades are recommended for specific applications.

The degree of size control which these machines are able to apply is severely limited and is dependent upon the grade and type of polymer, the physical size of the product and the operator. Consequently, the result may or may not include a wide range of sized granules and 'dust' and will have different characteristics in the feed hopper to those exhibited by the regular shaped virgin polymer. However, this type of machine is widely used as 'beside machine' granulators in locations where sprues and runners are chopped up directly and then separated from the product and fed back into the machine hopper. Under these conditions the degree of dilution is usually sufficient to avoid any adverse feeding effects.

Conversely, their use as a departmental machine is quite frequently to be found in extrusion factories where the granules are either blended in a specific proportion or else they are used solely for a specific application where reduced output efficiency is accepted as a premium against the use of substandard material. Granulators are also used as the first stage in more complex recovery procedures.

Distinct from the high impact chopping action, there are granulators which have a rasping or shaving action, and are specifically designed to receive lengths of extrusions. These machines are marginally less noisy than the rotary granulators although both produce a very high noise level.

The cable industry is one instance of special adaptions being made to cut electric wire into short sections which

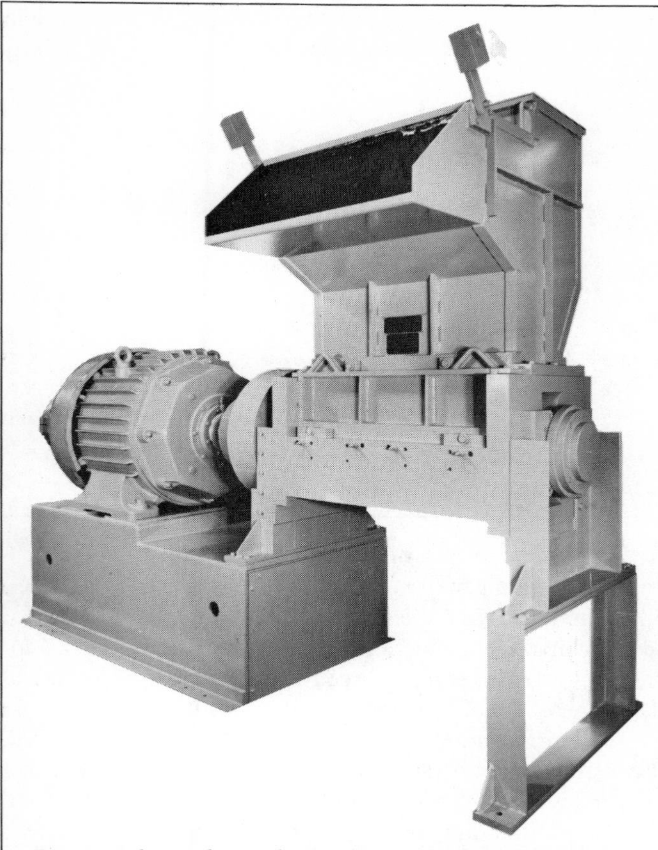

Fig. 2. A large throat, heavy duty, granulator suitable for central granulation of all kinds of material sizes and types. (Courtesy of USI Engineering Ltd.)

can subsequently be vibrated to separate plastics from conductors. Plastics film also needs special attention. Here machines are fitted with means of agglomerating by pressure or heat, or both, before material is fed to the granulator.

Difficulty is experienced when attempting the reduction of expanded/cellular forms of plastics and many attempts have been made to deal with them, either by direct compression (by feeding into an extruder having enlarged throat and a rapid compression via a conical screw) or by agglomerating by heat and pressure. Products of granulators can be further reduced to powder form by grinding at low temperatures, sometimes requiring the use of liquid nitrogen or CO_2.

Re-granulating does not enhance the physical qualities. Processes aimed at recovering some of the materials stability involve the use of heat and melt the polymer. It is subsequently extruded through multiple ribbon dies and chopped into regular cubes or cylinders either whilst hot at the extruder or after cooling in machines which bear a resemblance to the rotary granulators. This sort of process is the one frequently adopted by trade reprocessors who undertake re-blending of colours and additions to the formulation. The mixers and blenders employed are the standard machines used in polymer formulating—high speed paddle mixers or, more usually, internal mixers. Where pelletising is carried out at the extruder, vibrating chutes are used to aerate and cool the polymer before bagging.

Chemical equipment

The nature of thermoplastics is such that they can, by heating in controlled conditions, be broken down into the solvents, gases, *etc*, from which they were originally derived. Detailed explanation of these methods is beyond the scope of this chapter, however it should be noted that where clarity is of vital importance these techniques must be adopted.

Similarly, recovery of expanded plastics, and their disposal, demands a process whereby the volume is considerably reduced. Chemical means are available whereby expanded polystyrene can be recovered and converted into useful sheet material.

MATERIALS AND DESIGNS

Reprocessing entails the use of additional energy on the polymer and the consequent generation of heat. Thermoplastic polymers are able to withstand re-working in this way to varying extents before, eventually, degradation of physical properties prevents any further use. Table 2 places thermoplastics in some order of reprocessability, although it must be recognised that the decision as to its ability to withstand a further re-cycling is somewhat arbitrary and will necessarily vary with company policy.

Physical deterioration or degradation is usually preceded by a darkening of colour. Reprocessing of light and natural coloured polymers is thus more critical, although they can sometimes be re-coloured or blended for use in darker products. However, polymers can be compounded

Table 1. Typical granulator size/power relationship. (Courtesy of USI Engineering Ltd.)

Throat opening (inches)	5 × 4	6 × 12	10 × 12	9 × 20	12 × 20	18 × 28	16 × 30	16 × 43	20 × 50
Length of knives (inches)	5	12	12	20	20	28	30	43	50
Light duty (horsepower)	3		10		20	50–100			50–150
Heavy duty (horsepower)	5½	20		50–100			75–200	100–300	

Note. 1 inch = 25.4 mm; 1 horsepower = 0.7457 kW

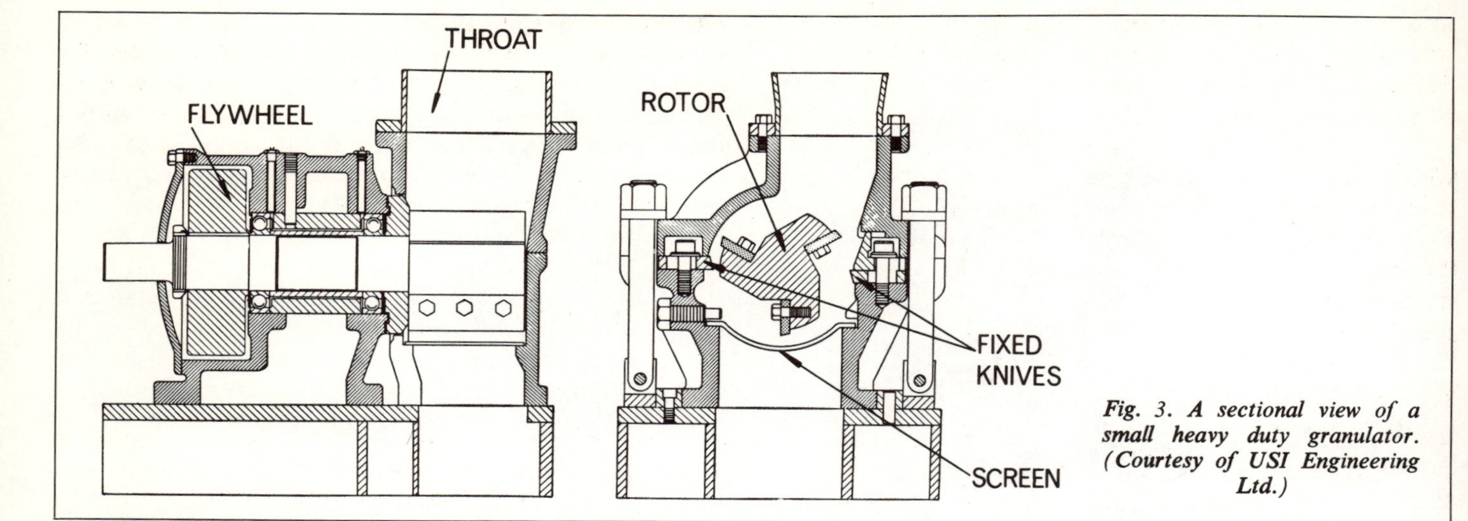

Fig. 3. A sectional view of a small heavy duty granulator. (Courtesy of USI Engineering Ltd.)

(usually at an increased cost) to withstand an increased number of cycles. Choice of material and design, therefore, plays an important role in its subsequent ability to be reprocessed. Along with material and design the intended reprocessing operation must be considered together with the product's characteristics.

It is inevitable that some processing differences are found between the virgin polymer and its reprocessed counterpart. More of the heat history has been utilised. These effects, however, are not always agreed between experts, who therefore adopt various techniques to re-use the material.

Reprocessed material is often blended with virgin polymers and re-used in a wide range of proportions; it is also used by itself. The controlling factor in this decision is often based upon the quantity available. Experience will show how wide a range of proportions can be satisfactorily tolerated if necessity demands.

Table 2. A guide to the reprocessability of plastics.

Material	Effect
Polystyrene Polyethylene	Very small changes
Plasticised PVC Cellulose Acetate PMMA Impact Polystyrene Polypropylene	Slight degradation, physical property loss or moisture absorption
ABS Nylon Polycarbonate	As above, but changes more marked
UPVC Polyacetal	Significant degradation and property loss
Coated fabrics and paper Co-extrudates of e.g. Polyvinylidene Chloride/PE/Cellulose Disimilar materials Lacquered materials	Polymer separation virtually impossible
Thermosets	Useful only as fillers

Accumulating effects of heat are first visible in the processing machinery, resulting in different output rates, different melt characteristics and a consequent need for adjustments to production specifications and work standards. For these reasons alone there is much to recommend the segregation of reprocessed material. It is the only way of maintaining control over the history of the material. Blends have the disadvantage that it is impossible to know how many times some of the mixture has been re-cycled and hence they are likely give rise to spasmodic failures which are impossible to logically eradicate. There is also cause to consider the use of re-ground material as a different grade material and use it entirely on other machinery under conditions which have been found to be satisfactory by experiment. Some companies use it to make products which have a less demanding specification, instead of purchasing separate grades of polymer. It can be seen that by careful examination and grading of a product range, a situation can be induced where virgin, 1st regrind, 2nd regrind, *etc*, are allocated to products permanently.

Whichever policy is adopted it is essential that its implications are appreciated and rigidly applied. Reprocessed materials used haphazardly can be the cause of chaotic production and lose more money than their direct disposal. However, intelligent re-use can be the difference between profit and loss.

ECONOMICS

Technical viability and a moral desire to achieve maximum material utilisation are not the sole justification for reprocessing. A polymer which costs x when purchased as virgin polymer will cost $x + y$ when used the second time as 'reprocessed', and possibly somewhat inferior, material. The subject of cost can become complex and abstract, but must be considered before a reprocessing policy is adopted.

Consider a polymer purchased for a moulding application where 70% will be saleable product and the remainder available for reprocessing. By simply disposing of the 30% the polymer price has increased by a factor of 1.43, whereas to reprocess the 30% will cost x pence and only 70% of it will be used at the second attempt. The cost will then be increased by the cost of reprocessing, but 91% of the original material will be used.

Clearly, an expression can be derived which will relate the fraction of material used and the reprocessing cost, to show the savings which can be obtained and, conversely, the true cost of polymer 'actually used' if no reprocessing is practiced. Allowance can also be made for changes in processing rates.

REPROCESSING PRACTICE

The enemies of reprocessing, apart from the deterioration of the materials inherent qualities, are dust and careless storage. It has been shown that reprocessing can be profitable, but it will only remain so if subsequent production is free from spasmodic difficulties which dust and quality variation cause. Material for reprocessing must receive equal, if not greater, care as finished products, for it is the raw ingredient for second stage polymer.

Cleanliness is often the consideration which prompts regrinding machines to be sited by the side of injection moulding machines or to be built into a blow moulding machine.

Reprocessing after extrusion operations presents a more difficult handling problem. Material for reprocessing arises at periods when the operator is intent on establishing (or re-establishing) the machine's stable working conditions, and he is therefore relatively unconcerned at that moment about the waste, apart from clearing it from machine. The prime need is therefore a tidy work area, a systematic working method, and suitable containers for segregating the waste. Purgings which fall on the floor or on the machine frame are almost certainly useless. Waste polymer should be removed from the production area and transferred to the reprocessing area as quickly as is convenient.

The reprocessing department

Reprocessing is a dusty operation. It is often a spasmodic operation. It is always noisy. These factors, and the need for cleanliness, make the planning of a reprocessing operation a matter of importance if consistent products and good labour relations are to be maintained. Three work areas are required:

(a) Reception area
(b) Recovery area
(c) Recovered materials storage

Reception area. The planning of a reception area for reprocessing is usually the least considered aspect, although it is probably the sector which affects subsequent production to the greatest extent. Materials arriving can be in a wide variety of shapes and will arrive when it suits the production departments. The first requirement is, therefore, a clear space. The other equally important need is for a plan of storage, first of similar polymers and, second, of similar products, in a manner such that they will not accumulate dust and dirt.

Recovery area. Because of the presence of noise and dust the machinery should be located in an enclosed area, acoustically insulated and with good ventilation. Anti-vibration mountings should be used for all equipment and protection provided for nose, ears and throat of all personnel who have to be in the environment. It is also advisable for some form of metal detection and separation equipment to be included in the recovery cycle.

Storage of recovered material. Once recovered, the material should be stored, in the same manner as virgin polymer, within the factory materials store. However, during the processing and immediately afterwards it is necessary to have a foolproof system of identification so that polymers remain segregated and uncontaminated.

Chapter 21

Disposal of Plastics

Professor G Scott *BA BSc*
The University of Aston in Birmingham

Plastics are being increasingly used in applications where a limited life is required. These applications range from non-returnable packaging containers, of which plastics sacks, heavy gauge bags and drums utilise a substantial proportion of the tonnage of polyethylene now manufactured, to products of longer life that will be used over a period of time and will, therefore, require to have a period of useful life before being finally disposed of. In this category may be considered jerry cans, mulching films for agricultural use and ropes and twines for packaging purposes.

This chapter will not be seriously concerned with engineering components and plastics used in the automotive or building industries since it will be some considerable time before plastics play an important enough part in this type of application to present a serious problem in disposal. However, examination of motor car bodies which have been exposed to the environment for many years shows that, even when the metallic parts of the chassis are essentially broken down by rusting, the rubber and plastics components remain relatively unchanged. This may well, therefore, produce a long-term problem for the environment since, unlike the all-steel chassis which can be compressed and the scrap metal reprocessed, plastics consist of a wide variety of chemical entities which cannot be usefully reprocessed *in toto*. Separation of plastics and rubbers into their individual chemical species on the basis of present technology presents an insurmountable problem from the economic point of view and, at present, incineration (see later) seems to be the only viable alternative in spite of its associated atmospheric pollution hazard.

At the present time, however, by far the greatest potential hazard to the environment lies with one-way packaging containers or wrappings. These are largely superseding traditional materials such as cardboard, glass and metals, and offer a number of inherent advantages over these materials.

Economic advantages of plastics

As the cost of metals and other traditional materials increases, so the cost of the cheaper thermoplastics become increasingly competitive. Even more important is the relative simplicity of the injection moulding, film blowing or extrusion operation which, to a large extent, determines the cost of the fabricated article. As further improvements in the productivity of plastics fabrication equipment continue to be made it seems likely that, assuming the problem of disposal can be overcome, plastics will become increasingly dominant in the packaging industry in the future.

Technical advantages of plastics

Plastics packages are considerably lighter than their metal or glass counterparts. In competition with the glass bottle, plastics therefore offer the dual advantage of lightness coupled with ability to withstand shock. Glass bottles break with the associated hazard in public places. For this and economic reasons they have traditionally been recycled. The one-way plastics container therefore offers an obvious advantage over glass and to a lesser extent over metals.

The outstanding advantage the thermoplastics have over cellulose-based packaging materials is their very good barrier properties. The advantage of polyethylene in food-wrapping is without parallel over traditional wrapping materials, such as paper or metal foil, due to its resistance to moisture and mechanical stress.

Plastics properties create problems in disposal

It is the very advantages of plastics in the packaging field which are the cause of their potential environmental hazard. Cellulose, because it is hydrophilic and is swollen by water, is subject to rapid degradation by water-borne bacteria. Paper and cardboard therefore disappear rapidly when exposed to the environment. Metal cans and drums, similarly, are rapidly reduced to rust when exposed to the environment, although ironically there is evidence here that surface coatings can extend the lifetime of a metal can beyond that which is acceptable.

Plastics are not readily wetted by water and do not absorb moisture to any appreciable extent. Bacteria cannot therefore attack them and the more stable ones, which include polyethylene and PVC, persist in the environment for many years almost unchanged.

Ironically, many polymers are not inherently stable, particularly under the high temperature and high shear conditions experienced in the barrel of an injection moulding machine. The polymer fabricator is not always aware of this fact since, particularly in the case of the polyolefin thermoplastics, the polymer manufacturer is normally concerned to ensure that his product will stand up to the most severe conditions that the fabricator is likely to impose upon it. To do this the manufacturer has normally introduced small amounts of antioxidants and stabilisers into his polymer which will ensure that the melt viscosity of the polymer remains sensibly constant during the processing operation. In the case of PVC, the position has traditionally been somewhat different. Here the problem is rather an aesthetic one in that the high temperatures employed during the injection moulding of rigid PVC bottles causes an intense discolouration of the plastics with associated elimination of the highly corrosive hydrogen chloride gas. PVC fabricators normally prefer to compound to their own formulation to achieve minimum discolouration and maximum clarity of the final product.

In all cases, however, the additives which are incorporated to produce polymer stability during the processing operation are also responsible for the virtual indestructibility of the polymer product under ambient conditions. It can be shown that, if the degradation of polymers obey the normal laws of chemical reactions (and all the evidence suggests that they do), then to achieve a satisfactory processing stability at 200°C, the resulting fabricated article would have a life of 10–20 years at ambient temperatures, assuming that no other degradative factors were also operative.

The chemical reactions leading to high temperature degradation and breakdown of the polymer chain involve the incorporation of oxygen in a free radical chain reaction with subsequent breakdown of the highly unstable hydroperoxide with polymer chain scission.

$$\mathrm{ROOH} \longrightarrow \mathrm{RO\cdot} + \mathrm{\cdot OH}$$

$$\underset{(\mathrm{R})}{-\mathrm{CH_2}-\mathrm{CH_2}-\mathrm{CH_2}-} \xrightarrow{\mathrm{RO\cdot}} \underset{(\mathrm{R\cdot})}{-\mathrm{CH_2}-\dot{\mathrm{C}}\mathrm{H}-\mathrm{CH_2}-} \quad \xrightarrow{\mathrm{O_2}} \quad (1)$$

$$\mathrm{R\cdot} + -\mathrm{CH_2}-\overset{\mathrm{OOH}}{\overset{|}{\mathrm{CH}}}-\mathrm{CH_2}- \xleftarrow{\mathrm{RH}} -\mathrm{CH_2}-\overset{\mathrm{OO\cdot}}{\overset{|}{\mathrm{CH}}}-\mathrm{CH_2}-\ (\mathrm{ROO\cdot})$$

$$-\mathrm{CH_2}-\mathrm{CHO} + \mathrm{\cdot OH} + \mathrm{\cdot CH_2}-$$

The chemical sequence is shown above for polyethylene ($-CH_2-CH_2-CH_2-$; RH) and is common to all the hydrocarbon polymers, including polypropylene and polystyrene. The unstable hydroperoxide (ROOH) acts as a source of initiation of the chemical reaction and a direct consequence of this is that degradation begins slowly and auto-accelerates as the hydroperoxide concentration builds up in the polymer. There is initially, therefore, a slow rate of change in the melt flow index of the polymer which is a sensitive indication of the reduction in molecular weight, but this rapidly begins to change as the hydroperoxide is formed and breaks down (see Fig. 1). The higher the processing temperature, the faster is the rate of hydroperoxide formation and hence the shorter is the safe processing time of the polymer. Other factors, like the degree of shear of the polymer in the extruder and the amount of oxygen which has been occluded in the polymer, are also important and in practice it is almost impossible to exclude oxygen completely during compounding and fabrication.

The effect of antioxidants and stabilisers is to delay the onset of the rapid change in melt flow index which occurs as a result of hydroperoxide build up. A typical effect on melt flow index is also shown in Fig. 1, and normally the polymer manufacturer uses enough antioxidant to take care of any eventuality which may arise during fabrication (*eg* machine breakdown which results in the polymer being held at high temperature for a considerable period of time). Two main types of antioxidant are used in hydrocarbon polymers. These act by different mechanisms and are often used in combination to give synergistic effects, *ie* a reinforcement of the effect of one antioxidant by another so that the effect achieved is greater than what might have been expected on the basis of the sum of the effects of the individual antioxidants[1a].

The first acts by removing the active free radicals which are involved in the cyclical chain reaction described in (1). Normally the alkylperoxy radical (ROO·) is the reaction species present in highest concentration and this reacts readily with phenolic and arylamine antioxidants (AH) to give a new stable free radical species which cannot continue the chemical chain reaction (2a). The absorption of oxygen and the rapid formation of hydroperoxide is therefore slowed down:

$$\mathrm{ROO\cdot + AH \xrightarrow{(a)} ROOH + A\cdot}\ \text{Inert radical} \qquad (2)$$
$$\mathrm{ROOH} \xrightarrow{(b)} \text{Non-radical products.}$$

The second mechanism involves the removal of the unstable hydroperoxide which is the main source of new free radicals in the system, a reaction which does not involve radical formation (2b). Many sulphur and phosphorus compounds act in this way and are used in synergistic combination with phenols and aryl amines[1–4].

Effect of processing and fabrication on environmental stability of plastics

The environmental stabilities of thermoplastics depend on their ability to resist the same kind of oxidative process involving molecular oxygen of the atmosphere as that described in (1). As has already been pointed out, the rate of this reaction at ambient temperatures is very much slower than it is at fabrication temperatures and without the intervention of other chemical reactions the lifetime of most plastics products would be very long indeed. However, hydroperoxides formed during the processing stage of the products life are profoundly unstable to ultra-violet (UV) light and even to the shorter wavelengths of the sun's spectrum which is at the lower frequency end of the UV spectrum. Also some of the breakdown products which are formed by breakdown of hydroperoxides (notably carbonyl compounds) are also readily photoactivated to highly reactive radical species (3).

$$\mathrm{ROOH \xrightarrow{h\nu} RO\cdot + \cdot OH}$$
$$\mathrm{RO\cdot} \rightarrow \mathrm{-C(=O)-} \xrightarrow{h\nu} \mathrm{-\dot{C}(-\dot{O})-} \qquad (3)$$

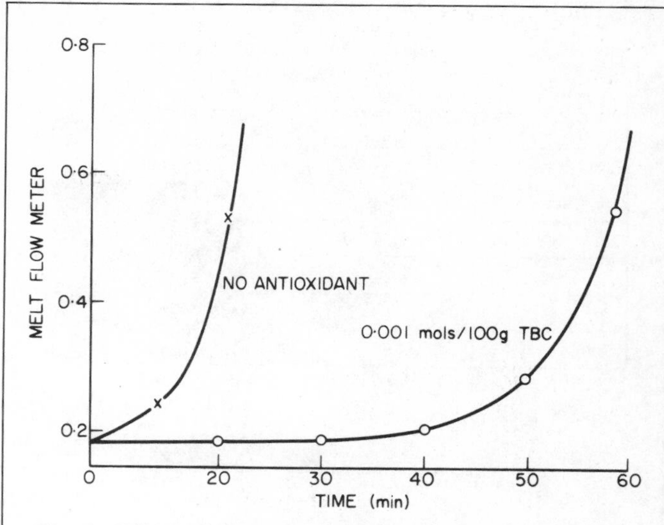

Fig. 1. Effect of an antioxidant (TBC) on the melt flow index of unstabilised polypropylene at 165°C.

This means, therefore, that UV light could be an effective photo-initiating process for degradation if there were hydroperoxides present in the polymer at the end of the processing operation. Unfortunately this is the very situation the polymer manufacturer has to avoid if his process is not to operate on a 'knife-edge'. There is no doubt, however, that the more severe the processing operation in terms of time-temperature and shear, the more light sensitive will the fabricated product be[14]. In general terms this is due to the depletion of stabilisers and the build-up of hydroperoxides and other photo-sensitising groups in the polymer.

All antioxidants and stabilisers have some effect on UV stability and some are much more effective than others. This is particularly true of the peroxide decomposers (2b) some of which are very powerful UV stabilisers for polymers and are widely used in combination with UV screening agents in applications where very good outdoor stability is required[5,6].

The commercially available packaging polymers are, for the above reasons, relatively stable to the outdoor environment even without the deliberate introduction of UV stabilising agents. Fig. 2 compares the stability of the common packaging materials in an accelerated exposure test. It can be seen that by far the most stable is PVC, which is one of the major polymers used for non-returnable bottles. The least stable is polypropylene which is at present more expensive than PVC and does not give the same clarity in the injection moulded bottle. Low density polyethylene, which is the most common film wrapping material is next in stability to PVC, and high density polyethylene falls between polypropylene and low density polyethylene. This latter fact is somewhat surprising since low density polyethylene, due to its branched chain structure, tends to oxidise more rapidly than the high density polymer [1c], but the reason for this appears to be in the morphology of the polymer rather than in its chemical structure[7].

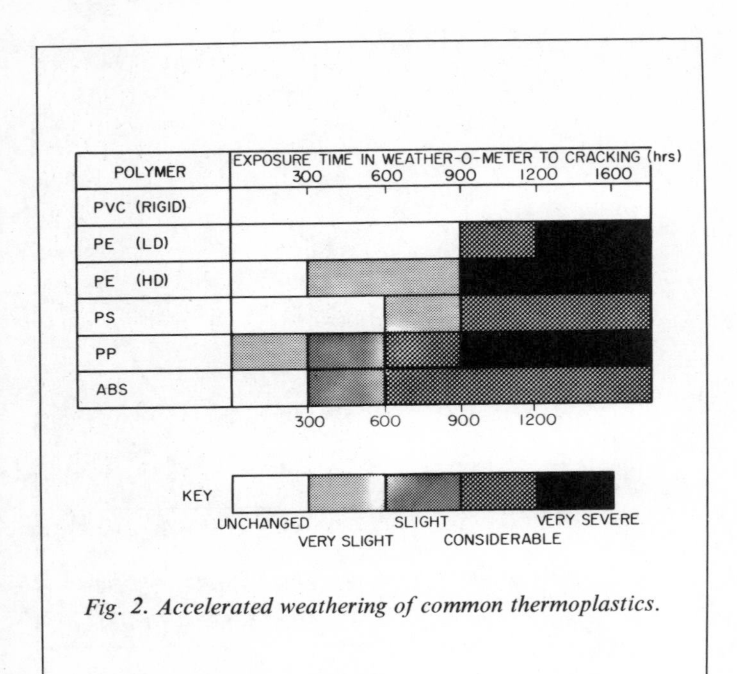

Fig. 2. Accelerated weathering of common thermoplastics.

Reprocessing
It has been shown that individual plastics can be reprocessed to give fabricated products of satisfactory quality. Here again the main cost is in separating one type of plastics from another. There is a very real possibility that further research will lead to economically viable automatic separation processes and this long-term solution is being investigated[14].

Tipping and land-fill
The most important means of plastics disposal at present is by transporting in admixture with other refuse and using as land-fill material[15,16]. The two major disadvantages of plastics in open tipping are firstly that plastics come to the surface of the tip as the organic matter compacts around them and secondly that plastics that are effectively buried cause 'sponginess' in the ground which makes it unsuitable for building purposes. It has been claimed that there are enough open holes in the ground to accommodate city refuse for many years to come, but almost certainly transport costs would become high in some urban areas as the volume of plastics waste increases.

ALTERNATIVE DISPOSAL TECHNIQUES

In view of the present persistence of packaging plastics in the environment a number of complementary disposal techniques have to be considered. For plastics articles which can be readily collected for central disposal the following techniques, all of which are being operated on a development or full scale basis, should be considered.

Incineration
Most plastics burn in a plentiful air supply to produce heat which can in turn be used to provide power. An exception is PVC which does not burn readily, although it can be incinerated in combination with other waste[8]. Some plastics, notably PVC, give off obnoxious gases which not only constitute a toxicity hazard[8] but also reduce the life of the incineration plant[9,10]. There is evidence that both effluent and corrosion problems can be overcome but at a considerable cost. No useful products besides heat are produced by incineration and, as the amount of plastics in waste increases and public pressure for improvement in atmospheric pollution also increases, it must be considered to be of doubtful future viability on economic grounds.

Pyrolysis
Plastics can be decomposed at high temperatures in the absence of oxygen to give potentially useful chemical compounds. In some cases (*eg* polystyrene) the parent monomer can be recovered[11]. But unless some prior separation process is performed in order to carry out individual pyrolysis of each type of plastics, the cost of separating the organic chemicals produced would probably be prohibitive. There is also a similar corrosion problem with PVC to that occurring during incineration. The most favourable application of pyrolysis is probably in tyres, from which useful chemicals have been recovered[12,13].

BUILT-IN DISPOSABILITY

The techniques discussed so far apply only where collection of plastics waste is feasible. It does not apply to litter which is becoming increasingly obnoxious particularly in the countryside, both from an aesthetic point of view and because of its ecological effects. The cost of collecting this type of plastics waste is quite out of the question and some alternative long-term solution must be found. As has already been pointed out, paper sacks, cardboard packages and metal drums all disappear on exposure to the environment and an objective of current research is to produce plastics packaging with a similar limited lifetime. Two main methods are being investigated.

Biodegradable plastics
The first is to make plastics as biodegradable as cellulosic materials. There are, in principle, two alternative ways of doing this. The first is to synthesise new polymers which will be more susceptible to attack by bacteria than the present materials[17]. This, of necessity, means that bacteria must be able to enter into the structure of the plastics to destroy it, which in turn means that new plastics must be synthesised which are hydrophilic. In so doing, however, the useful moisture resistant properties of the plastics would be destroyed. In addition, the cost of any such specialised plastics would probably be prohibitive for general use. The second approach is to develop new bacteria which are capable of attacking massive plastics more rapidly. It is already known from study of bio-degradation of detergents that bacteria can accommodate to hydrocarbon feedstocks. However, there is no evidence that there is such a development in the offing in the plastics field and even if it did prove possible the ability to control such new rapacious bacteria in the environment is open to question.

The probability is that they would also attack plastics products with intended long life as well as waste materials.

Photodegradable plastics

The second approach, which has received some attention during the past two years, is the development of UV sensitive plastics which will be relatively stable indoors but will degrade rapidly under the influence of sunlight out-of-doors. Again, two different approaches have been taken.

The first is to incorporate photosensitive groupings in the polymer backbone. The introduction of carbonyl, the product normally produced by hydroperoxide breakdown has been investigated by Guillet and his co-workers[18] and some success has been reported. The same economic arguments apply to a tailor-made photodegradable plastics as applies to a biodegradable plastics; that is, it will cost very much more than the common packaging materials.

The second approach is to incorporate an additive during processing that does not interfere with the melt stability at this stage but which accelerates photo-oxidation during environmental exposure[20]. Several chemical compounds have been evaluated which accelerate the degradation of the common thermoplastic packaging polymers by factors varying between 3 and 7 times. The polyolefins and polystyrene (including high impact polystyrene) are particularly susceptible to this process. PVC is less readily degraded, due to the very considerable loading of stabilisers introduced to restrain colour formation during processing but which are also powerful UV stabilisers.

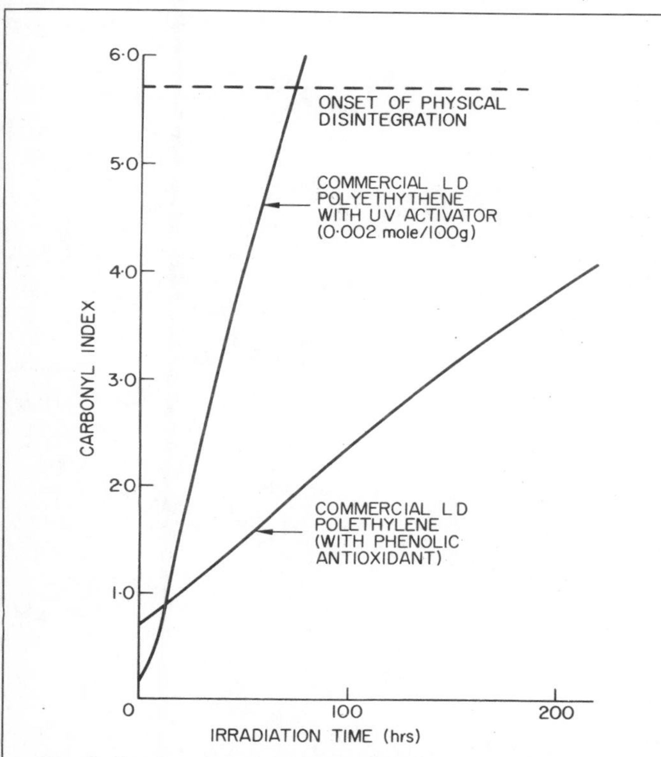

Fig. 3. Comparison of UV degradability of LD polyethylene with and without UV activator additive.

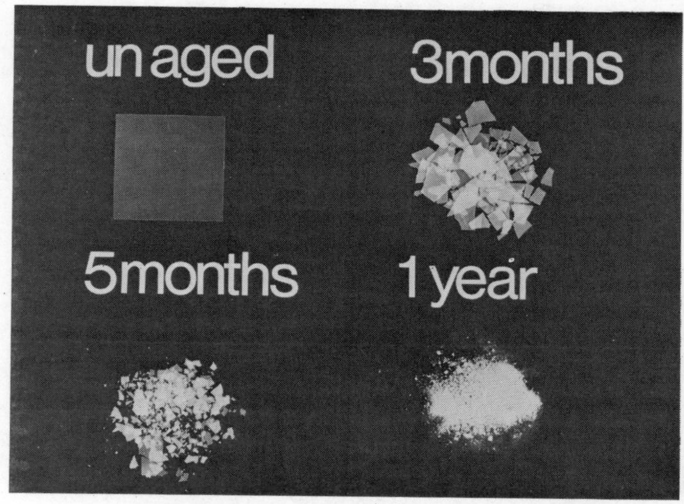

Fig. 4. Embrittlement and fragmentation of polymer under the influence of wind and weather as a result of carbonyl formation.

The effect of such an additive on the UV degradation of polyethylene is shown in Fig. 3 which shows carbonyl formation in the polymer with time in the presence and absence of additive. It can be shown that carbonyl formation correlates directly with the physical properties of the polymer which rapidly loses strength and embrittles. Embrittlement and fragmentation are particularly rapid in the case of the more crystalline polymers which crumble even under the influence of wind and weather (see Fig. 4). The polymer particles so formed are chemically modified by oxidation to be now subject to attack by normal bacterial action and it is anticipated that these will quickly become part of the soil.

The application of photo-bio-degradation principles should add little to the cost of fabricated products since there is no interference with the processing operation. Although the initial application will almost certainly be to the problem of litter, where the environment for decay is most favourable, there is also the possibility of applying it to rubbish dumps and even to sewage by appropriate separation and treatment[19]. It seems likely that in the future varying degrees of degradability will be built into a range of plastics products and, although certain aspects require further research (*eg* a built-in early warning system for degradation), these problems are soluble in principle and there exists a real prospect that the growth of plastics in packaging will not be limited by the problem of plastics litter.

REFERENCES

1. G Scott, Atmospheric Oxidation and Antioxidants, *Elsevier,* London, 1965. (a) p.203 et seq. (b) p.295; 299; (c) p.273.
2. G Scott, Mechanisms of Reactions of Sulfur Compounds, **4**, 99, (1969).
3. G Scott and P A Shearn, *J. Appl. Poly. Sci.,* **13**, 1329, (1969).
4. G Scott, *Europ. Poly. J.,* Supplement, 189 (1969).

5. J E Bonkowski, *Textile Res. J.*, 243 (1969).
6. S L Fritten, R N Haward and G R Williamson, *Brit. Poly. J.*, **2**, 217, (1970).
7. E Turi, L G Roldan, F Rahl and H J Oswald, *A.C.S. Polymer Reprints*, **5**, (No. 2), 558 (1964).
8. Anon. *Environmental Science and Technology,* **2**, 89, (1968).
9. Anon, *Chem. and Eng. News,* Jan. 8, 13 (1968).
10. Anon, *Chem. Eng.*, June 15 (1970), p.88.
11. A S Wood, *Modern Plastics,* March (1970), p.50.
12. Anon, *Chem. Eng.*, Dec. 2, 53 (1968).
13. Anon. *Environmental Science and Technology*, **3** (2), 119, (1969).
14. G Scott, Unpublished work.
15. T J Sorg and H L Hickman, Sanitary Landfill Facts, *U.S. Dept. of Health, Education and Welfare* (1968).
16. R L Steiner and R Keintz, Sanitary Landfill, A Bibliography. U.S. Government Printing Office, Washington D.C., *U.S. Public Health Service Publication No. 1819* (1968).
17. R Lewin, *New Scientist and Science Journal,* Feb. 25, 440 (1971).
18. F J Golemba and J E Guillet, *S.P.E. Journal,* 26 (1970). See also, *Plastics and Rubber Weekly,* 327, 10, (May 15, 1970).
19. G Scott, *Plastics, Rubbers and Textiles,* Sept. 26, 361 (1970).
20. G Scott, *New Scientist*, 47, 293 (1970).
See also, R A Dobson, *Maintenance Engineering,* 50, April (1971).

Chapter 22

Designing for Plastics

V E Yarsley *MSc DSc(Tech) CEng FRIC MIChemE FPI*
Yarsley Research Laboratories Ltd

It is fairly true to say that apart from public prejudice, which was more intense in Britain perhaps than elsewhere, no single factor has inhibited the progress and use of plastics as has design, or more accurately, the lack of good design. Apart from wartime uses, which were generally well specified and could afford to be in some degree speculative, the application of plastics was for many years limited to the production of articles well known in traditional metals and woods: in short, plastics were merely substitutes in every sense of the word. Quite apart from the fact that in those pioneer days little was known of the new synthetics, either as to their broad physical properties or consistency of quality, little if any attempt was made to adjust either the materials used, or the processes applied, to the properties and functionality of the end products. When the then often risky decision had been taken to use plastics in place of traditional materials, it was customary to follow the general shape and form of the metal prototype, with complete disregard of the fact that not only did plastics differ fundamentally from metals for example, but the methods evolved over long years for the manipulation of metals could not of necessity be applied to plastics. In many cases the approach was *ad hoc* in the extreme; so long as the plastics article broadly resembled the prototype that was regarded as sufficient, and little attempt was made to take advantage of the special properties plastics could offer other than their colour versatility. Not surprisingly the results were frequently disastrous, and with such failures the good name of plastics (if indeed it existed at all at the time) was further damaged. Even more unfortunately, little attempt was made to profit by such failures, since much of the data so collected was of a qualitative nature, indeed it was the early and almost complete lack of performance data which deterred many of the early potential users of plastics, and it is small wonder that they refused to risk the use of plastics in preference to wood or metals with which they had years of experience, and for which performance data extending over many years was usually available. It was only when manufacturers appreciated that they frequently had to forget their past experience with traditional materials, that they accepted that the use of plastics demanded that they should design both in and for the new materials. To design in plastics meant not only that they would select the correct material with properties to meet the needs of their proposed end product, but that the method of producing such an end product should capitalise the process advantages which stemmed from the unique properties of plastics, namely their long-chain structure, in which they differed so radically from metals. As this became realised and applied, so the quality of the plastics products gradually improved, and almost *pro rata* came a growth of public confidence in the new materials.

Some measure of the progress which has been made, and indeed of the magnitude of the problems which faced the replacement of traditional materials by plastics is given in the BS4618: Introduction: 1970, "Recommendations for the Presentation of Plastics Design Data". Here it is pointed out that plastics are more sensitive to environmental changes, and that their properties vary widely with temperature, humidity, service conditions, and with time. It is stressed equally that the performance of plastics products is usually markedly affected by the method and conditions of their manufacture. These 'Recommendations' deal with the presentation of data and not with design procedures, and are intended rather to guide the designer to the selection of the correct material for a particular use. The five parts (with some sub-sections still in course of preparation) comprising the specification deal respectively with mechanical properties; structural properties; thermal properties; environmental and chemical effects; other properties; and is almost unique among formal publications of this type, in that it states not only what is required, but also gives details of the tests to be carried out and in many instances explains their significance. This British Standards Institution publication is perhaps symptomatic of the change in approach to design in plastics, which has taken place during the past few years. From the position a decade ago when design signified in the main attention to the outward form and general functionality of an article, and was (with certain notable exceptions) little more than an art which was at best at the 'have-a-go' level, today the danger is rather the other way round. The science of designing has become so sophisticated, with its own terminology which would be almost unknown even to the specialist of a few years ago, that there is now some danger of lack of adequate communication between the practical man concerned with production, and the theoretical polymer physicist who talks in terms of viscoelastic functions, isochronous stress/strain curves, isometric stress/time curves, and the Laplace Transformation. Such consideration is beyond the terms of the present survey, but those interested to pursue the subject more deeply would do well to read the paper by P C Powell and S Turner[1] presented at the Plastics Institute Conference at Cranfield early in 1971. Further references for more detailed study are collected in the Bibliography.

Despite all this sophistication and the amazing amount of performance data on plastics now available to the designer, the fundamentals remain the same—what materials can be used to meet a specific need, and how can these be converted into the desired end product?

THE SELECTION OF THE CORRECT PLASTICS MATERIAL

Today, with the proliferation of plastics types and grades, selection of the best plastics material to use is frequently not a simple problem. In the main this has to take into account not only the functionality of the end product, but the possible life span of the article to be produced and its material and production cost. Even more fundamental is of course the basic question, why plastics? However, it must be assumed that for very good reasons the designer has satisfied himself that on all these counts plastics can offer advantages over traditional materials.

The plastics materials available to the designer fall broadly into four groups, the thermoplastics, the thermosets, the composites or reinforced materials, and the cellular or expanded materials. Which he will choose will largely depend on his cost allowance, which, together with processing costs and the size of his expected market, sum up the economics of his problem. Of almost equal importance is the nature of his end product, including the functional nature and the expected duration of its service. Summed up briefly the main problem facing the designer is to meet functional requirements at an economic cost. All too frequently in the past, as was demonstrated in early plastics toys and household articles, performance was sacrificed to material costs, and this was aggravated in many cases by poor design. Good design can help, but it cannot make up for poor or cheap material.

In deciding which plastics material to use the designer must assess the fundamental requirements of his end product, and optimise these from the properties which plastics can offer, such as mechanical strength in relation to weight, electrical properties, chemical resistance, dimensional stability, heat and moisture resistance, and average these out against what he expects will be the requirements of his end product. Quite obviously no single material is likely to combine all the properties he needs, and it may well be that to obtain what he considers a reasonable average on functionality he may be forced to sacrifice some cost advantage, and the lower volume costs of some plastics is often a critical factor in deciding for plastics versus metal. For example, relatively costly plastics such as acetals and nylon have successfully competed with brass where good corrosion resistance is required. Although of considerable importance, the weight factor is by no means the only one in favour of plastics, it can also materially influence the decision as to which plastics material to use. For example, where high rigidity is required in the end product, the plastics material with the highest modulus need not of necessity be selected. Compared with nylon 66 the modulus of polypropylene is only about half, so that for equal stiffness a polypropylene section would expectedly have to be significantly greater than that of nylon. In actual fact, when corrected for density, the weights required in these materials for equal performance are practically identical.

Of the four groups of plastics materials referred to above, thermoplastics and thermosets present the designer with very much the same problems, with the significant difference that in the latter the material has not only to be moulded to the required shape, but it has to undergo the chemical transformation usually styled 'curing'. Thus where thermosets are to be used the problem of mould design is radically different, involving as it does the rapid transfer of heat to and from the plastics material, an operation which differs considerably with the material used. On the other hand, providing the mould has been designed to allow for dimensional changes during the curing stage, and that exit from the mould is not hindered by undercuts and the like, the problem of ejecting a moulded thermoset is relatively speaking easier than that of a thermoplastics unit. In both cases, however, as will be shown later, the thermoforming of plastics no matter what the type and what the method of manipulation used, depends for its economic operation and ultimate functional success very much on the skill of the designer and that of the actual moulder, and also increasingly in these days of sophisticated machine control, on the designer of the actual moulding unit.

Less onus falls perhaps on the designer in the case of the third group of plastics, the composites or laminates. These are mainly used in the form of sheets, rods and tubes, which are applied very much as are conventional materials in these physical forms. Orientation of the reinforcing materials plays a large part, particularly in the case of profiles which are frequently 'laid up' by hand to ensure that the reinforcing fibre or fabric is placed at those points which have to sustain maximum stress during ultimate use, or which are likely to suffer dimensional change in processing. The design problem in such cases is perhaps intensified by the fact that structures of this type are usually quite large, and as many of them are destined for the chemical industry, they have to withstand what is usually classified as 'use in adverse conditions'. Composites incorporating carbon fibres as reinforcement have opened up entirely new concepts in design. Here it is not so much a question of doing something better, as doing something which no other materials can do.

Though materials of the fourth group remained relatively insignificant for many years under the general but quite descriptive title of 'expanded' plastics, they have rocketed to prominence and importance in the last few years under the now more generally accepted designation of cellular materials. They include almost all thermoplastics and thermosets, although only a limited number (especially of the latter) have attained commercial importance. The potential of cellular materials was recently extended by the new ICI process in which a sandwich material can be produced to any reasonable profile, having a cellular core enclosed within a solid outer 'skin'. As may be expected, cellular plastics offer a remarkable stiffness/weight ratio, and their commercial availability has added almost a new dimension to the designer in plastics. From small and humble beginnings twenty years ago when 'FUF' (foamed urea formaldehyde) was offered as a packaging material for florists, cellular plastics have now advanced as structural materials offering entirely novel design possibilities. These have already been applied in the manufacture of furniture and in automobile fittings, and in both have created a mild but quite definite revolution in design. It is certain that as designers become more acquainted with these quite unique materials, they will be widely applied in what may be classified as heavy-duty applications, many of them in usage areas hitherto considered impossible.

DESIGN IN RELATION TO THE MANIPULATION OF PLASTICS

Under this heading of 'manipulation' is included any process whereby the solid plastics material, usually in powder or 'nib' form, is converted to the finished end product. It embraces moulding by compression, transfer, injection or extrusion, blow moulding, rotational moulding, and any process whereby a layer of comminuted plastics material is heat-sintered into a finished unit. In all of them the product designer has to work very closely with the manufacturer not only of the plastics material, but also with the mould maker and to an ever-increasing extent with the maker of the moulding machine.

In some cases, especially where large units (such as sink drain units) are required, several methods may be available, and the ultimate choice will depend on the 'number-off' required. In the main, the problem is one of moulding, either by compression but more usually by injection, and it is here that the designer of the product may well be also the mould designer, or at any rate he must work in very close collaboration with one. Whatever the circumstances the product designer must know what the machine manufacturer can offer him, and it may be well to consider this aspect of the problem before that of the mould which is directly associated with the outward and visible shape of the end product.

Production design in relation to the plant used for plastics manipulation

As has already been stated, there are a number of methods available to the designer, but in the main the question is one of moulding the plastics material by the application of heat and pressure. This is an area in which considerable progress has been and is being made, and this has stemmed from the realisation that moulding is indeed something more than applying heat and pressure, and that the accurate control of both could materially affect the physical properties of the end product. Contrary to earlier practice, when it was assumed that the essence of good moulding was to avoid undue flash and 'shorts', it is now realised that how the plastics materials flow into and in the mould, as also the rate and uniformity of filling, can have a considerable influence on the quality of end products.

It is not surprising therefore that current design trends in the manufacture of moulding machines, particularly for injection and extrusion, are mainly in the direction of precise control of temperature, pressure and material flow. The result is in many cases that the outward appearance of the machines is changing, with the control ancillaries forming separate sections, and strategically placed transducers and thermocouples enable the operator to control pressure and temperature at critical points during the actual moulding operation. In many cases the data so collected are continuously 'logged', so that in the case of difficulty, trouble-shooting is reduced to a minimum, and the operator knows such important facts as the rate at which the fluidised plastics material flows into the mould. A considerable amount of research and development work is currently in hand which will link the rheology of the plastics material with the mechanical functioning of the injection machines. Since this inevitably means a closer collaboration between the men who design the molecular structure of the polymers, and those who build the moulding machines, it is becoming even more essential for the product designer to keep in close touch with both, and to co-ordinate the results of their efforts in what is the focal point of the whole operation, the product mould.

Fig. 1. A very successful design of chair made from an injection moulding of ABS.

Mould design for performance and appearance

Some interesting general observations on the question were recently made in a documented interview by Martyn Rowlands[2], one of Britain's leading plastics designers. He stressed the fact that it is essential to design round the product, and at the same time to be fully aware of the fabrication process to be used, in order to design according to aesthetics. He saw 'the difference between the stylist and the industrial designer as an appreciation of how to use the properties of the materials in the finished products'.

The first essential the designer has to consider is of course to reproduce in his mould the general form and functionality of the required end product, but as Martyn Rowlands' stressed, with reference to the set of cutlery he designed for BRS, this need not be traditional in appearance. In doing this he must provide the essential means for the introduction of the plastics material, positioning the necessary sprues, runners and gates, so that not only is the mould most easily and rapidly filled, but that the process of flow shall so orientate the material to use to the best advantage its long-chain molecular structure. As far as possible the juxtaposition of thin and thick sections should be avoided, since the prevalence of the latter is likely to cause 'sinks' due to unequal contraction through varied rate of cooling. Having successfully filled the mould with the fluidised plastics the designer has also to take account of its easy removal, so that the positioning of ejector pins, and the nature and location of the 'split' in the mould have to be taken into account. All these factors will in turn be influenced by size and 'number-off' of the mouldings required. He may have to decide for economic reasons between the use of a larger multi-impression mould which will be slower in operation, and a quick cycling mould with very few cavities. He will also have to take into account dimensional changes both in the mould and moulded unit during the moulding process, and also the possiblity of ultimate 'creep' when the product comes into use. This incidentally was a factor which for many years was lightly regarded or completely ignored, but which is currently being closely studied. Today end-product performance can indeed by predicted with a fair degree of accuracy[3].

What has been described so far in the design of the mould has been mainly concerned with its external features, *ie* the shape of the moulded unit, and the means whereby the mould is filled and the moulded unit is removed (particularly the latter) depends on the rate at which the moulding can be cooled. From the viewpoint of cost, quite naturally the more rapid the solidification and ejection, the greater the throughput per machine/man hour for a given unit. Though this is true it has to be balanced against the danger that too rapid cooling may cause stress areas in the moulding, and too quick or inefficient ejection from the mould may cause deformation. Thus, heat transfer both of the plastics material used and of the mould itself is a matter of considerable importance and again it is one with which designers are currently very much concerned. As C E Waters[4] pointed out recently, one of the most critical and often rate-limiting aspects of injection moulding is the cooling cycle, it is often the key both to higher production and better quality products. The provision of the heating/cooling channels within the mould varies widely with the type of moulding, compression for thermosets and injection for thermoplastics. There are of course the special cases, such as the thermosetting alkyd 'Injak' recently introduced by BIP (Chemicals) for injection moulding[5]. Generally speaking for the injection moulding of the polyolefins, polystyrene and some of their co-polymers, a fairly low temperature (below 100°F—40°C) will need the use of some degree of refrigeration. The cellulosics, PVC, acrylics, ABS, polyvinylidene fluoride and PPO/PS copolymers, require temperatures between 100–200°F (40–95°C), and cooling can be accomplished with a cooling tower, except in the case of high-speed production of fairly thin walled units. On the other hand, temperatures above 200°F (95°C) are required for nylon, polycarbonates, acetals and fluorinated polyolefins. Not only is it essential to have the correct temperature according to the type of plastics used, but it is equally important to have this uniformly throughout the mould. This may mean that in moulding units of varying section, the cooling must be concentrated at the points of thickest section. Failure to do this may mean that the unit leaves the mould unevenly cooled, with the result that stresses develop and 'sink' marks are visible: in such cases surface finish may also be affected. Whilst the above observations have been made with the main reference to injection moulding, different problems are presented in extrusion and blow moulding, and of course in the compression moulding of thermosets where, though important, they are possibly least critical. In all moulding operations the design and selection of the cooling system is of considerable importance, the aim of the designer in all cases being to minimise the cooling requirements as far as possible.

PLASTICS DESIGN AND THE FUTURE

Looking back to 1945 when the institution of the Horner's award by the British Plastics Federation first brought the spotlight of attention on the significance of good design in plastics, progress has been rapid and considerable. The special needs for design in and for plastics are now generally accepted, and the volume of literature available on the subject is so vast as to bewilder any but the most experienced designers. The future success lies in the complete collaboration of the mould designer with the manufacturer of the plastics materials and of the plant for their manipulation. At present the process is still to some degree empirical, but that it is rapidly graduating from this stage is evident from the recent announcement that Olivetti has developed a computer-based design system known as OPTAL (Olivetti Press Tool Automation Language)[6]. This system reduces any press or injection tool to a series of standard components, about 20 in common use, in various sizes. The new development fills the requirements of a data processing system, in that it is simple enough even for the untrained to use yet comprehensive enough to carry out sophisticated work. It is of course in line with the sequential control of machines now in operation, and as such carries out the 'instructions' of the designer as programmed. It is unlikely that we shall see design for plastics further impersonalised in the foreseeable future, but it is indicative of the extent to which design in this, as in many other areas, has graduated from an art to a science.

The increasingly significant role which plastics will play in influencing the design of consumer goods, was recognised some years ago by the Plastics Institute in the formation of a Design Committee, the membership of which includes professional designers and representatives both of the academic and industrial side of the industry. Its aim is to further in the broadest possible sense education in design in plastics, to inaugurate educational courses to this end, and furthermore to ensure that our educational system at all levels includes not only the technology of plastics, that is a knowledge of the plastics materials themselves, but also what was in the earlier days of the industry rather vaguely referred to as 'aesthetics'.

Having regard to the wide publicity recently given both in the lay and technical press to the contribution which plastics make to the growing hazard of pollution, it is of topical interest that the Plastics Institute is convening a working party on designing for disposability. It is certainly more logical that it might be easier to design plastics units to facilitate their after-use disposal, than to expect to attain this by chemical decomposition as has been alternatively suggested.

REFERENCES

1. 'The Transformation of Research Results into Design Practice', P C Powell and S Turner, *Plastics Institute Conference,* Cranfield, January 1971.
2. 'Design in Europe', *British Plastics.* 1971, **44**, (5), 92–93.
3. 'Prediction of end-product performance: the present position'. P C Powell, *Plastics & Polymers,* 1971, **39**, (139), 43–51.
4. 'Good Planning needed for effective Mould Cooling', C E Waters. *Plastics Technology,* 1970, **16**, (12), 43.
5. 'Injak', BIP (Chemicals) *Financial Times,* 4.6.71, page 10.
6. 'Tool Design by Computer', *Financial Times,* 4.6.71, page 10.

BIBLIOGRAPHY

In addition to the references cited, the following items are recommended for further reading:

1. 'Plastics Engineers Data Handbook'. A B Glanvill. *The Machinery Publishing Co. Ltd.* London, 1971.
2. 'Design in Plastics'. *Council of Industrial Design,* 1970. Case history of typical design problems in Plastics.
3. 'Impact Tests and Service Performance of Thermoplastics'. P I Vincent. *Plastics Institute,* London, 1971.
4. 'Standardising Design Data for Plastics'. W F Ratclyffe. *Plastex* 1968.
5. 'Engineering Properties of Thermoplastics'. Edited by R M Ogorkiewicz. *Wiley–Interscience,* 1970.
6. 'The Economic Design of Plastics Mouldings'. S Joisten, U Knipp, H Schulze, A Trauzow and O Walter. *Kunststoffe,* 1971, **61**, 325–331.

Product Guide

THERMOSET MATERIALS

Epoxide	A
Melamine formaldehyde	B
Phenol formaldehyde	C
Polyester	D
Urea formaldehyde	E
Silicones	F

Anchor Chemical Developments Ltd	C
Aquitane Organico (UK) Ltd	A
Arcode Ltd	C
Arinor Ltd	D
Ault & Wiborg Ltd	D
Automobile Plastics Co Ltd	A D
BASF United Kingdom Ltd	B D
B & K Moulds Ltd	A D
BIP Chemicals Ltd	B C E
BP Chemicals International Ltd	B C D E
Bakelite Xylonite Ltd (BXL)	A B C D
Bayer Chemicals Ltd	A D
Berk Ltd	C
Bexford Ltd	D
Bibby Chemicals Ltd	D
Blackburn & Oliver Ltd	B C D
British Chemical Products & Colours Ltd	D
British Steel Corporation	D
British Traders & Shippers	B
Bush Beech & Segner Bayley Ltd	A B C E
CIBA-GEIGY (UK) Ltd	A B C E
Campbell (Rex) & Co Ltd	D
Carborundum Co Ltd (The)	A
Catalin Ltd	C
Chemical & Insulating Co Ltd (The)	A C
Chemical Trading Co Ltd	D
Chemoplast (B & I) Ltd	B
Clarke (H) & Co (Manchester) Ltd	D
Cornbrook Resin Co Ltd	C
Cornelius Chemical Co Ltd	F
Cray Valley Products Ltd	C D E
Cromwell (E M) & Co Ltd	D
Devcon Ltd	A
Dow Chemical Co (UK) Ltd	A
Dryburgh (D) & Co Ltd	C
Du Pont Co (United Kingdom) Ltd	D
Durham Raw Materials Ltd	D
Du Vergier (E) & Co Ltd	D
Featly Products Ltd	B C D E
Ferguson (James) & Sons Ltd	C E
Ferro (Great Britain) Ltd	D
Fibre Form Ltd	C
Fordath Ltd	C
Freeman Chemicals Ltd	D
Gansolite Ltd	D
Greef (R W) & Co Ltd	A B E F
Guest Industrials Ltd	B
Hadrian Plastics Ltd	A D
Halmatic Ltd	A C D
Hardy (M W) (Mercantile) Ltd	C D
Hercules Powder Co Ltd	B C
Hermetite Products Ltd	A D
Hoechst UK Ltd	A C D
Honeywell-Atlas Ltd	D
Huls (UK) Ltd	D
Hysol Sterling Ltd	A
Imperial Chemical Industries Ltd	A B C D E F
Insulating Components & Materials Ltd	A C D
Island Plastics Ltd	A D
Jacobson van den Berg & Co (UK) Ltd	F
KW Chemicals Ltd	D
Kingsley & Keith (Chemicals) Ltd	A C
Laporte-Synres Ltd	B C D E
Mica & Micanite Supplies Ltd	A C D F
Micanite & Insulators Ltd	A C D F
Middlesex Oil & Chemical Works Ltd	D F
Midland Silicones Ltd	F
Minnesota Mining & Manufacturing Co Ltd	A D
Mining & Chemical Products Ltd	E
Mitchell (W A) & Smith Ltd	C D
Mosses & Mitchell Ltd	A C
Moulding Powders Ltd	C
Mycalex & Rosite Ltd	D
Novadel Ltd	D
Peter Plastics Ltd	A
Pinchin Johnson & Associates Ltd	D
Plastanol Ltd	D
Plastic Coatings Ltd	A
Plastotype Ltd	C
Plessey & Co Ltd (Components Group) (The)	C
Re-fab Ltd	D
Resinous Chemicals	B C D E
Schenectady-Midland Ltd	A C
Scott Bader Co Ltd	D
Shell Chemicals (UK) Ltd (Industrial Chemicals)	A
Smith (Wilfrid) Ltd	C
Spicer Cowan Ltd	D
Sterling Moulding Materials Ltd	C
Sterling Varnish Co Ltd (The)	AD
Surface Coating Synthetics Ltd	A D
Swales (William A) & Co Ltd	D
Sylpon Manufacturing Co Ltd	D
Synthetic Resins Ltd	A B C D E
TAC Construction Materials Ltd	A C D
Tar Residuals Ltd	B E
Tufnol Ltd	A C
Turner Brothers Asbestos Co Ltd	C
Union Carbide UK Ltd (Chemicals Division)	A F
VMG Plastics Ltd	BC
Varnish Industries Ltd	C
Weil (Joseph) & Son Ltd	B C E
Witco Chemical Co Ltd	D
Yorkshire Dyeware & Chemical Co Ltd (The)	C

THERMOPLASTIC MATERIALS

ABS	A
Acrylics	B
Fluorocarbons (PCTFE)	C
Fluorocarbons (PTFE)	D
Polyacetals	E
Polyamides (Nylon)	F
Polycarbonates	G
Polypropylenes	H
Polystyrenes	I
Polyethylenes	J
PVC	K

ADI Plastics Ltd	B
Albis Plastic Co (Great Britain) Ltd	A I J K
Alpine Insulations Ltd	I
Ansonite Manufacturing Co	F
Archway Plastics Ltd	A B H I
Arcode Ltd	A I J K
Associated Resin Compounds	I J
BASF United Kingdom Ltd	A F I J
BIP Chemicals Ltd	F
BP Chemicals International Ltd	B I J K
Bakelite Xylonite Ltd (BXL) (Polyethylene Division)	J
Bakelite Xylonite Ltd (BXL) (PVC Division)	H I J K
Bayer Chemicals Ltd	A F G
Begg & Co (Thermoplastics) Ltd	I J
Beldam Asbestos Ltd	D
Belgrave Plastic Developments Ltd	I J K
Bibby Chemicals Ltd	F
Bricell Plastics Ltd	I K
British Celanese Ltd	H I J
British Chemical Products & Colours Ltd	F
British Traders & Shippers	A F J K
Bunzl & Biach (British) Ltd	F H I J
Bush Beech & Segner Bayley Ltd	A K
Chemoplast B & I Ltd	A B K
Clarke (H) & Co (Manchester) Ltd	E
Coin Insulation Ltd	K
Cole (R H) Ltd	A I J K B
Cole Plastics Ltd	A I J
Cole Polymers Ltd	B
Collins & Harmer Plastics Ltd	A I
Commercial Plastics Ltd	J K
Compounded Polymers Ltd	D
Connor (Herbert) Ltd	I J
Cork Growers Ltd	I
Cornelius Chemical Co Ltd	B
Cow (P B) (Plastics) Ltd	K
Crane Packing Ltd	D
Cray Valley Products Ltd	B
Croda Chemicals Ltd	H
Cromwell (E M) & Co Ltd	B
Curtis (Neville) & Co Ltd	A
Curtis Plastics Ltd	I

Cyanamid of Great Britain Ltd	B
D O & E Industries Ltd	A B F I H J K
Dohm Industrial Ltd	J
Dow Chemical Co (UK) Ltd	I J K
Dugdale (D) & Son & Co Ltd	K
Du Pont Co (United Kingdom) Ltd	B D E F J
Durham Raw Materials Ltd	B
Du Vergier (E) & Co Ltd	B G
East Anglia Plastics Ltd	H I J K
Elson & Robbins Ltd	K
Engineering Polymers	G
Ferro (Great Britain) Ltd	B E F H I J
Fluorocarbon Co Ltd	C D
Freeman Chemicals Ltd	B
Genrista Ltd	H
Grant Plastics Co Ltd	A I J
Grilon & Plastic Machinery Ltd	F
Guest Industrials Ltd	K
Hardy (M W) (Mercantile) Ltd	J
Hellyar (John) & Co Ltd	H I J
Hoechst UK Ltd	C D E F H I J K
Holliday (G J) & Co	J
Huls (UK) Ltd	F I J K
Hydralon Ltd	F
Imperial Chemical Industries Ltd (ICI)	B D E F H J K
Industrial Polymers Ltd	I
Industrial Reels Ltd	I
Insulating Components & Materials Ltd	B D E G H I J K
Intrama Ltd	I J
KW Chemicals Ltd	D J K
Kautex Ltd	J
Kaylis Chemical Ltd	A H I J
Kelcoat Engineering Plastics Ltd	F
Kingsley & Keith (Chemicals) Ltd	C D F J K
Lanplas	I J
Lennig Chemicals Ltd	B
Lincoln Plastics Ltd	J
London Varnish & Enamel Co Ltd	B
Luxus Ltd	F I J
Mackins & Wood Ltd	A I J K
Marbon UK Ltd	A F
May & Baker Ltd	D F
Middlesex Oil & Chemical Works Ltd	F
Minnesota Mining & Manufacturing Co Ltd	C K
Monopol Plastics Ltd	F I J
Monsanto Chemicals Ltd	A I J
Mosses & Mitchell Ltd	D F H J
Muehlstein-Northwestern Ltd	H I J
Mundet Cork & Plastics Ltd	I
National Adhesives & Resins Ltd	B
Norsk Hydro (UK) Ltd	K
North Hill Plastics Ltd	B
PDI Ltd	G
PRM (London) Ltd	I
Page (Charles) & Co Ltd	I
Permanoid Ltd	K
Peter Plastics Ltd	F
Phillips Petroleum UK Ltd	J K
Phoenix Rubber Ltd	H J K
Plastanol Ltd	B
Plasticable Ltd	K
Plastic Coating Equipment Ltd	D F J K
Plastic Coatings Ltd	C
Plastic Extruders Ltd	J
Plucknett (C J) & Co Ltd	B
Polymon Developments Ltd	A H I J
Polypenco Ltd	D E F I J
Polyservices (Bedford) Ltd	C
Poron Insulation Ltd	I
Preston (J) Ltd	H K
Rubert & Co Ltd	B
SIC Plastics Ltd	A I
Scott Bader & Co Ltd	B K
Schulman (A) Inc. Ltd	J
Shawinigan Ltd	K
Shell Chemicals (UK) Ltd (Industrial Chemicals)	H I J K
Siegrist Orel Ltd	D K
Sim (L A) & Co Ltd	A J
Smith (Stanley) & Co	H J K
Southern Industries Agency	H I J K
Spicer Cowan Ltd	J K
Splintex Ltd	B
Sterling Moulding Materials Ltd	A I
Stevens (Michael S) Ltd	A H I J
Stuart, Kinney & Co Ltd	I J
Sylpon Manufacturing Co Ltd	B
TAC Construction Materials Ltd	B
Tar Residuals Ltd	A I J K
Tekta Packaging Ltd	I
Telcon Plastics Ltd	F
Trig Engineering Ltd	C
Troviplast Ltd	F
Fullis (John) & Son Ltd	D F H J K
US Industrial Chemicals Co	J
Uniroyal Ltd	A
VMG Plastics Ltd	I K
Vactite Wire Co Ltd	D
Victoria International Plastics Ltd	I J
Victor Plastics (Manchester) Ltd	H I J
Vinablend Ltd	K
Vinatex Ltd	J K
Vinyl Compositions Ltd	K
Vinyl Reprocessors Ltd	K
Vinyl Products Ltd	B I K
Visijar Laboratories Ltd	F
Wandex Distributors (London) Ltd	D I J
Weil (Joseph) & Son Ltd	B D F I
Wilson (J) & Co Ltd	I J
Witco Chemical Co Ltd	B
Wokingham Plastics Ltd	B

PVC SHEETING SUPPLIERS

Film	A
Flexible	B
Rigid	C
Not defined	D

APV Kestner	C
Alpa Steel & Plastics Ltd	D
Anderson (D) & Son Ltd	D
Armoride Ltd	D
Armshire Reinforced Plastics Ltd	B C
Ascol Products Ltd	B C
Ashton (Thomas A) Ltd	B C
Attewell (B) & Sons Ltd	B C
Attwater & Sons Ltd	C
BIP Reinforced Products Ltd	C
Bakelite Xylonite Ltd (BXL) (Industrial Prod. Div)	B C
Baxenden Chemical Co Ltd	C
Benker Co Ltd	C
Birmingham Mica Ltd	B C
Borst Bros Ltd	B C
Bourne Plastics Ltd	B C
Bradford Glass Co Ltd	C
Bricell Plastics Ltd	C
Brigtown Industries Ltd	B
British Celanese Ltd	A B C
British Celophane Ltd	A B
British Visqueen Ltd	A B
Burns (J) Ltd	C
CSK Plastics Ltd	A B C
Caligen Foam Ltd	B
Cape Universal Building Products Ltd	C
Cellgrave Ltd	B C
Central Manufacturing & Trading Co (Dudley) Ltd	A B C
Clear View Ltd	A
Clements (R A) (London) Ltd	D
Coates (James) Bros Ltd	C
Coes (Derby) Ltd	A B C
Coin Insulation Ltd	B C
Commercial Plastics Ltd	A B C
Comoy Components Ltd	C
Cork Growers Ltd	B C
Coubro & Scrutton (M & I) Ltd	B C
Cox (William J) (Sales) Ltd	C
Crossley (Henry) (Packings) Ltd	B C
Cyroma Plastics Ltd	D
DO & E Industries Ltd	A B C
Davidson & Co (Coachbuilders) Ltd	B C
Davis (F J) (Motor Accessories) Ltd	D
Dixopak Ltd	C
D-Mac Ltd	C
Douthwaite (T N) Ltd	A B C
Duraplex (Plastics)	A B C
East Coast Plastics Ltd	D
Elliott (E) Ltd	B
Envopak Ltd	D
Essex Mica Co Ltd	B C
Ewart Plastics Ltd	D
Export Packing Service Group	B
Flexo Plywood Industries Ltd	C
Foam Engineers Ltd	B
Forbes (Kenneth) (Manchester) Ltd	B C
Frost (C B) & Co Ltd	B C
General Celluloid Co Ltd	D
General Developments Co (Glasgow) Ltd	B C
General Foam Products Ltd	B C
Gilchrist & Fisher Ltd	D
Giplex Ltd	D
Goudies of Bothwell Ltd	D
Graham & Wylie Ltd	A B C
HJB Plastics Ltd	A B C
Hadrian Plastics Ltd	B C
Hall & Co (South East) Ltd	A C
Halmatic Ltd	C
Harrison & Jones (Flexible Foam) Ltd	B
Heaven Dowsett & Co Ltd	B C
Homogeneous Plastics Ltd	D
Imperial Chemical Industries	A B C
Insulating Components & Materials Ltd	B C
Insulating Materials Ltd	C
Insulation Equipments Ltd	B C
Kalle UK Ltd	A B C
Kay-Metzeler Ltd	B
Lambert Bros (Walsall) Ltd	C
Lane (P H) Ltd	D
Langley (London) Ltd	D
Latter (A) & Co Ltd	A
Lintafoam Ltd	B
Lloyds Cartons Ltd	B C
McCrae & Drew Ltd	B
Mardle (G B) Ltd	A
Marley Floor Tile Co Ltd	A B
Mathews & Yates Ltd	B C
Mellowes Orb Engineering Ltd	A B C
Metal Box Co Ltd	A B
Micanite & Insulators Ltd	C
Mill Plastics Ltd	C
Minnesota Mining & Manufacturing Co Ltd	A
Mitchell (Andrew) & Co Ltd	D
Modern Veneering Co Ltd	C
Moon Aircraft Ltd	C
Morane Plastic Co Ltd	A
Mosses & Mitchell Ltd	A B C
Nairn Coated Products Ltd	B
Nairn Floors	B
Nairn International	B
Newalls Insulation & Chemical Co Ltd	C
Newmor Plastics Ltd	B

Nickwood Plastics Ltd	D
Nylonic Engineering Co Ltd	A
P & B Plastics Ltd	A B
PDI Ltd	C
Permafence Ltd	A
Permali Ltd	C
Phipps Plastic Products Ltd	B C
Phoenix Rubber Ltd	B
Plastic Constructions Ltd	A B C
Plastic Weldings Ltd	D
Plastics Design & Engineering Ltd	C
Plastics Fabrication & Printing Ltd	A B
Plastics Reinforced (Liverpool) Ltd	C
Plastics Marketing Co Ltd	D
Plasti-Shapes Ltd	D
Polypenco Ltd	A B C
Poron Insulation Ltd	B C
Power Plastics Ltd	C
Precision Products (Leeds) Ltd	B C
Prodorite Ltd	B C
Railko Ltd	C
Rector (M R) Ltd	B
Rubbarite Ltd	B
SOM Plastics Ltd	A
Salon (Nelson) Ltd	C
Saro Products Ltd	C
Singleton Flint & Co Ltd	B
Smith (Stanley) & Co	A B C
Snow (J E) Plastics Ltd	B C
Splintex Ltd	C
Stephens (Plastics) Ltd	A B
Stewarts & Lloyds Plastics	D
Storey Bros & Co Ltd	A B
Strebor Plastics	A B C
Suntex Safety Glass Industries Ltd	B C
Technovac	C
Telcon Plastics Ltd	B C
Transatlantic Plastics Ltd	A B C
Triplex Foundries Group	B C
Tuckers (Sheffield) Ltd	A B C
Tufnol Ltd	C
Tullis (John) & Son Ltd	A B C
Turner & Hughes Ltd	D
Uniroyal Ltd	C
Vacuum Formers Ltd	D
Venesta International-Vencel Ltd	B
Viskase Ltd	A
WB Industrial Plastics Ltd	A
Wallington Weston & Co Ltd	B
Wardle (Bernard) (Everflex) Ltd	A B C
Wilson (F E) (Plastics) Ltd	C
Zephyr Plastic Products Ltd	A B C

COMPRESSION & TRANSFER MOULDERS

AEI Moulded Products Division
AI Paramount Plastics
Acme Showcard & Sign Co Ltd
Armshire Reinforced Plastics Ltd
Attewell (B) & Sons Ltd
Ball Plastics Ltd
Bakelite Xylonite (BXL) Ltd
Banarse (S) & Co Ltd
Basildon Moulding Co Ltd
Bendix & Herbert Ltd
Bettix Ltd
Birmingham Mica Ltd
Birmingham Plastics Ltd
Bluemel Bros Ltd
Blue Peter Urethene Products Ltd
Boddington (W H) & Co Ltd
Bonnella (D H) & Son Ltd
Borst Bros Ltd
Bristol Aerojet Ltd
British Steel Springs Ltd
Brookes & Adams Ltd
Bushing Co Ltd
Byron Jardine Ltd
Byson Appliance Co Ltd
CSK Plastics Ltd
Camden Optical Co Ltd
Charlesworth (L G G) Ltd
Clarke (H) & Co (Manchester) Ltd
Coes (Derby) Ltd
Combined Optical Industries Ltd
Constructional Products (Chaderton) Ltd
Co-ordinators Service Engineering Ltd
Cornercroft Plastics Ltd
Cosmocord Ltd
Coventry Motor & Sundries Ltd
Cox (William) (Sales) Ltd
Crabtree Engineering (Rochdale) Ltd
Crane Packing Ltd
Crystalate Mouldings Ltd
Dale (John) Ltd
Denton & Co (Rubber Engineers) Ltd
Derwent Plastics Ltd
Dohm Holdings Ltd
Dunlop Co Ltd (Precision Rubbers Division)
Dura Co Ltd
Ebonestos Industries Ltd
Ekco Plastics Ltd
Elliott (E) Ltd
Elm Plastics Ltd
English Electric Plastics Division
Essex Mica Co Ltd
Evans (Frederick W) Ltd
Evered & Co (Holdings) Ltd
Ewarts Ltd
FPT Industries Ltd
Ferguson Shires Ltd
Fibre Form Ltd
Fothergill & Harvey Ltd
Fourfold Mouldings Ltd
Fraser & Glass Ltd
Freeman (William) & Co Ltd
Frost (C B) & Co Ltd
GC Plastics Ltd
Glass Fibre Developments Ltd
Gloster Engineering (Cheltenham) Ltd
Greenwood Rawlins & Co Ltd
Hadrian Plastics Ltd
Hallite Plastics Ltd
Harrison (A T) & Co (Mouldings) Ltd
Harrison Bros (Plastics) Ltd
Harwin Engineers Ltd
Healey Mouldings Ltd
Henry & Thomas Ltd
Holloid Plastics Ltd
Hornflowa Ltd
Humpage & Moseley Ltd
Hyde Plastics Ltd
ICL Plastics Ltd
Imco Plastics Ltd
Industrial Mouldings Ltd
Ingram (Thomas A) & Co Ltd
Initial Plastics Ltd
Injection Moulded Plastics Ltd
Insulated Components & Materials Ltd
Insulators Ltd
International Computers Ltd
Invenit Industrial Plastics Ltd
Jobling (James A) & Co Ltd
Keil (E) & Co Ltd
Kent Mouldings
Lancasters Plastics Ltd
Lindsay & Williams Ltd
Litholite Mouldings Ltd
London Artid Plastics Ltd
Long & Hambly Ltd
Lovell Bros & Stapley
Lukely Engineering & Moulding Co Ltd
MP Plastics (Maurice Powell Ltd)
Mackie (James) & Sons Ltd
McMurdo Instrument Co Ltd
Manuplastics Ltd
Marton Plastics Ltd
Mawson Taylor Ltd
Mellows Orb Engineering Ltd
Mercury Signs Ltd
Merriott Mouldings Ltd
Metropolitan Flexible Products Ltd
Mica & Micanite Supplies Ltd
Mica Products Ltd
Minerva Mouldings Ltd
Motools Ltd
Mundel Cork & Plastics Ltd
Mycalex & Rosite Ltd
NB Mouldings Ltd
National Plastics Ltd
Neosid Ltd
Nettle Accessories Ltd
Owen (Henry) & Sons Ltd
P & S (Cirencester) Ltd
Paramount Plastics Products Ltd
Peerless Plastics Ltd
Perkins (E) & Co Ltd
Permali Ltd
Plasmic Ltd
Plastic Engineers Ltd
Plastics (Wednesbury) Ltd
Plessey Co Ltd (Components Group) (The)
Plutec Ltd
Precision Mouldings (Plastics) Ltd
Precision Plastics Ltd
Precision Products (Leeds) Ltd
Pye of Cambridge Ltd
Radiamp Co Ltd
Rialko Ltd
Ranton & Co Ltd
Ray Engineering Co Ltd
Ray Mouldings Ltd
Reinforced Plastics Applications (Swansea) Ltd
Rendar Instruments Ltd
Roanoid Ltd
Rolls-Royce (Composite Materials) Ltd
Rootes Mouldings Ltd
STC (Standard Telephones & Cables) Ltd
Sales Aids (Sedlescombe)
Salon (Nelson) Ltd
Seaforth Plastics
Serk (R & D)
Sima Plastics Ltd
Smith (Stanley) & Co
Souplex Ltd
Spencer Knight & Co Ltd
Stadium Ltd
Stechford Mouldings Ltd
Streetly Manufacturing Co Ltd (The)
Super Oil Seals & Gaskets Ltd
Sure Form Plastics Ltd
Swift (S M) (Exeter) Ltd
TAC Construction Materials Ltd
Tekta Packaging Ltd
Terminal Insulators Ltd
Therm-Plast Ltd
Thermo Plastics Ltd
Thetford Moulded Products Ltd
Tomkinson (F W) Ltd
Torpey (Sylvester) & Sons Ltd
Townstal Products Ltd
Triton Plastics Ltd
Tufnol Ltd
UECL Ward Brook
Ultra Electronics (Components) Ltd
Ulster Plastics Ltd
United Moulders Ltd
Unitex Ltd
Vacmobile Ltd
Vale (J S)
Viking Industrial Plastics Ltd
WCB Containers Ltd
Ward & Goldstone Ltd
Wheatley (A) Ltd

Whiteley Electrical Radio Co Ltd
Witton Moulded Plastics Ltd
Wolley (Frederick) Ltd
Xenit Products Ltd

INJECTION MOULDERS

AC Plastic Industries Ltd
ACW Ltd
AEI Moulded Products Division
AP Industries (Nelson) Ltd
Acrow Plastics Ltd
Advanced Moulding Developments Ltd
Alexander (Silverthorne) Ltd
Ansorite Manufacturing Co
Armitage Plastics Ltd
Astex Ltd
Atlas Plastics Ltd
Attewell (B) & Sons Ltd
Aubit Plastics Ltd
B & B Plastics Ltd
BEF Products (Essex) Ltd
B & G Plastics Ltd
Bakelite Xylonite Ltd (BXL)
(Cascelloid Division)
Bakelite Xylonite Ltd (BXL)
(Scintellex Products)
Ball Plastics Ltd
Ball Plastics (Clacton) Ltd
Barclay Stuart (Plastics) Ltd
Bardex (Plastics) Ltd
Basildon Moulding Co Ltd
Bendix & Herbert Ltd
Bettix Ltd
Bibby (R) (Engineers) Ltd
Birchware Ltd
Birkbys Ltd
Birmingham Mica Co Ltd
Birmingham Plastics Ltd
Blewis & Shaw (PVC) Ltd
Bluebell Dolls Ltd
Bluemel Bros Ltd
Blue Peter Urethene Products Ltd
Boardman (A) Ltd
Boddington (W H) & Co Ltd
Bonnella (D H) & Son Ltd
Borst Bros Ltd
Bourne (H & W) Ltd
Brookes & Adams Ltd
Byron Jardine Ltd
CSK Plastics Ltd
Camden Optical Co Ltd
Capon Heaton & Co Ltd
Carter Bros (Billingshurst) Ltd
Cashmore Plastics Ltd
Cassidy Bros Ltd
Castle Products Ltd
Celmac Plasclip Ltd
Chadburn Darwen Ltd
Charlesworth (L G G) Ltd
Chemical Pipe & Vessel Co Ltd
Chemidus Plastics Ltd
Cisterns Ltd
Clarke (H) & Co (Manchester) Ltd
Clarke (James) & Sons (Stoke-on-Trent) Ltd
Clayton Wright (Howard) Ltd
Coinmechs Ltd
Collard (S A) Ltd
Columbia Products Co Ltd
Colwyn Plastics Ltd
Combined Optical Industries Ltd
Constructional Products (Chadderton) Ltd
Cope Allman Plastics Ltd
Coral Plastics Ltd
Cornercroft (Plastics) Ltd
Coventry Motor & Sundries Co Ltd
Crawford Plastics Ltd
Creators Ltd
Critchley Bros Ltd
Crittall-Hope Ltd
Crystalate (Mouldings) Ltd
Daleman (Richard) Ltd
Davall (S) & Sons Ltd
Denton & Co (Rubber Engineers) Ltd
Derwent Plastics Ltd
Dines Plastics Ltd
Dobson (John) (Milnthorpe) Ltd
Dormex Industrial Plastics Ltd
Dowty Seals Ltd
Dunlop Co Ltd
(Polymer Engineering Division)
Durapipe & Fittings Ltd
Ebonestos Industries Ltd
Edge (Wilfred) & Sons Ltd
Eeto Plastics Ltd
Efi-Astex Ltd
Egatube Ltd
Ekco Plastics Ltd
Elco Plastics Ltd
Elford Plastics Ltd
Elliott (E) Ltd
Elm Plastics Ltd
Elta Plastics Ltd
Enalon Plastics Ltd
Enfield Plastics Ltd
English Electric Plastics Division
Essex Mica Co Ltd
Eurobung Ltd
Evans (Frederick W) Ltd
Evered & Co (Holdings) Ltd
Ever Ready Plastic Co Ltd
FWH Plastics Ltd
Falconcraft Ltd
Ferguson Shiers Ltd
Fern Plastic Products Ltd
Fibrenyle Ltd
Fisher (Harold) (Plastics) Ltd
Forbes (Kenneth) (Plastics) Ltd
Fordham Pressings Ltd
Forest Row Plastics Ltd
Foster Bros Plastics Ltd
Fourfold Mouldings Ltd
Fraser & Glass Ltd
Frost (C B) Ltd
Frys Diecastings Ltd
Futters Ltd
GC Plastics Ltd
GFW Plastics Ltd
GKN Sankey (Plastics Division)
GKN Screws & Fasteners Ltd
Gazelle Plastics Ltd
Girdlestone Electronics Ltd
Gloster Engineering (Cheltenham) Ltd
Glover Plastics Ltd
Glyndon Plastics Ltd
Glynwed Plastics Ltd
Goldring Manufacturing Co Ltd
Goodburn Plastics Ltd
Goodfellow Metals Ltd
Greenwood Rawlins & Co Ltd
Grover (A F) & Co Ltd
Hallite Plastics Ltd
Hall Mark Plastics Ltd
Hampshire Plastics & Engineering Co Ltd
Harrison Bros (Plastics) Ltd
Harwin Engineers Ltd
Healey Mouldings Ltd
Healy Injection Mouldings Ltd
Heathrod Ardwyn & Co Ltd
Hellerman Plastics
Henley Plastics Ltd
Hettich & Walls Plastics Ltd
Henry & Thomas Ltd
Hills Precision Die Castings Ltd
Hilton Plastics Ltd
Holmbush Plastics Ltd
Holpak Ltd
Hozelock Ltd
Humpage & Moseley Ltd
Hupfield Plastics Ltd
Hutchings (F A) (Kingston) Ltd
ICL Plastics Ltd
ID Precision Plastics
I to I Plastics Ltd
IMI Developments Ltd
Illingworths (Plastics) Ltd
Illsen (H) Ltd
Imco Plastics Ltd
Industrial Mouldings Ltd
Industrial Reels Ltd
Ingram (Thomas A) & Co Ltd
Initial Plastics Industries Ltd
Injection Moulded Plastics Ltd
Insulators Ltd
International Computers Ltd
Invenit Industrial Plastics Ltd
Invicta Plastics Ltd
Jagger (Albert) Ltd
Jay Plastics Ltd
Jeffries (P R) (Guernsey) Ltd
Jepson & Co Ltd
Jig Tools (Pentyrch) Ltd
Jobling (James A) & Co Ltd
Johnsen & Jorgensen (Plastics) Ltd
Judge International Ltd
Judge International Housewares Ltd
KP Plastics (Bletchley) Ltd
Keil (E) & Co Ltd
Kent Mouldings
Kent Plastics UK Ltd
Kenure (J F) Ltd
Kerr (J & W) (Plastics) Ltd
Ketch Plastics Ltd
Kingston Plastics Ltd
LB (Plastics) Ltd
LGS Ltd
L & P Plastics Ltd
Lettercast Ltd
Lincoln Plastics Ltd
Link 51 Ltd
Litholite Mouldings Ltd
London Artid Plastics Ltd
London Association for the Blind
London Bankside Products Ltd
Lorival Ltd
Lovell Bros & Stapley
Lucent Plastics Ltd
MCM Tools Ltd
Mackie (James) & Sons Ltd
McMurdo Instrument Co Ltd
Manuplastics Ltd
Marley Extrusions Ltd
Marley Floor Tile Co Ltd
Marshall (C F) & Son Ltd
Martin-Golod Products Ltd
Marton Plastics
Measom Freer & Co Ltd
Mellowes Orb Engineering Ltd
Mentmore Manufacturing Co Ltd
Merriott Mouldings Ltd
Merrivale Manufacturing Co Ltd
Metal Box Co Ltd
Metal & Plastic Products Co
(Balsall Common) Ltd (The)
Metropolitan Flexible Products Ltd
Metropolitan Plastics Ltd
Mica Products Ltd
Minerva Mouldings Ltd
Mobbs Miller Ltd
Monk (Anthony) (Engineering) Ltd
Montgomery (Daniel) & Son Ltd
Moorside Machining Co Ltd
Morning (Croydon) Ltd
Moss (Robert) Ltd

Moulded Fasteners Ltd
Moulded Plastics (Birmingham) Ltd
Moulding Service Ltd
Mundel Cork & Plastics Ltd
NB Mouldings Ltd
National Plastics Ltd
Neosid Ltd
Nettle Accessories Ltd
Nico Manufacturing Co Ltd
Norbert Plastics Ltd
North (Robert) & Sons Ltd
North West Plastics Ltd
Nutshell Cap Co Ltd
Nylon Plastics Co Ltd
Nylonic Engineering Co Ltd
Oates Ltd
Olympic Plastics Ltd
Omega Plastics Ltd
Optical Products Ltd
Owen (Henry) & Sons Ltd
Osprey Plastics Ltd
PH Plastics Ltd
Paragon Plastics Ltd
Paramount Plastics Products Ltd
Park Lane Plastics (Aldridge) Ltd
Parkinson Cowan Ltd
Parsons Bros Ltd
Paton, Clavert & Co Ltd
Peel Engineering Co Ltd
Peerless Plastics Ltd
Pendry (Plastics) Ltd
Performance Plastics Ltd
Perkins (E) & Co Ltd
Perry (E S) Ltd
Pershore Mouldings Ltd
Pharmaceutical Plastics Ltd
Pickup (John) Ltd
Pioneer Plastic Containers Ltd
Pillar Plastics Ltd
Pioneer Oilsealing & Moulding Co Ltd
Plasmic Ltd
Plastic Engineers Ltd
Plasticisers Ltd
Plastics Manchester Ltd
Plastics Wednesbury Ltd
Plastiglide Products Ltd
Plessey Co Ltd (Components Group) (The)
Plutec Ltd
Polytubes (Nylaflex) Ltd
Polythene Drums Ltd
Polytop Plastics Ltd
Poplar Playthings Ltd
Poppe Rubber & Tyre Co Ltd
Portex Ltd
Powell (E & F) Ltd
Precision Mouldings (Plastics) Ltd
Precision Plastics Ltd
Preci-Spark Ltd
Prior (John) Engineering Ltd
Prior (John) Plastics Ltd
Pritchard (Plastics) Ltd
Protected Conductors Ltd
Punfield & Barstow (Mouldings) Ltd
Pye of Cambridge Ltd
QD Plastics Ltd
QED Design & Development Ltd
Quinton (T H) Ltd
Radiamp Co Ltd
Rainbow (S H) Ltd
Raleigh Industries Ltd
Ranton & Co Ltd
Ray Engineering Co Ltd
Raychem Ltd
Rediweld Ltd
Regina Industries Ltd
Rendar Instruments Ltd
Resoid Ltd
Roanoid Ltd
Roanoid Plastics Ltd
Rock Electrical Accessories Ltd
Rolinx Ltd
Rollite Products (Bridlington) Ltd
Rolls-Royce (Composite Materials) Ltd
Rootes Plastics Ltd
Rosedale Associated Manufacturers Ltd
Rotalac Co Ltd
Rowley (Plastics) Ltd
Rubberplas (Birmingham) Ltd
Rutland Plastics Ltd
Ryford Ltd
STC (Standard Telephones & Cables) Ltd
SOM Plastic Mouldings Ltd
Salon (Nelson) Ltd
Salter Packaging Ltd
Salter Plastics Ltd
Screenprints (Vacuum Formers) Ltd
Securos Ltd
Selcol Products Ltd
Sendale (Plastics) Ltd
Serk (R & D)
Severn Plastics Ltd
Shaw Munster Ltd
Silleck Engineering Ltd
Simco Plastics Ltd
Snow (J E) Plastics Ltd
Souplex Ltd
Spa Plastics Ltd
Spencer Knight & Co Ltd
Speglestein (S) & Son Ltd
Stadium Ltd
Stanley (Alfred) & Sons Ltd
Stantone Plastics Ltd
Stechford Mouldings Ltd
Stephenson Blake & Co Ltd
Sterilin Ltd
Stewart Plastics Ltd
Stones Plating Co Ltd
Strebor Plastics
Streetly Manufacturing Co Ltd
Studio Plastic & Metal Components Ltd
Submarine Products Ltd
Sundt Plastics Ltd
Sunnytoys (Distributors) Ltd
Sunstan Industrial Mouldings Ltd
Supa-ware (Southend) Ltd
Super Oil Seals & Gaskets Ltd
Sure Form Plastics Ltd
Swift (S M) (Exeter) Ltd
TAC Construction Materials Ltd
TPT Ltd
TT Containers Ltd
TT Mouldings Ltd
Tallon Plastics Ltd
Techno-Plast Ltd
Technovac
Terminal Insulators Ltd
Textile Mouldings Ltd
Thermoplastics Ltd
Thetford Moulded Products Ltd
Thomas & Vines Ltd
Tinker Grayson Ltd
Tomkinson (F W) Ltd
Toone Plastics Ltd
Torpey (Sylvester) & Sons Ltd
Townstal Products Ltd
Transatlantic Plastics Ltd
Tratt Plastics Ltd
Tresco Plastics Ltd
Truform Plastics Ltd
Tullis (John) & Son Ltd
Turbro Ltd
Tyne Plastics Ltd
UG Closures & Plastics Ltd
Ulster Plastics Ltd
Ultra Electronics (Components) Ltd
United Moulders Ltd
United-Carr Ltd
Vacmobile Ltd
Vale (J S)
Vanguard Plastics Ltd
Vero Precision Engineering Ltd
Victoria Plating Co Ltd
Viking Industrial Plastics Ltd
Vitamol Precision Ltd
WCB Containers Ltd
Walker Litherland Plastic Ltd
Wall & Leigh Thermoplastics Ltd
Ward & Goldstone Ltd
Waterside Plastics Ltd
Wells-Hinton Plastics Ltd
Welsh Trust (Rhigos) Ltd
West Country Converters Ltd
Western Pressings Ltd
Westinghouse Brake & Signal Co Ltd
Westway Models Ltd
Wheatley (A) Ltd
Wheway Frames Ltd
Whiteley Electrical Radio Co Ltd
Willamot Industrial Mouldings Ltd
Willow Plastics Engineering Co Ltd
Wilmot Breeden Ltd
Witton Moulded Plastics Ltd
Woollen & Co Ltd
Wolley (Frederick) Ltd
Wragby Plastics Ltd
Wye Plastics Ltd
Wyllie-Young Ltd
Xenit Products Ltd
Xlon Products Ltd

EXTRUDED SECTION SUPPLIERS

AI Paramount Plastics
Airfix Industries Ltd
Airfix Plastics Ltd
Allen (George) (Plastics) Ltd
Anderson (D) & Son Ltd
Anson-Bassan Plastic Rods Ltd
Anson Plastics Ltd
Anorite Manufacturing Co
Ardray Manufacturing Co Ltd
Arrow Plastics Ltd
Astex Ltd
Aubit Plastics Ltd
BP Chemicals International Ltd
BTL Mould & Die Co Ltd
Ball Plastics Ltd
Bardex (Plastics) Ltd
Berkshire Mouldings Ltd
Bibby (R) (Engineers) Ltd
Bluebell Dolls Ltd
Bluemel Bros Ltd
Boyriven Ltd
Brighouse Plastics Ltd
British Celanese Ltd
British Ropes Ltd
British Steel Springs Ltd
British Visqueen Ltd
Brodie Dowmex Ltd
CSK Plastics Ltd
Capon Heaton Ltd
Carter Bros (Billingshurst) Ltd
Celmac Plasclip Ltd
Central Manufacturing & Trading Co (Dudley) Ltd
Chemical Pipe & Vessel Co Ltd
Chemidus Plastics Ltd
Cimid Ltd
Cisterns Ltd
Clayton-Wright (Howard) Ltd
Clear Span Ltd
Coba Plastics Ltd

Colgate (D J) Plastics Ltd
Commercial Plastics Ltd
Compoflex Ltd
Contemporary Furnishing Co Ltd
Co-ordinators Service Engineering Ltd
Coventry Motor & Sundries Co Ltd
Cow (P B) (Plastics) Ltd
Crabtree Engineering (Rochdale) Ltd
Crane Packing Ltd
Critchley Bros Ltd
Crittall-Hope Ltd
Crossley (Henry) Packings) Ltd
Davidson Industries
Dormex Industrial Plastics Ltd
Dunlop Co Ltd
(Polymer Engineering Division)
Dunlop Co Ltd
(Precision Rubbers Division)
Durapipe & Fittings Ltd
Duraplex Plastics
Duratube & Wire Ltd
Econopack Ltd
Ekco Plastics Ltd
Egatube Ltd
Ecco Plastics Ltd
Elm Plastics Ltd
English Electric Reinforced
Plastics Division
Enkothene Ltd
Espinasse Ltd
Evered & Co (Holdings) Ltd
Extruda Ltd
FPT Industries Ltd
Fibrenyle Ltd
Flowmaster Ltd
Forbes (Kenneth) (Manchester) Ltd
Forbes (Kenneth) (Plastics) Ltd
Formica Ltd
Foster Bros Plastics Ltd
Foster Plastics (Rainford) Ltd
Freeman (William) & Co Ltd
Frost (C B) & Co Ltd
Giplex Ltd
Glynwed Plastics Ltd
Goodfellow Metals Ltd
Graham & Wylie Ltd
Griflex Products Ltd
HJB Plastics Ltd
Hall & Co (South East) Ltd
Hartley Baird Ltd
Heaven Dowsett & Co Ltd
High Peak Plastics Ltd
Hispeed Plastics Ltd
Holmbush Plastics Ltd
Hygienex Industries Ltd
Impetus Building Components Ltd
Industrial Plastic Extrusions Ltd
Insulating Components & Materials Ltd
Judge International Ltd
Judge International Housewares Ltd
Jupiter Plastics Ltd
KW Chemicals Ltd
Kalle (UK) Ltd
LB (Plastics) Ltd
L & P Plastics Ltd
Lamb (N) (Scrap) Plastics
Lambert Bros (Walsall) Ltd
Liermann Ltd
Lincoln Plastics Ltd
Lorival Ltd
Low & Bonar (Textiles & Packaging) Ltd
Lucent Plastics Ltd
MAM Rubber Manufacturing Co Ltd
Marley Extrusions Ltd
Marley Floor Tile Co Ltd
Marshall (C & C) Ltd
Marshall (C F) & Son Ltd
Mathews & Yates Ltd
Meldrum (A W) Thermoplastics Ltd
Metal Box Co Ltd
Middleton Plastics Ltd
Mobbs Miller Ltd
Moon Aircraft Ltd
Moseley (David) & Sons Ltd
Multicraft Ltd
Muntz Plastics
National Plastics Ltd
Negri Bossi Evans Ltd
Netlon Ltd
Norseman (Cables & Extrusions) Ltd
Nylonic Engineering Co Ltd
Olympic Plastics Ltd
Ormond Brassfoundry Ltd
Osma Plastics Ltd
PDI Ltd
Paragon Plastics Ltd
Paramount Plastics Products Ltd
Perkins (E) & Co Ltd
Permanoid Ltd
Phipps Plastic Products Ltd
Pharmaceutical Plastics Ltd
Pillar Plastics Ltd
Plastestrip Ltd
Plastic Constructions Ltd
Plasticable Ltd
Plastic Engineers Ltd
Plastic Extruders Ltd
Plastic Tube & Conduit Ltd
Plasticisers Ltd
Plex (Engineering) Ltd
Plextrude Ltd
Poet Plastics Ltd
Polygrow Plastics
Polypenco Ltd
Polytubes (Nylaflex) Ltd
Portex Ltd
Precision Mouldings (Plastics) Ltd
Pye of Cambridge Ltd
Rainbow (S H) Ltd
Range Valley Engineering Ltd
Ravenscroft Plastics Ltd
Reddiplex Ltd
Rehau-Plastics Ltd
Roanoid Ltd
Roanoid Plastics Ltd
Roberts (Alfred) & Sons Ltd
Rosedale Associated Manufacturers Ltd
Ross Thermoplastics Ltd
Rotalac Co Ltd (The)
Rowley Plastics Ltd
Rustless Curtain Rod Co Ltd
SOM Plastic Mouldings Ltd
Sefton Mills Ltd
Smith (Stanley) & Co Ltd
Snow (J E) Plastics Ltd
Stewart (P B) Ltd
Studio Plastic & Metal Components Ltd
Sundt Plastics Ltd
Surfleet (Plastics) Products
TAC Construction Materials Ltd
TPT Ltd
Tecalemit (Engineering) Ltd
Telcon Plastics Ltd
Temple Building Products Ltd
Tenaplas Ltd
Thermo Plastics Ltd
Transatlantic Plastics Ltd
Tuckers (Sheffield) Ltd
Tullis (John) & Son Ltd
USI Engineering Ltd
Unitex
Varicol Ltd
Victor Plastics (Manchester) Ltd
Wandleside Warren Wire Co Ltd
Ward & Goldstone Ltd
Wardle (Bernard) (Everflex) Ltd
Warne (William) & Co Ltd
Wetty Oates & Co Ltd
Windshields of Worcester Ltd
Wragby Plastics Ltd

BLOW MOULDERS

Al Paramount Plastics
Able Developments Co
Acme Showcard & Sign Co Ltd
Airfix Industries Ltd
Airfix Plastics Ltd
Anchor Chemical Developments Ltd
Atlas Plastics Ltd
Autodevices (1960) Ltd
Bakelite Xylonite (BXL) Ltd
(Cascelloid Division)
Barclay-Stuart (Plastics) Ltd
Bettix Ltd
Blagden & Noakes (Holdings) Ltd
Blewis & Shaw (VC) Ltd
Blow Mouldings Ltd
Bluebell Dolls Ltd
Carter Bros (Billingshurst) Ltd
Combex Ltd
Corfield (Reginald) Ltd
Cox (William J) (Sales) Ltd
Dale (John) Ltd
Daleman (Richard) Ltd
Dines Plastics Ltd
Elm Plastics Ltd
Fibrenyle Ltd
Fordham Pressings Ltd
Foster Plastics (Rainford) Ltd
Fraser & Borthwick Ltd
Freeman (William) & Co Ltd
Hadrian Plastics Ltd
Hampshire Plastics & Engineering Co Ltd
Harcostar Ltd
Hispeed Plastics Ltd
Holpak Ltd
I to I Plastics Ltd
Ingram (Thomas A) & Co Ltd
Jobling (James A) & Co Ltd
Lacrinoid Products Ltd
Lax & Shaw Ltd
Lovel Bros & Stapley
Mica Products Ltd
Middleton Plastics Ltd
Montgomery (Daniel) & Son Ltd
Northern (Neon) Lights Blackpool Ltd
Paklite Ltd
Paramount Plastic Products Ltd
Peel Engineering Co Ltd
Pershore Mouldings Ltd
Pillar Plastics Ltd
Plysu Products Ltd
Polygrow Plastics
Polythene Drums Ltd
Poplar Playthings Ltd
Precision Products (Leeds) Ltd
Priestly Studios Ltd
QD Plastics (Glasgow) Ltd
RF Development Co Ltd
Raleigh Industries Ltd
Regis Machinery (Sussex) Ltd
Rexpak
Rollite Products (Bridlington) Ltd
Rosedale Associated Manufacturers Ltd
Ross Chemical & Storage (Plastics) Ltd
SOM Plastic Mouldings Ltd
Selcol Products Ltd
Sharna-Ware (Manufacturing) Ltd
Silvaflame Co Ltd
Studio Plastic & Metal Components Ltd
Supa-Ware (Southend) Ltd
TT Containers Ltd
Tennant (Charles) & Co Ltd
Transatlantic Plastics Ltd
Tuckers (Sheffield) Ltd
UG Closures & Plastics Ltd
WCB Containers Ltd
Ward Adams Co Ltd (The)

Waterside Plastics Ltd
Willamott Industrial Mouldings Ltd
Xlon Products Ltd

THERMOFORMING (VACUUM)

Al Paramount Plastics
Acme Showcard & Sign Co Ltd
Acrow Plastics Ltd
Advance Plastic Products Ltd
Ainsworth Marsland & Co Ltd
Airfix Plastics Ltd
Alpine Plastics Ltd
Aluminium & Plastic Products Ltd
Armitage Plastics Ltd
Arrow Plastics Ltd
Attewell (B) & Sons Ltd
Ball Plastics Ltd
Blow (J J) Ltd
Bluemel Bros Ltd
Boardman (A) Ltd
Boxmakers (Manchester) Ltd
Rigtown Industries Ltd
British Needle Co Ltd
British Steel Springs Ltd
British Vita Co Ltd
Burns (J) Ltd
Bushing Co Ltd
CSK Plastics Ltd
Carville Ltd
Cellgrave Ltd
Chadburn-Darwen Ltd
Clayton Wright (Howard) Ltd
Clearex Products Ltd
Coes (Derby) Ltd
Combex Ltd
Comoy Components Ltd
Compact Cases Ltd
Courtaulds Engineering Ltd
Coventry Motor & Sundries Ltd
Cox (William J) (Sales) Ltd
Crystalate Mouldings Ltd
Cyroma Plastics Ltd
Daleman (Richard) Ltd
Dewey Waters & Co Ltd
Dormex Industrial Plastics Ltd
Drelco Ltd
Driclad Ltd
Eastwood Plastics Ltd
Econopack Ltd
Egatube Ltd
Elm Plastics Ltd
Enalon Plastics Ltd
Ewart Plastics Ltd
Falconcraft Ltd
Fasweld Ltd
Ferguson Shiers Ltd
Fibrebond Ltd
Fibre Form Ltd
Finlay (William) (Belfast) Ltd
Flight Equipment & Engineering Ltd
Forbes (Kenneth) (Manchester) Ltd
Forbes (Kenneth) (Plastics) Ltd
Fordham Pressings Ltd
Formed Plastic Containers Ltd
Formold Ltd
Formpak Ltd
GC Plastics Ltd
GFW Plastics Ltd
Gazelle Plastics Ltd
General Developments Co (Glasgow) Ltd
General Celluloid Co Ltd
Gilchrist & Fisher
Gina Plastics Ltd
Giplex Ltd
Gloster Engineering (Cheltenham) Ltd
Glynwed Plastics Ltd
Goldring Manufacturing Co (Great Britain) Ltd
Goodall (R A) & Co Ltd
Greenwood Rawlins & Co Ltd
HJB Plastics Ltd
Hadrian Plastics Ltd
Hartley Baird Ltd
Hills Precision Die Castings Ltd
Hindmarsh Smith Plastics Ltd
Hispeed Plastics Ltd
Huntley Boorne & Stevens Ltd
Hyde Plastics Ltd
ICL Plastics Ltd
Initial Plastics Ltd
Invicta Plastics Ltd
Irvine, Martin Plastics Ltd
Jennings Bros (Harringay) Ltd
Jupiter Plastics Ltd
Kabi (Electrical & Plastics) Ltd
Kedo Plastics Ltd
King Packaging Ltd
LB (Plastics) Ltd
Lambert Bros (Walsall) Ltd
Lane (P H) Ltd
Liermann Ltd
Lincoln Plastics Ltd
Linecross Plastics Co Ltd
Link Associates (RPB) Ltd
Lintafoam Ltd
Lloyds Cartons Ltd
London Bankside Products Ltd
Longwell Green Coachworks Ltd
Lovell Bros & Stapley
Lynester Products Ltd
MacEchern Ltd
Manhattan Equipment Ltd
Marley Extrusions Ltd
Martin's Products Ltd
Mercury Signs Ltd
Metal Box Co Ltd
Metal & Plastic Products Co (Balsall Common) Ltd (The)
Metropolitan Flexible Products Ltd
Midland Industries Ltd
Mill Plastics Ltd
Modern Fittings Ltd
Moon Aircraft Ltd
Moore (Harold) & Son Ltd
Moulton Mouldings Ltd
Nickwood Plastics Ltd
Nightingale Signs (Blackburn) Ltd
North West Plastics Ltd
Northern (Neon) Lights (Blackpool) Ltd
Notreme Engineering Developments Ltd
Olympic Plastics Ltd
Omega Plastics Ltd
Orbex Ltd
PDI Ltd
P & S (Cirencester) Ltd
Paramount Plastics Products Ltd
Parnall & Sons Ltd
Peel Engineering Co Ltd
Perks (Mark) Ltd
Pharmaceutical Plastics Ltd
Phipps Plastic Products Ltd
Plastestrip Ltd
Plastic Display Units (Strand) Ltd
Plastics (Manchester) Ltd
Plastifilm Products Ltd
Polygrow Plastics
Precision Products (Leeds) Ltd
Priestly Studios Ltd
Pye of Cambridge Ltd
QD Plastics (Glasgow) Ltd
QED Design & Development Ltd
Raleigh Industries Ltd
Reinforced & Microwave Plastics Group (BAC) Ltd
Ridout (D H) & Son Ltd
Robinson Thermoforming (E S & A Robinson Ltd)
Rutherford Plastics & Engineering Ltd
Salon (Nelson) Ltd
Screenprints (Vacuum Formers) Ltd
Silvaflame Co Ltd
Silver (E) (B S) Ltd
Soplaril (Great Britain) Ltd
Splintex Ltd
Suntex Safety Glass Industries Ltd
Swansea Plastics & Engineering Ltd
Taylor (C F) (Plastics) Ltd
Technichrome Products (Plastics) Ltd
Technovac
Telcon Plastics Ltd
Tenaplas Ltd
Tennänt (Charles) & Co Ltd
Transatlantic Plastics Ltd
Triplex Foundries Group Ltd
Tuckers (Sheffield) Ltd
Uniroyal Ltd
Unit Buying Services Ltd
Vacuum Formers Ltd
Varicol Signs Ltd
Victoria Plating Co Ltd (Plastics Division)
Vitafoam Ltd
WCB Containers Ltd
Walkers of Manchester
West Country Converters Ltd
Westford Plastics & Engineering Ltd
Westway Models Ltd
Wilson (F E) (Plastics) Ltd
Windshields of Worcester Ltd
Wokingham Plastics Ltd

ROTATIONAL MOULDERS

Armshire Reinforced Plastics Ltd
Ash Plastic Products Ltd
Coes (Derby) Ltd
DM Industrial Plastics Ltd
Dip Mouldings Ltd
Dohm Industrial Ltd
Elm Plastics Ltd
Elson & Robbins Ltd
Fecher (M J) Ltd
Griflex Products Ltd
Hadrian Plastics Ltd
Hennessey Products (Hampshire) Ltd
Hispeed Plastics Ltd
Initial Plastics Industries Ltd
Invicta Plastics Ltd
Marley Extrusions Ltd
Marley Foam Ltd
North West Plastics Ltd
PD (Technical Mouldings) Ltd
Phoenix Rotational Mouldings Ltd
Plastic Rotational Mouldings Ltd
Plastic Dip Mouldings Ltd
Polythene Drums Ltd
Regina Industries Ltd
Rotary Castings Ltd
SOM Plastic Mouldings Ltd
Stress Free Plastics Ltd
Thermo Plastics Ltd
WCB Containers Ltd
Xlon Products Ltd

FUSION BONDERS (DIP COATING)

Airfix Industries Ltd
Airfix Plastics Ltd
Aquitaine-Organico (UK) Ltd
Ardray Manufacturing Co Ltd
Arlington Plastics Development Ltd

Armshire Reinforced Plastics Ltd
Auriol (Guildford) Ltd
Avon Rubber Co Ltd
B & M Mouldings Ltd
Bluemel Bros Ltd
Borden Chemical Co (UK) Ltd
Bordesley Rubber Co Ltd
British Steel Springs Ltd
Bunting Electric Co Ltd
Bushing Co Ltd
Cashmore Plastics Ltd
Clyde Paper Co Ltd
Coates (James) Bros Ltd
Coes (Derby) Ltd
Coin Insulation Ltd
Corrosion Control Ltd
Courtaulds Engineering Ltd
Crabtree Engineering (Rochdale) Ltd
Crane Packing Ltd
DM Industrial Plastics Ltd
Denton & Co (Rubber Engineers) Ltd
Ferguson Shiers Ltd
Fluorocarbon Ltd
Frost (C B) & Co Ltd
General Developments Co (Glasgow) Ltd
Greenbank Engineering Co Ltd
Hadrian Plastics Ltd
Heaven Dowsett & Co Ltd
High Peak Plastics Ltd
Hydralon Ltd
ICL Plastics Ltd
Insulcap Ltd
Kelcoat Engineering Plastics Ltd
Ketch Plastics Ltd
LB Plastics Ltd
Litholite Mouldings Ltd
Loyne (Belfast) Ltd
Loyne Ltd
Merlin Reinforced Plastics Ltd
Metal Box Co Ltd
Metal Pressings Ltd
Newmore Plastics Ltd
Norris Cheshire Ltd
Parglas Ltd
Plastic Coating Equipment Ltd
Plastic Coatings Ltd
Plastic Constructions Ltd
Plastic Dipping Co Ltd
Plastic Dip Mouldings Ltd
QD Plastics (Glasgow) Ltd
Reinforced Plastics Applications (Swansea) Ltd
Resistoid Ltd
Robinson Waxed Paper Co Ltd
Rudic Products Ltd
Siegrist Orel Ltd
Stelling (G H) & Sons Ltd
Sundt Plastics Ltd
Tanerone Ltd
Tennant (Charles) & Co Ltd
Thermoplastic Coatings Ltd
Transatlantic Plastics Ltd
Tullis (John) & Son Ltd
Valley Plastic Coaters Ltd
Ward & Goldstone Ltd
Wheway Frames Ltd

FABRICATORS & WELDERS

Chemical Industry Fabrication	A
General Fabrication	B
Film Fabrication & Sealing	C
Welding Service	D

Company	Services
AC Plastic Industries Ltd	B D
APV Kestner	A B
Acalor (1948) Ltd	A B D
Acrilite Ltd	B
Acrow (Plastics) Ltd	B C
Advance Plastic Products Ltd	B D
Airborne Upholstery Ltd	B D
Ainsworth Marsland & Co Ltd	B D
Airothene Ltd	B
Airpak Ltd	B D
Akrus Ltd	B D
Alexander (S M) Plastics Ltd	B D
Alpa Steel & Plastics Ltd	B D
Amalgamated Plastics (ERS) Ltd	B C
Amplex Appliances (Kent) Ltd	C
Anglo-American Vulcanised Fibre Co Ltd	B
Aqua-Dynamics Ltd	B D
Ardleigh Laminated Plastics Co Ltd	A B
Armitage Plastics Ltd	B
Armshire Reinforced Plastics Ltd	A B C
Attewell (B) & Sons Ltd	B
Attwater & Sons Ltd	B
BP Chemicals International Ltd	B
Bakelite Xylonite Ltd (BXL) (Flexible Packaging Division)	C
Bakelite Xylonite Ltd (BXL) (PVC Division)	A B
Bakelite Xylonite Ltd (BXL) (Scintillex Products)	B D
Bassett (H S) & Son Ltd	B
Birmingham Mica Co Ltd	B
Blight & White Ltd	B
Bondlite Ltd	B
Booth (William) & Co (Metal Work) Ltd	A B
Borden Chemical Co (UK) Ltd	A B
British Steel Corporation (Chemicals Division)	B
British Steel Springs Ltd	B
British Visqueen Ltd	C
Burns (James) (Loose Leaf Bindings) Ltd	B C D
Bushboard Co Ltd	B
CSK Plastics Ltd	A B
Cago Ltd	B
Caligen Foam Ltd	B
Capper Neill Plastics Ltd	A
Chadburn Darwen Ltd	B
Charlesworth (L G G) Ltd	A B
Chemical Pipe & Vessel Co Ltd	A
Chemical Plastic Fabrications Ltd	A
Coates (James) Bros Ltd	A B C
Coes (Derby) Ltd	A B D
Coin Insulation Ltd	B
Colgate (D J) Plastics Ltd	C
Comoy Components Ltd	A B
Cope Allman Plastics Ltd	B
Coplastix Ltd	B D
Corfield (Reginald) Ltd	B D
Corrosion Control Ltd	A B D
Courtaulds Engineering Ltd	A B D
Coventry Motor & Sundries Ltd	B D
Cow (P B) (Plastics) Ltd	B
Crabtree Engineering (Rochdale) Ltd	A B
Crane Packing Ltd	B
Cravens Homalloy (Preston) Ltd	B
Creators Ltd	B D
Crossley (Henry) (Packings) Ltd	A B
Cyroma Plastics Ltd	B C D
DM Industrial Plastics Ltd	B
Davidson & Co (Coachbuilders) Ltd	B
Davis (F J) (Motor Accessories) Ltd	B D
Denco (Clacton) Ltd	B
Denton & Co (Rubber Engineers) Ltd	B D
Dewey Waters & Co Ltd	A B D
Dobson (John) (Milnthorpe) Ltd	B D
Dorman & Smith Ltd	A B
Dormex Industrial Plastics	A B D
Douthwaite (T N) Ltd	A B D
Drix Plastics Ltd	C D
Du Vergier (E) & Co Ltd	B
Dunbee Ltd	B
Dunlop Co Ltd (Polymer Engineering Division)	B
Durapipe & Fittings	B
Duraplex (Plastics)	B D
Duraweld Ltd	D
East Coast Plastics Ltd	D
Eastwood Plastics Ltd	B D
Elfi-Astex Ltd	B D
Egatube Ltd	B D
Elco Plastics Ltd	D
Electra Plastic Welders Ltd	D
Elk & Co Ltd	D
Elta Plastics Ltd	B D
Enalon Plastics Ltd	B
English Electric Reinforced Plastics Division	B
Ensecote Ltd	A
Envopak Ltd	B C
Export Packing Service Group Ltd	B D
FPT Industries Ltd	B D
Falconcraft Ltd	B
Fasweld Ltd	C
Ferguson Shiers Ltd	B
Fisons Scientific Apparatus Ltd	D
Flexo Plywood Industries Ltd	B
Fluorocarbon Co Ltd	A B
Forbes (Kenneth) (Manchester) Ltd	A B D
Forbes (Kenneth) (Plastics) Ltd	A B D
Forest Engineering (Notts) Ltd	B
Formatrend Ltd	D
Frost (C B) & Co Ltd	B
Gazelle Plastics Ltd	B
General Celluloid Co Ltd	C D
General Developments Co (Glasgow) Ltd	A B
Giplex Ltd	D
Gilchrist & Fisher Ltd	B D
Gina Plastics Ltd	B D
Giplex Ltd	C
Glass Fibre Fabrications Ltd	A B D
Godfrey Insulation Ltd	B
Goudies of Bothwell Ltd	B
Graham & Wylie Ltd	B
Graham (H G) & Son Ltd	B
Greenbank Engineering Co Ltd	B D
Griffiths (A E) (Smethwick) Ltd	A B
Griflex Products Ltd	B
HJB Plastics Ltd	B D
Hadrian Plastics Ltd	A B
Haines & Sherman Ltd	B D
Halmatic Ltd	B
Harrison & Jones (Flexible Foam) Ltd	B
Harvey (J J) (Manchester) Ltd	A B
Hartley Baird Ltd	A B
Henry & Thomas Ltd	B
High Peak Plastics Ltd	D
Hilma Ltd	D
Hispeed Plastics Ltd	B D
Hozelock Ltd	B
ICL Plastics Ltd	D
Impetus Building Components Ltd	B D
Ingram (Thomas A) & Co Ltd	B
Initial Plastics Industries Ltd	B
Injection Moulded Plastics Ltd	D
Insulating Components & Materials Ltd	B
Insulation Equipments Ltd	A B
Intaplas Mouldings Ltd	D
Invicta Plastics Ltd	B
Jagger (Albert) Ltd	B
Jennings Bros (Harringay) Ltd	B
Jupiter Plastics Ltd	B D
Kay-Metzeler Ltd	B D
Kings Plastics Ltd	D
Kingston Plastics Ltd	D
LB (Plastics) Ltd	B
Lambert Bros (Walsall) Ltd	A B D
Lane (P H) Ltd	C D
Lawrence (I J) & Son Ltd	D
London Association for the Blind	C

London Bankside Products Ltd B D
Longwell Green Coachworks Ltd B
Low (Gordon) (Plastics) Ltd C D
Lynester Products Ltd D
MP Plastics (Maurice Powell Ltd) A
Maison Fittings Ltd B
Mardle (G B) Ltd B D
Marshall (C & C) Ltd B
Marshall (C F) & Sons Ltd D
Martins Birmingham Ltd B D
Mathews & Yates Ltd A B D
Mellowes Orb Engineering Ltd B
Mercury Signs Ltd B
Merlin Reinforced Plastics Ltd A D
Mica & Micanite Supplies Ltd A B
Micanite & Insulators Ltd B
Mickleover Transport Ltd B
Mill Plastics Ltd B
Mitchell (Andrew) & Co D
Mitchenall Bros Ltd A
Modern Veneering Co Ltd B
Monks (M) & Co Ltd A
Montgomery (Daniel) & Son Ltd A
Moon Aircraft Ltd A B
Moore (Harold) & Son Ltd A B D
Morgan & Grundy Ltd A B
Mosses & Mitchell Ltd B D
Moulded Fasteners Ltd D
Nickwood Plastics Ltd D
Norris (Cheshire) Ltd B
North East Glass Fibre Works Ltd A
North West Plastics Ltd B
Norton Plastics Ltd D
Novoplas Ltd B
Nutt & Co Ltd D
Olympic Plastics Ltd B D
Omega Plastics Ltd B
Osma Plastics Ltd B
P & B Plastics Ltd B
PDI Ltd A B
Parglas Ltd B D
Peerless Plastics Ltd B
Pendred (Norman) & Co Ltd B D
Performance Plastics Ltd D
Permafence Ltd B
Permali Ltd A B
Phoenix Timber Co Ltd B
Pickup (John) Ltd B
Plasro Plastics Ltd D
Plastic Constructions Ltd A B D
Plastic Dip Mouldings Ltd B
Plastic Supplies (Bristol) Ltd D
Plastic Weldings Ltd D
Plastics Design & Engineering Ltd A B C
Plastics Fabrications Co Ltd B
Plastics Fabrication & Printing Ltd B D
Plastics Reinforced (Liverpool) Ltd A B
Plastifilm Products Ltd D
Plysu Products Ltd B
Polybags Ltd C
Polygrow Plastics D
Portobello Fabrications Ltd B
Power Plastics Ltd B
Pressure Sealed Plastics Ltd D
Prodorite Ltd A B
Pylon Developments Ltd B
QD Plastics (Glasgow) Ltd A B D
RF Development Co Ltd D
Rainbow (S H) Ltd D
Raleigh Industries Ltd B
Rayburn Plastics Ltd A B D
Rector (M R) Ltd D
Reddiplex Ltd C
Rediweld Ltd A B D
Redland Quilting Ltd D
Regent Plastics Ltd B
Reinforced Plastics Applications (Swansea) Ltd A B D
Resistant Equipment Ltd B D
Rolinx Ltd D
Ronder Plastic Assemblies Ltd A B D
Rotafoil Ltd B
Screenprints (Vacuum Formers) Ltd B
Serk (R & D) A B D
Sherborne G R P Mouldings A B
Shrewsbury Plastics & Engineering Ltd A B
Snow (J E) Plastics Ltd B
Solport Bros Ltd D
South Western Plastics Ltd D
Splintex Ltd B
Standard Engineering Plastics Ltd B D
Stephens (Plastics) Ltd B D
Sundt Plastics Ltd B D
Swansea Plastics & Engineering Ltd B
Symbol Signs B
TAC Construction Materials Ltd D
TT Mouldings Ltd D
Taylor (C F) (Plastics) Ltd A B
Taylor Industries Ltd A B
Technical Treatments Ltd B D
Technichrome Products (Plastics) Ltd D
Tenaplas Ltd A B D
Thermoplastics Ltd A B
Thames Estuary Plastics Ltd C
Transatlantic Plastics Ltd A B D
Triplex Foundries Group Ltd B
Tropical Packers (Cheshire) Ltd B
Tuckers (Sheffield) Ltd A B D
Tufnol Ltd B
Tullis (John) & Son Ltd B
Turner & Hughes Ltd C
Tye (John) & Sons (Packages) Ltd D
Tygadure (Division of Fothergill & Harvey Ltd) B
UK Plastics Ltd B D
Uniroyal Ltd B
Unit Buying Services Ltd D
United Moulders Ltd D
Vacmobile (Ireland) Ltd B
Varicol Signs Ltd B
Vicsons Ltd B D
Visijar Laboratories Ltd B
Vitafoam (Wraystone) Ltd B
Vitafoam (Bradford) Ltd B
Vogue Plastics Ltd C
WB Industrial Plastics Ltd A B
Walker Litherland Plastic Ltd B
Ward (Thomas W) Ltd A
Warren (F R) & Co Ltd B D
Wear Ventilator & General Sheet Metal Co Ltd A B D
Weldex Plastics Ltd D
Weltonhurst Ltd A B
West Country Converters Ltd B C
Westfield Engineering Co (Marine) Ltd B
Westinghouse Brake & Signal Co Ltd B
Whaley Welding Co Ltd A B
Wheway Frames Ltd D
Wilson (F E) Plastics Ltd B D
Wilson Sheriff Ltd A B
Wokingham Plastics Ltd B
Xlon Products Ltd B D

HIGH PRESSURE LAMINATE FABRICATORS

Facing sheets	A
Machineable Sections	B
Bonding service	C

AVP Industries Ltd A
Aaronson Bros Ltd A B
Acalor (1948) Ltd C
Airscrew Weyroc Ltd A
Aladdin Components Ltd B
Alpa Steel & Plastics Ltd B
Anderson (C F) & Son Ltd A B
Anglo American Vulcanised Fibre Co Ltd A B
Arborite Ltd A
Ardleigh Laminated Plastics Co Ltd C
Ardray Manufacturing Co Ltd B
Arlington Plastics Development Ltd B
Armabord Ltd A
Armoride Ltd B
Armshire Reinforced Plastics Ltd A B C
Ascol Products Ltd A B C
Attwater & Sons Ltd B
Avlam Ltd A B C
BPB Industries Ltd A B
Bakelite Xylonite Ltd (BXL) (Ind. Prod. Div.) A B
Bassett (H S) & Son Ltd A C
Birmingham Mica Co Ltd B
Bonded Laminates Ltd A C
Borden Chemical Co (UK) Ltd B
Borst Bros Ltd A
Bourne Plastics Ltd B
Bribond Signs Ltd C
British Gypsum Ltd A B
British Steel Springs Ltd A B
British Transfer Printing Co Ltd C
Burbridge (H) & Son Ltd B
Burns (J) Ltd A B
Bushboard Co Ltd A B C
Bushing Co Ltd (The) B
Cambell (Malcolm) (Plastics) Ltd B
Canvac Products Ltd B
Catalin Ltd A
Cellgrave Ltd C
Central Manufacturing & Trading (Dudley) Ltd A
Chamberlain Phipps Ltd C
Chester Electroplating Co C
Clark (H) & Co (Manchester) Ltd B
Clyde Paper Co Ltd C
Coated Specialities Ltd C
Coates (James) Bros Ltd B C
Coes (Derby) Ltd A B
Coin Insulation Ltd A B C
Commercial Plastics Ltd A B
Comoy Components Ltd A B
Compra Plastics Ltd A
Concargo Ltd C
Corrosion Control Ltd B
Courtaulds Engineering Ltd B
Cox (William J) (Sales) Ltd B
Cravens Homalloy (Sheffield) Ltd B
Cromwell (E M) & Co Ltd B C
Crossley (Henry) (Packings) Ltd B
D O & E Industries Ltd B C
Davall (S) & Sons Ltd B
Davidson & Co (Coachbuilders) Ltd A B C
Dean & Furbisher (Darrington) Ltd A
Denton & Co (Rubber Engineers) Ltd C
Dewey Waters & Co Ltd A B C
Dorman & Smith Ltd A
Dormex Industrial Plastics Ltd A
Douthwaite (T N) Ltd B C
Du Vergier (E) & Co Ltd A
Dunbee Ltd B
Duraplex (Plastics) B
Efi-Astex Ltd C
Elliott (E) Ltd C
Elson & Robbins Ltd A B
Elta Plastics Ltd C
Enalon Plastics Ltd B
Ensecote Ltd B
Essex Mica Co Ltd B
Evans (Frederick W) Ltd C
FPT Industries Ltd B
Ferguson Shiers Ltd B C
Finlock Group Ltd A

Company	Class
Fitzgibbon & Murray Ltd	B
Flexible Reinforcements Ltd	B
Flexo Plywood Industries Ltd	A
Forbes (Kenneth) (Manchester) Ltd	B
Forbes (Kenneth) (Plastics) Ltd	B
Formica Ltd	A
Formwood Ltd	A
Foster (W H) & Sons Ltd	B C
Gazelle Plastics Ltd	B
General Developments Co (Glasgow) Ltd	B
Getalit Ltd	A B
Graham & Wylie Ltd	A B C
Griflex Products Ltd	B
Hadrian Plastics Ltd	A B
Hall & Co (South East) Ltd	A B
Halmatic Ltd	B
Heaven Dowsett & Co Ltd	B
Hedges Reinforced Plastics Ltd	B
Industrial Mouldings Ltd	A
Ingram (Thomas A) & Co Ltd	B
Insulating Components & Materials Ltd	B
Insulating Materials Ltd	B
Insulation Equipments Ltd	A B C
Isere Nord	A B
JBC Plastics (Dewsbury) Ltd	B
Kay-Metzeler Ltd	C
Kew Laminates Ltd	A B
La Brecque Engineering Co Ltd	B
Laconite Ltd	A
Langley (London) Ltd	C
Latter (A) & Co Ltd	B
Lintafoam Ltd	C
Litholite Mouldings Ltd	B
Loftheath Plastics Ltd	A C
Longwell Green Coachworks Ltd	B
MP Plastics (Maurice Powell Ltd)	B C
MacEchern Ltd	B
Marglass Ltd	C
Marley Floor Tile Co Ltd	B
Mackie (James) & Sons Ltd	B
Marston Excelsior Ltd	B
Mathews & Yates Ltd	A B C
Mercury Signs Ltd	A
Merlin Reinforced Plastics Ltd	B
Metal Box Co Ltd	B
Mica & Micanite Supplies Ltd	B
Micanite & Insulators Ltd	B
Mica Products Ltd	B
Mickleover Transport Ltd	B
Minerva Mouldings Ltd	B
Minnesota Mining & Manufacturing Co Ltd (MMM)	B
Modern Veneering Co Ltd	A C
Moon Aircraft Ltd	C
Morane Plastic Co Ltd	C
Mosses & Mitchell Ltd	B
Moulded Fasteners Ltd	B
Nairn Floors	A
Newmor (Plastics) Ltd	A
North East Glass Fibre Works	A
Pancolite Plastics Ltd	B
Parglas Ltd	A B
Peel Engineering Co Ltd	B
Peerless Plastics Ltd	A
Perkins (E) & Co Ltd	A B
Permali Ltd	B
Performance Plastics Ltd	B
Phoenix Timber Co Ltd	A C
Plastic Constructions Ltd	B C
Plastics Design & Engineering Ltd	B C
Plastic Fabrication Co Ltd	A B C
Plastics Reinforced (Liverpool) Ltd	B C
Poron Insulation Ltd	B C
Precision Insulation Products Ltd	B
Precision Products (Leeds) Ltd	B
Precision Units (Dorset) Ltd	C
Priestly Studios Ltd	A
Prodorite Ltd	B
Railko Ltd	B
Rayburn Plastics Ltd	B
Raylite Supplies Ltd	A
Rector (M R) Ltd	C
Reinforced Plastics Applications (Swansea) Ltd	A B C
Rolls-Royce (Composite Materials) Ltd	B
SBD Construction Products Ltd	A B
Sanenwood Products Ltd	B
Serk (R & D)	B
Shadbolt (F R) & Sons Ltd	A
Shaw Hathernware	B
Short Bros & Harland Ltd	B
Shrewsbury Plastics & Engineering Ltd	A B
Sindall Concrete Products Ltd	B
Smith (Stanley) & Co	A
Stainless Steel Fabricators Ltd	B
Stanton (A E) Plastics (Southern) Ltd	B
Storey Bros & Co Ltd	B
Superide Ltd	B
Swales (William A) & Co Ltd	C
Symbol Signs	B
Taylor (C F) (Plastics) Ltd	B
Telcon Plastics Ltd	B
Tenoosa Ltd	A
Textile Bonding Ltd	C
Thames Estuary Plastics Ltd	B
Trade Laminators Ltd	A C
Transatlantic Plastics Ltd	B C
Tryka Ltd	A
Tuckers (Sheffield) Ltd	B C
Tufnol Ltd	B
Turbro Ltd	B
Turner & Hughes Ltd	B
Vacuum Research Ltd	C
Venesta International-Vencel Ltd	A
Vitalam (Mandleberg) Ltd	B
Ward & Goldstone Ltd	A
Ward (Thomas W) Ltd	B
Wardle (Bernard) (Everflex) Ltd	B
Wear Ventilator & General Sheet Metal Co Ltd	C
Westinghouse Brake & Signal Co Ltd	B C
Westpole Products Ltd	B
Willmott Son & Phillips Ltd	B
Zephyr Plastic Products Ltd	B

GLASS REINFORCED PLASTICS MOULDERS

AC Plastic Industries Ltd
APV Kestner Ltd
Abbott (Malcolm) Ltd
Acalor (1948) Ltd
Acrow Plastics Ltd
Acrow (Engineers) Ltd
Anderson (D) & Son Ltd
Anglo American Vulcanised Fibre Co Ltd
Anmac Ltd
Applied Plastics (Scunthorpe) Ltd
Ardleigh Laminated Plastics Co Ltd
Armitage Plastics Ltd
Armshire Reinforced Plastics Ltd
Arnold Designs Ltd
Attwater & Sons Ltd
BIP Reinforced Products Ltd
B & K Moulds Ltd
BS Equipment Ltd
BTR Reinforced Plastics Ltd
Bakelite Xylonite Ltd (BXL) (Industrial Products Division)
Basildon Moulding Co Ltd
Birkbys Ltd
Birmingham Mica Co Ltd
Birmingham Plastics Ltd
Blight & White Ltd
Blythe (William) & Co Ltd
Bondlite Ltd
Borden Chemical Co (UK) Ltd
Bourne Plastics Ltd
Brensal Plastics Ltd
Bristol Aerojet Ltd
British Steel Springs Ltd
Buckingham (J F) Ltd
Burns (J) Ltd
Bushing Co Ltd (The)
Cammell Laird (Anglesey) Ltd
Cape Universal Building Products Ltd
Capper Neill Plastics Ltd
Central Manufacturing & Trading Co (Dudley) Ltd
Clarke (H) & Co (Manchester) Ltd
Clover Leaf (Products) Ltd
Coin (Glass Fabric) Mouldings Ltd
Coin Insulation Ltd
Commercial Plastics Ltd
Comoy Components Ltd
Con Cargo Ltd
Corrosion Control Ltd
Cortina Plastics Ltd
Coubro & Scrutton (M & I) Ltd
Courtaulds Engineering Ltd
Crane Packing Ltd
Cravens Homalloy (Preston) Ltd
Cravens Homalloy (Sheffield) Ltd
Cromwell (E M) & Co Ltd
Crossley (Henry) (Packings) Ltd
DB Plastics (Waltham Abbey) Ltd
DM Industrial Plastics Ltd
Darpress Industrial Ltd
Davidson & Co (Coachbuilders) Ltd
Dewey Waters & Co Ltd
Dormex Industrial Plastics Ltd
Douthwaite (T N) Ltd
Duple Motor Bodies (Northern) Ltd
Duple Group Sales Ltd
ERF Engineering Ltd (Plastics Division)
Enalon Plastics Ltd
Engineering & Glassfibre Developments Ltd
English Electric Reinforced Plastics Division
Ensecote Ltd
Essex Mica Co Ltd
Fi-Glass Developments Ltd
Finlock Group Ltd
Flexo Plywood Industries Ltd
Forbes (Kenneth) (Manchester) Ltd
Forbes (Kenneth) (Plastics) Ltd
Fothergill & Harvey Ltd
GC Plastics Ltd
General Developments Co (Glasgow) Ltd
Gilbern Sports Car (Components) Ltd
Glascade (G R P) Ltd
Glasdon Ltd
Glass Fibre Developments Ltd
Glassfibre Fabrications Ltd
Glass Fibre Laminates Ltd
Glass Fibre Engineering Ltd
Gloster Engineering (Cheltenham) Ltd
Greenbank Engineering Co Ltd
Griffiths (A E) (Smethwick) Ltd
Hadrian Plastics Ltd
Hall & Co (South East) Ltd
Halmatic Ltd
Heaven Dowsett & Co Ltd
Hedges Reinforced Plastics Ltd
Helmets Ltd
Hills Precision Die Castings Ltd
Humpage & Moseley Ltd
ICL Plastics Ltd
Imco Plastics Ltd
Imperial Chemical Industries Ltd (ICI)
Insulating Components & Materials Ltd
International Computers Ltd
Island Plastics Ltd
JBC Plastics (Dewsbury) Ltd
Kenney Plastics Co Ltd
Kirton Kayaks

LG Plastics
La Breque Engineering Co Ltd
Lambert Bros (Walsall) Ltd
Langley (London) Ltd
Litholite Mouldings Ltd
Loftheath Plastics Ltd
Longwell Green Coachworks Ltd
Lynester Products Ltd
MP Plastics (Maurice Powell Ltd)
MacEchern & Co Ltd
Marshall of Cambridge (Engineering) Ltd
Marston Excelsior Ltd
Mellowes Orb Engineering Ltd
Merlin Reinforced Plastics Ltd
Mica & Micanite Supplies Ltd
Micanite & Insulators Ltd
Mickleover Transport Ltd
Microplas Ltd
Minerva Mouldings Ltd
Mitchenall Bros Ltd
Mitra (Plastics) Ltd
Modern Plastics Ltd
Monks (M) & Co Ltd
Morgan, Berkeley & Co Ltd
Mosses & Mitchell Ltd
Moulded Fastners Ltd
Moulded Plastics (Birmingham) Ltd
Mycalex & Rosite Ltd
Newtown Industries
North Manchester Plastics Ltd
Norton Plastics Ltd
Novoplas Ltd
Osma Plastics Ltd
P & S (Cirencester) Ltd
Pancolite Plastics Ltd
Parglas Ltd
Pearce Road Signs Ltd
Peel Engineering Co Ltd
Peerless Plastics Ltd
Perkins (E) & Co Ltd
Permali Ltd
Petitdemange (R) Ltd
Phosco Ltd
Plastic Constructions Ltd
Plastic Development
Plastic Dip Mouldings Ltd
Plastics Design & Engineering Ltd
Plastics Reinforced (Liverpool) Ltd
Plastona (John Waddington) Ltd
Plessey Co Ltd (Components Group) (The)
Portobello Fabrications Ltd
Prodorite Ltd
Pyke (M C) & Associates
Pylon Developments
Radnall (F A) & Co Ltd
Rayburn Plastics Ltd
Rediweld Ltd
Redland Pipes Ltd
Regina Industries Ltd
Reinforced & Microwave Plastics Group
Reinforced Plastics Applications
(Swansea) Ltd
Resinform Ltd
Resistant Equipment Ltd
Rock Electrical Accessories Ltd
Rollite Products Ltd
Rolls-Royce (Composite Materials) Ltd
Ronder Plastic Assemblies Co
Rootes Mouldings Ltd
Rotafoil Ltd
Rubberplas (Birmingham) Ltd
Rydgeway Plastics Ltd
SBD Construction Products Ltd
Sailcraft Ltd
Sales Aids (Sedlescombe)
Salter Plastics Ltd
Serk (R & D)
Shaw Hathernware
Sherborne G R P Mouldings
Short Bros & Harland Ltd
Shrewsbury Plastics & Engineering Ltd
Sindall Concrete Products Ltd
Specialised Mouldings Ltd
Stainless Steel Fabricators Ltd
Stechford Mouldings Ltd
Storey Bros & Co Ltd
Strebor Plastics
Straco Ltd
Swansea Plastics & Engineering Ltd
Swift (S M) (Exeter) Ltd
Symbol Signs
TAC Construction Materials Ltd
TPT Ltd
Taylor (C F) (Plastics) Ltd
Temple Building Products Ltd
Thetford Moulded Products Ltd
Thomson (T) Sons & Co (Barrhead) Ltd
Tod (W & J) Ltd
Transatlantic Plastics Ltd
Triton Plastics Ltd
Tuckers (Sheffield) Ltd
Tufnol Ltd
Tullis (John) & Son Ltd
Unitex
Vectis Laminates Ltd
Vitamol (Clyde) Ltd
Vitesta Ltd
WCB Containers Ltd
Waterside Plastics Ltd
Watson Industrial Plastics Ltd
Welsh Mouldings (1964) Ltd
Westbrook Marine Co Ltd
Western Laminates Ltd
Westfield Engineering Co (Marine) Ltd
Westford Plastics & Engineering Ltd
Westpole Products Ltd
Whaley Welding Co Ltd
Whitson (James) & Co Ltd
Widnes Foundry & Engineering Co Ltd
Wincanton Transport & Engineering Co Ltd
Willmott Son & Phillips Ltd
Woodland Plastics
Zonex Ltd

EXPANDED POLYSTYRENE MOULDERS

Airfix Industries Ltd
Airfix Plastics Ltd
Alpine Insulations Ltd
Arcol Thermoplastics Ltd
Armshire Reinforced Plastics Ltd
Astex Ltd
Ball Plastics Ltd
Basildon Moulding Co Ltd
Bourne (H & W) Ltd
Bricell Plastics Ltd
British Steel Springs Ltd
Brough (G & S) Ltd
Burns (J) Ltd
DM Industrial Plastics Ltd
Denton & Co (Rubber Engineers) Ltd
Conbox Co Ltd
Dohm Industrial Ltd
Elford Plastics Ltd
Elliott (E) Ltd
Foam Plastics Packaging Ltd
Foley Packaging & Insulation Ltd
Fraser & Glass Ltd
Frost (C B) & Co Ltd
Girdlestone Electronics Ltd
Invicta Plastics Ltd
Jackmore Ltd
Jobling (James A) & Co Ltd
Judge International Ltd
Latter (A) & Co Ltd
Levity Plastics Ltd
Long & Hambly Ltd
Plastic Dip Mouldings Ltd
Polythene Drums Ltd
Polytop Plastics Ltd
Pontefract Box Co Ltd
Poron Insulation Ltd
QED Design & Development Ltd
Reinforced Plastics Applications
(Swansea) Ltd
Ridout (D H) & Son Ltd
Rock Electrical Accessories Ltd
Rondopack Plastics Ltd
Ross Warmafoam Ltd
STC (Standard Telephones & Cables) Ltd
Siegrist Orel Ltd
Tekta Packaging Ltd
Transatlantic Plastics Ltd
Thyne (William) (Plastics) Ltd
United Moulders Ltd

FOAMED PLASTICS FABRICATORS

Acrow (Plastics) Ltd
Airfix Industries Ltd
Arcol Thermoplastics Ltd
Arlington Plastics Development Ltd
Armshire Reinforced Plastics Ltd
Avon Rubber Co Ltd
Astex Ltd
Bakelite Xylonite Ltd (BXL)
Baxenden Chemical Co Ltd
Boston Foam Mouldings Ltd
Bourne (H & W) Ltd
Bowater Scott Corporation Ltd
Bricell Plastics Ltd
Brigtown Industries Ltd
British Vita Co Ltd
Caligen Foam Ltd
Central Manufacturing & Trading Co
(Dudley) Ltd
Clifford Covering Co Ltd
Coes (Derby) Ltd
Commercial Plastics Ltd
Coplastix Ltd
Cork Growers Ltd
Cyroma Plastics Ltd
DM Industrial Plastics Ltd
Delcon Foam Plastics Ltd
Donbox Co Ltd
Driclad Ltd
Elson & Robbins Ltd
Ferguson Shiers Ltd
Fomair Ltd
Foam Plastics Packaging Ltd
Foley Packaging & Insulation Ltd
Freeman (William) & Co Ltd
Frost (C B) & Co Ltd
General Celluloid Co Ltd
Hadrian Plastics Ltd
Harrison & Jones (Flexible Foam) Ltd
Hills Precision Die Castings Ltd
Hispeed Plastics Ltd
Hyman (I & J) Ltd
Imperial Chemical Industries Ltd (ICI)
JBC Plastics (Dewsbury) Ltd
Kay-Metzeler Ltd
Lane (P H) Ltd
Levity Plastics Ltd
Lintafoam Ltd
Marley Foam Ltd
Mayfair Plastics Ltd
Montgomery (Daniel) Ltd
Netlon Ltd

Newalls Insulation & Chemical Co Ltd
Non-Sag Seating Ltd
North West Plastics Ltd
Omega Plastics Ltd
Orr (William) (Foam) Ltd
Plastic Dip Mouldings Ltd
Plastikade (Manchester) Ltd
Plastics Fabrication Co Ltd
Plastics Fabrication & Printing Ltd
Pontefract Box Co Ltd
Poron Insulation Ltd

Pritex (Plastics) Ltd
QED Design & Development Ltd
Reinforced Plastics Applications (Swansea) Ltd
Rubbarite Ltd
Rubber Plastics Ltd
Screenprints (Vacuum Formers) Ltd
Serk (R & D)
Silleck Engineering
Superide Ltd
Technovac Ltd

Tekta Packaging Ltd
Tesa Tapes Ltd
Thermalon Ltd
Transatlantic Plastics Ltd
Tuckers (Sheffield) Ltd
Unitex
Venesta International-Vencel Ltd
Vitafoam Ltd
Vitafoam Wrayston Ltd
Vitafoam Bradford Ltd
Waterside Plastics Ltd

ADDRESSES

A C Plastic Industries Ltd
Prospect Works
216 Sydenham Road
Croydon CRO 2EB Surrey
01-689 1555

A C W Ltd
40 Hutcheon Street
Aberdeen Scotland
Aberdeen 26357

A D I Plastics Ltd
Preston New Road
Blackpool FY3 9TP Lancs
Blackpool 61301

AEI Moulded Products Division
Oak Mill
Skipton Road
Colne Lancs
Colne 3838

A I Paramount Plastics
5–9 Annerley Station Road
London SE20
01-778 2201

A P Industries (Nelson) Ltd
Hollin Bank
Brier Field
Nelson Lancs
Nelson 65217

A P V Kestner Ltd
Greenhithe Kent
Greenhithe 3281

A V P Industries Ltd
Lea Valley Trading Estate
London N18 3HS
01-807 0411

Aaronson Bros Ltd
Town Wharf
Rickmansworth Herts
Rickmansworth 76655

Abbott (Malcolm) Ltd
59 West End
Redruth Cornwall
Redruth 5601

Able Developments Co
Tenbury Wells Worcs
Tenbury Wells 600

Acalor (1948) Ltd
Kelvin Way
Crawley Sussex
Crawley 23271

Acme Showcard & Sign Co Ltd
Green Street
Brimsdown
Enfield Middlesex
01-804 1651

Acrilite Ltd
The Old Mill
Standon
Ware Herts
Ware 821551

Advanced Moulding Developments Ltd
Navigation Street
Wolverhampton
Wolverhampton 24854

Advance Plastic Products Ltd
White Lund Estate
Morecambe Lancs
Lancaster 67401

Airborne Upholstery Ltd
Arterial Road
Southend-on-Sea
Southend-on-Sea 525265

Airfix Industries Ltd
Haldane Place
Garratt Lane
London SW18
01-870 0151

Airfix Plastics Ltd
Windmill Road
Sunbury-on-Thames
Sunbury-on-Thames Middx
Sunbury-on-Thames 85131

Ainsworth Marsland & Co Ltd
38–40 Denham Road
Sheffield 11
Sheffield 60375

Airlite Plastics Ltd
Church Street
Dorking Surrey
Dorking 4696

Airothene Ltd
56 Beethoven Street
London W10
01-969 4999

Airpak Ltd
Overton
Overton Hants
Overton 441

Akrus Ltd
2 Gorleston Road
London N15
01-800 4827

Airscrew Weyroc Ltd
Weybridge Surrey
Weybridge 45599

Albis Plastic Co (Great Britain) Ltd
York House
Empire Way
Wembley Middx
01-903 2147

Aladdin Components Ltd
Western Avenue
Greenford Middx
01-578 2300

Alexander (S M) Plastics Ltd
Cromwell Road
St Neots Hunts
St Neots 3140

Alexander (Silverthorne) Ltd
Silverthorne Lane
Cradley Heath
Warley Worcs
Cradley Heath 66325

Allen George (Plastics) Ltd
Northway House
High Road
London N20
01-446 0061

Allen Plastics Ltd
303–305 Portswood Road
Southampton SO2 1LD
Southampton 57595

Alpa Steel & Plastics Ltd
Maybank Road
London E18 IEY
01-504 9211

Alpine Insulations Ltd
Alpine Works
Newton Road
Crawley Sussex
Crawley 26326

Alpine Plastics Ltd
49 Bowlers Crofts
Basildon Essex
Basildon 20281

Altrincham Plastics Ltd
76a Stamford New Road
Altrincham Cheshire
061-928 2803

Aluminium & Plastic Products Ltd
70 Dock Street
Newport Mon
Newport 59188

Amalgamated Plastics (E R S) Ltd
30–31 Station Close
Potters Bar Herts
Potters Bar 50771

Amplex Appliances (Kent) Ltd
6a Tylney Road
Bromley BR1 2RL Kent
01-640 5531

Amtico Flooring Ltd
Foleshill Road
Coventry CV6 5AG
Coventry 88771

Anchor Chemical Developments Ltd
Birch Vale
Stockport Lancs
New Mills 3233

Anderson (C F) & Son Ltd
Islington Green
London N1 2XJ
01-226 1212

Anderson (D) & Son Ltd
Stretford
Manchester M32 OYL
061-865 4444

Anglo-American Vulcanised Fibre Co Ltd
21 Bath Street
London EC1V 9EP
01-253 8484

Anmac Ltd
Daleside House
Daleside Road
Nottingham NG2 3GG
Nottingham 89661

Anson Plastics Ltd
133 Blyth Road
Hayes Middx
01-573-5626

Ansorite Manufacturing Co
Highbury Grove
London N5
01-226 0909

Applied Plastics (Scunthorpe) Ltd
Butherwick Road
Althorpe
Scunthorpe Lincs
Scunthorpe 74381

Aqua-Dynamics Ltd
16 Berkeley Street
London W1
01-493 2766

Aquitaine-Organico (UK) Ltd
Colthrop Lane
Thatcham
Newbury Berks
Kennet Bridge 456

Arborite Ltd
Factory T4
Norham Road
North Shields Northumberland
North Shields 70241

Arcol Thermoplastics Ltd
Newlandsfield Works
Riverford Road
Glasgow S3
041-649 2831

Archway Plastics Ltd
Causton Road
London NE6
01-340 4971

Arcode Ltd
Plantation House
Mincing Lane
London EC3
01-623 7074

Ardleigh Laminated Plastics Co Ltd
Halstead Essex
Halstead 3434

Ardray Manufacturing Co Ltd
3–5 Chesterfield Mews
London N4
01-800 4652

Arlington Plastics Development Ltd
South Road
Harlow Essex
Harlow 24611

Arinor Ltd
Stockbury House
Church Street
Storrington Sussex
Storrington 3355

Armabord Ltd
Peel Mill
Foulridge
Colne Lancs
Colne 1422

Armitage Plastics Ltd
Highlands Road
Shirley
Solihull Warwicks
021-705 6381

Armoride Ltd
Grove Mill
Earby
Colne Lancs
Earby 2205

Armshire Reinforced Plastics Ltd
Ipswich Road
Slough Bucks
Slough 26665

Arnold Designs Ltd
Chalford
Stroud GL6 8NR Glos
Brimscombe 2310

Arrow Plastics Ltd
Arrow Works
Hampden Road
Kingston-upon-Thames Surrey
01-546 6258

Ascol Products Ltd
Aycliffe Industrial Estate
Darlington Co Durham
Aycliffe 2622

Ash Plastic Products Ltd
P O Box 16
Rea South Street
Birmingham B5 6LD
021-643 9182

Ashton (Thomas A) Ltd
Speedwell Works
Sidney Street
Sheffield S1 3QB
Sheffield 25211

Associated Resin Compounds Ltd
Jury Lane Works
Sidlesham
Nr Chichester Sussex
Sidlesham 484

Astex Ltd
54 Crewys Road
London NW2
01-455 6722

Atlas Plastics Ltd
Brook Road
Wood Green
London N22 6TR
01-888 4944

Attewell (B) & Sons Ltd
Ridgeway Trading Estate
Iver Bucks
Iver 2030

Attwater & Sons Ltd
Hopwood Street Mills
Preston PR1 1TA
Preston 58245

Aubit Plastics Ltd
Harrowbrook Industrial Estate
Hinckley Leics
Hinckley 2023

Ault & Wiborg Ltd
Standen Road
Southfields
London SW18
01-874 7244

Auriol (Guildford) Ltd
Trading Estate
Farnham Surrey
Farnham 3366

Autodevices (1960) Ltd
Foundry Lane
Horsham Sussex
Horsham 60191

Automobile Plastics Co Ltd
Autoplax House
7 Henry Road
New Barnet Herts
01-449 9147

Avlam Ltd
20 Angel Factory Colony
London N18
01-807 1040

Avon Rubber Co Ltd
Industrial Products Division
Bradford-on-Avon Wilts
Melksham 3101

B A S F United Kingdom Ltd
(Plastics Division)
Knightsbridge House
197 Knightsbridge
London SW7
01-584 5080

B & B Plastics Ltd
Poyle Trading Estate
Colnbrook
Slough Bucks
Colnbrook 2254

B D J (England) Ltd
Maple Cross Industrial Estate
North Orbital Road
Rickmansworth Herts
Rickmansworth 79959

B E F Products (Essex) Ltd
1133 London Road
Leigh-on-Sea Essex
Southend-on-Sea 77771

B & G Plastics Ltd
25 Bath Road
Cheltenham GL53 7HG Glos
Cheltenham 53808

B I P Reinforced Products Ltd
Sutton Coldfield Warwicks
021-353 2411

B I P Chemicals Ltd
(Subsidiary of British Industrial Plastics Ltd)
PO Box 6
Popes Lane
Oldbury
Warley Worcs
021-552 1551

B & K Moulds Ltd
130 Verdant Lane
Catford
London SE6
01-698 1954

B & M Mouldings Ltd
167 Capstone Road
Chatham Kent
Medway 45492

B P B Industries Ltd
Ferguson House
15-17 Maryelbone Road
London NW1
01-935 8246

BP Chemicals International Ltd
(Plastics Department, Polymers Division)
Devonshire House
Piccadilly
London W1X 6AY
01-629 8867

BP Chemicals International Ltd
Lightpill Mills
Stroud Glos
Stroud 3301

B S Equipment Ltd
Freeth Street
Meadow Lane Nottingham
Nottingham 81231

B T L Mould & Die Co Ltd
Cradock Road
Luton Beds
Luton 54271

B T R Reinforced Plastics Ltd
Rockingham Road
Uxbridge Middx
Uxbridge 32484

Bakelite Xylonite Ltd (BXL)
27 Blandford Street
London W1H 3AD
01-935 9211

(Cascelloid Division)
Abbey Lane Leicester
Leicester 61811

(Flexible Packaging Division)
Huddersfield Road
Darton
Barnsley Yorks
Darton 3388

(Industrial Products Division)
Manningtree Essex

(Polyethylene Division)
PO Box 25
Inchyra Road
Grangemouth Stirlingshire
Grangemouth 3400

(PVC Division & Thermosetting Division)
12-18 Grosvenor Gardens
London SW1
01-730 0898

Ball Plastics Ltd
Kennel Lane
Billericay Essex
Billericay 51283

Ball Plastics (Clacton) Ltd
Valley Bridge Road
Clacton-on-Sea Essex
Clacton-on-Sea 21731

Banarse (P S) & Co (Products)
34-36 Mozart Street
London W10
01-969 0239

Barclay Stuart (Plastics) Ltd
25-27 Brunswick Street
Luton Beds
Luton 26363

Bardex (Plastics) Ltd
119 Guildford Street
Chertsey Surrey
Chertsey 3484

Barpak Products Ltd
Alma Park
Grantham Lincs
Grantham 2323

Basildon Moulding Co Ltd
Chester Hall Lane
Basildon Essex
Basildon 3366

Basset (H S) & Son Ltd
Bathurst Street
Swansea SA1 3RZ
Swansea 55904

Baveystock (A) & Co Ltd
Cooks Road
London E15 2PL
01-534 7591

Baxenden Chemical Co Ltd
Paragon Works
Baxenden
Accrington BB5 2SL
Accrington 34631

Bayer Chemicals Ltd
Kingsway House
Richmond Surrey
01-940 6077

Beadel (James) & Co Ltd
Frodsham House
Edwards Lane
Speke
Liverpool L24 9HX
051-486 1395

Begg & Co (Thermoplastics) Ltd
Harness Road
Woolwich Industrial Estate
London SE18
01-854 1236

Beldam Asbestos Co Ltd
Lascar Works
Hounslow Middx
01-570 7722

Belgrave Plastic Developments Ltd
Cheshire Works
Stockport Road
Cheadle Heath
Stockport Cheshire
Stockport 6515

Bendix & Herbert Ltd
270 London Road
London E7
01-472 991

Benker & Co Ltd
Commercial Street
Todmorden Lancs
Todmorden 2457

Berk Ltd
Berk House
Basing View
Basingstoke Hants
Basingstoke 29292 (01-486 6688)

Berkshire Mouldings Ltd
Temple Mill
Passfield
Liphook Hants
Passfield 281

Bettix Ltd
Dunbar Road
New Malden Surrey
01-942 6691

Bexford Ltd
Manningtree Essex
Manningtree 2424

Bibby Chemicals Ltd
8 Stanley Street
Liverpool L1 6EZ

Bibby (R) (Engineers) Ltd
Moss Side Works
Tavistock Road
Callington PL17 7DY Cornwall

Birchware Ltd
North Farm Estate
Royal Tunbridge Wells Kent
Tunbridge Wells 24158

Birkbys Ltd
PO Box No 2
Liversedge Yorks
Heckmondwike 3721

Birmingham Mica Co Ltd
3 South Road
Hockley Birmingham B18 5NN
021-554 5631

Birmingham Plastics Ltd
86-88 Hospital Street
Birmingham 19
021-236 1944

Blackburn & Oliver Ltd
Lamberhead Industrial Estate
Pemberton
Wigan Lancs
Wigan 82252

Blagden & Noakes (Holdings) Ltd
Plantation House
Mincing Lane
London EC3M 3HP
01-626 0081

Blewis & Shaw (PVC) Ltd
Lower Ham Road
Kingston-upon-Thames Surrey
01-546 1161

Blight & White Ltd
Prince Rock
Plymouth Devon
Plymouth 65333

Blow (J J) Ltd
Oldfield Works
Chatsfield Road
Chesterfield Derbys
Chesterfield 76635

Blow Mouldings Ltd
Hampton Street
Birmingham 19
021-236 3866

Bluebell Dolls Ltd
Tulketh Street
Southport Lancs
Southport 56871

Bluemel Bros Ltd
The Works
Wolston
Coventry CV8 3FU
Coventry 352244

Blue Peter Urethane Products Ltd
Worting Village
Basingstoke Hants
Basingstoke 4404

Blythe (William) & Co Ltd
Holland Bank Works
Church
Accrington BB5 4PD Lancs
Accrington 37211

Boardman (A) Ltd
Douglas Industrial Estate
Douglas
Lanark Scotland
Douglas 398

Boddington (W H) & Co Ltd
Horsmonden
Nr Tonbridge Kent
Brenchley 2277

Bonded Laminates Ltd
Chisenhale Road
London E3 5QZ
01-980 2005

Bondlite Ltd
Balm Mill
Roberttown
Liversedge Yorks
Heckmondwile 2007

Bonnella (D H) & Son Ltd
West Hill
Hoddesdon Herts
Hoddesdon 64484

Booth (William) & Co (Metal Work) Ltd
Old Road Works
Warrington Lancs
Warrington 33569

Borden Chemical Co (UK) Ltd
North Baddesley
Southampton SO5 9ZB
Rownhams 2131

Bordesley Rubber Co Ltd
Houghton Street
West Bromwich Staffs
021-553 2225

Borst Bros Ltd
Taylor's Road
Stotfold
Hitchin Herts
Stotfold 277

Boston Foam Mouldings Ltd
3 Maple Road
Boston Lincs
Boston 3338

Bourne (H & W) Ltd
64 Spring Lane
London SE25
01-654 7777

Bourne Plastics Ltd
Harby Road
Langar Nottingham
Harby 621

Bowater Scott Corporation
Knightsbridge
London SW1
01-584 7070

Boxmakers (Manchester) Ltd
444 Ordsall Lane
Salford M5 3EP

Boyriven Ltd
Riven Works
Bridgewater Road
Wembley HA0 1AW Middx

Bradford Glass Co Ltd
Spring Mill Street
Bradford BD5 7DT Yorks
Bradford 33400

Brensal Plastics Ltd
Highbridge Somerset
Burnham-on-Sea 2393

Bribond Signs Ltd
Victoria Road
Burgess Hill Sussex
Burgess Hill 5611

Bricell Plastics Ltd
Worsley Road North
Worsley
Manchester M28 5QJ
Farnworth 72781

Brigtown Industries Ltd
Green Lane
Cannock Staffs
Cannock 3891

Brighouse Plastics Ltd
Atlas Mill Road
Brighouse HD6 1ES Yorks
Brighouse 2271

Bristol Aerojet Ltd
Banwell
Weston-super-Mare Somerset
Banwell 2251

British Celanese Ltd
22-23 Hanover Square
London W1A 1BS
01-629 8000

British Celophane Ltd
Regal House
Twickenham Middx
01-892 0125

British Chemical Products & Colours Ltd
Canova House
22 Buckingham Street
London WC2
01-839 3634

British Gypsum Ltd
Ferguson House
15-17 Marylebone Road
London NW1
01-486 1282

British Industrial Plastics Ltd
PO Box 6
Popes Lane
Oldbury
Warley Worcs
021-552 1551

British Needle Co Ltd
Argosy Works
Victoria Street
Redditch Worcs
Redditch 63205

British Picker Co Ltd
Sandholme Mills
Todmorden Yorks
Todmorden 2457

British Ropes Ltd
Condercum House
PO Box 1AE
171 West Road
Newcastle-upon-Tyne
Newcastle 39111

British Steel Corporation
(Chemicals Division)
Orgreave
Sheffield S13 9NJ
Sheffield 3211

British Steel Springs Ltd
Handsworth
Birmingham B21 9EW
021-554 9045

British Traders & Shippers Ltd
Stevinson House
155 Fenchurch Street
London EC3
01-623 1541

British Transfer Printing Co Ltd
Stoke
Coventry Warwicks
Coventry 53381

British Visqueen Ltd
6 Hills Way
Stevenage Herts
Stevenage 3400

British Vita Co Ltd
Middleton
Manchester M24 2DB
061-643 4301

Brodie Dowmex Ltd
The Mews
Hildreth Street
London SW12
01-673 7512

Brookes & Adams Ltd
Shady Lane
Kingstanding
Birmingham B44 9DX

Brough (G & S) Ltd
25-29 Commercial Street
Birmingham B1 1RU
021-643 3574

Brymor Ltd
The Tannery
East Peckham
Tonbridge Kent
East Peckham 384

Buckingham (J F) Ltd
Priory Road
Kenilworth CV8 1LE
Kenilworth 52351

Bunting Electric Co Ltd
Lion Works
69 Warstone Lane
Birmingham 18
021-236 5868

Bunzl & Biach (British) Ltd
(Plastics Department)
Friendly House
21-24 Chiswell Street
London EC1
01-606 9966

Burbridge (H) & Son Ltd
Coventry Warwicks
Coventry 72361

Burn (James) (Loose Leaf Bindings) Ltd
Douglas Road
Esher Surrey
Esher 63513

Burns (J) Ltd
Wangye Works
Chadwell Heath
Romford RM6 4EX Essex
01-599 0211

Bush Beach & Segner Bayley Ltd
175 Tottenham Court Road
London W1
01-580 8041

Bushboard Co Ltd
Princes Way
Team Valley Trading Estate
Gateshead
Low Fell 878484

Collingwood Street
Newcastle NE1 1JD
Newcastle 28842

Smithfold Lane
Walkden
Worsley
Manchester
061-790 4491

Bushing Co Ltd (The)
South Drive
Hebburn Co Durham
Hebburn 832241

Byron Jardine Ltd
New Basford
Nottingham NG7 7HR
Nottingham 77731

Byson Appliance Co Ltd
Woolfold
Bury BL8 1TD Lancs
061-764 4832

CIBA–GEIGY (UK)
Duxford
Cambridge CB2 4QA
Sawston 2121

Campbell (Rex) & Co Ltd
A M P House
Dingwall Road
Croydon CR9 3QU
Surrey

Cago Ltd
59-63 Burlington Street
Aston
Birmingham 6
021-359 2881

Cammell Laird (Anglesey) Ltd
Beaumaris
Anglesey North Wales
Beaumaris 431

Caligen Foam Ltd
Broad Oak
Accrington BB5 2BS Lancs
Accrington 32241

Camden Optical Co Ltd
42-44 Bowlers Croft
Basildon Essex
Basildon 20506

Campbell (Malcolm) (Plastics) Ltd
Byron Road
Addleston Surrey
Weybridge 48356

Cape Universal Building Products Ltd
Exchange Road
Watford Herts
Watford 34551

Capper Neill Plastics Ltd
Foundry Lane
Horsham Sussex
Horsham 60121

Capon Heaton & Co Ltd
Hazelwell Mills
Stirchley
Birmingham 30
021-458 3511

Carborundum Co Ltd (The)
Trafford Park
Manchester 17
061-872 2381

Carpenter (J H) & Son (London) Ltd
107 York Way
London N7
01-485 3917

Carter Bros (Billingshurst) Ltd
Reliance Works
Billingshurst Sussex
Wisborough Green 551

Carville Ltd
Station Road
Dorking Surrey
Dorking 81681

Cassidy Bros Ltd
Mitcham Road
Blackpool Lancs
Blackpool 62288

Castle Products Ltd
Clinton Road
Leominster Herts
Leominster 2831

Catalin Ltd
54 Farm Hill Road
Waltham Abbey Essex
Waltham Cross 23344

Cellgrave Ltd
Phillip Road
London SE15

Celmac Plasclip Ltd
Highbank Mill
Godley
Hyde SK14 2QE Cheshire
Hyde 3542

Central Manufacturing & Trading Co (Dudley) Ltd
Halesowen Road
Dudley Worcs
Cradley Heath 69434

Chadburn Darwen Ltd
The Green
Borough Road
Darwen Lancs
Darwen 71578

Chamberlain Phipps Ltd
Wood Street
Higham Ferrers
Wellingborough Northants
Rushden 3084

Charlesworth (L G G) Ltd
North Hill Works
Worcester Road
Malvern Worcs
Malvern 61831

Chemical & Insulating Co Ltd (The)
West Aukland Road
Darlington Co Durham
Darlington 3547

Chemical Pipe & Vessel Co Ltd
Frimley Road
Camberley Surrey
Camberley 4414

Chemical Plastic Fabrications Ltd
Satley Trading Estate
Birmingham BB 1BG
021-327 3372

Chemicals Trading Co Ltd
25 Berkeley Square
London W1
01-499 1246

Chemidus Plastics Ltd
Brunswick Road
Cobbs Wood
Ashford Kent
Ashford 22271

Chester Electroplating Co
City Road
Chester CH1 3AE
Chester 46606

Cimid Ltd
2-10 Commonside East
Mitcham Surrey
01-640 4141

Cisterns Ltd
Addingham
Ilkley Yorks
Addingham 444

Chemoplast, B & I Ltd
52 Haymarket
London SW1
01-930 4734

Clarke (H) & Co (Manchester) Ltd
Atlas Works
Patricroft
Manchester M30 0RR
061-789 5301

Clarke (James) & Sons
(Stoke-on-Trent) Ltd
Stoke-on-Trent ST6 1DX Staffs

Clayton-Wright (Howard) Ltd
Wellesbourne Warwick
Stratford-upon-Avon 4222

Clear Span Ltd
Wellington Road
Greenfield
Oldham Lancs
Saddleworth 3244

Clearex Products Ltd
Fence Houses
Houghton-le-Spring Co Durham
Fence Houses 2031

Clifford Covering Co Ltd
Spring Road
Hall Green
Birmingham B11 3DN
021-777 5261

Clements (R A) (London) Ltd
1 Taunton Place
Gloucester Place
London NW1
01-262 8403

Clyde Paper Co Ltd
Clyde Paper Mills
Rutherglen
Glasgow
041-647 4381

Coated Specialities Ltd
Chester Hall Lane
Basildon Essex
Basildon 3331

Coates (James) Bros Ltd
Townley Street
Middleton
Manchester
061-643 2653

Coba Plastics Ltd
New Road
Kibworth Leicester
Kibworth 2118

Coes (Derby) Ltd
Thirsk Place
Ascot Drive
Derby DE2 8JL
Derby 46301

Coin (Glass Fabric) Mouldings Ltd
Coin Works
St Johns Road
Woking Surrey
Woking 64121

Coin Insulation Ltd
Coin Works
St Johns Road
Woking Surrey
Woking 64121

Coinmechs Ltd
Pavor Road
Watcombe
Torquay Devon
Torquay 38661

Cole (R H) Ltd
7/15 Landsdowne Road
Croydon CR9 2HB Surrey
01-686 4411

Cole Plastics Ltd (Cole (RH) Ltd)
Batford Mill
Harpenden Herts
Harpenden 60161

Cole Polymers Ltd (Cole (R H) Ltd)
Empire Works
Mitcham Road
Croydon CR0 3AB
01-684 3431

Coleman Industrial Plastics Ltd
Fir Street
Heywood Lancs
Heywood 68641

Colgate (D J) Plastics Ltd
Station Road
Loudwater
High Wycombe Bucks
High Wycombe 27325

Collard (S A) Ltd
Wetherby Road
Derby DE2 8HL
Derby 48533

Collins & Harmer Plastics Ltd
129 Nathan Way
Woolwich Industrial Estate
London SE18
01-854 9712

Columbia Products Ltd
Binstead
Ryde Isle of Wight
Ryde 3761

Combex Ltd
Great Portland Street
London W1
01-580 3264

Combined Optical Industries Ltd
198/200 Bath Road
Slough SL 4DW Bucks
Slough 21292

Commercial Plastics Ltd
Berkeley Square House
London W1X 6AN
01-629 8030

Comoy Components Ltd
90/92 Pentonville Road
London N1 9H2
01-837 1089

Compact Cases Ltd
Pontygwindy Industrial Estate
Caerphilly Wales
Caerphilly 5811

Compoflex Ltd
Lumb Mill
Delph
Nr Oldham Lancs
Saddleworth 5511

Compounded Polymers Ltd
Ford Road Industrial Estate
Clacton-on-Sea Essex
Clacton-on-Sea 22391

Compra Plastics Ltd
Dicker Mill
Hertford
Hertford 4677

Con Cargo Ltd
Oldmixon Industrial Estate
Winterstoke Road
Weston-super-Mare Somerset
Weston-super-Mare 28221

Connor (Herbert) Ltd
120 Beaufort Park
London NW11
01-455 5023

Connor (Herbert) (Kent) Ltd
39 Coblers Bridge Road
Herne Bay Kent
Herne Bay 3361

Constructional Products (Chadderton) Ltd
Chadderton
Oldham Lancs
061-624 1999

Contemporary Furnishing Co Ltd
Glynn Works
New Lane
Havant Hants
Havant 2486

Coordinators Service Engineering Ltd
Shady Lane
Great Barr
Birmingham 22A
021-357 4237

Cope Allman Plastics Ltd
Fitzherbert Road
Farlington
Portsmouth PO6 1SD
Cosham 70102

Coplastix Ltd
Site 3
Park Farm Industrial Estate
Redditch Warwicks
Ipsley 2777

Coral Plastics Ltd
Worsley Road North
Worsley
Manchester M28 5PU
Farnworth 74711

Corfield (Reginald) Ltd
29/31 Holmethorpe Avenue
Redhill Surrey
Redhill 64641

Cork Growers Ltd
Vulcan Street
Bootle Lancs
051-922 1917

Cornbrook Resin Co Ltd
Clough House Mill
Wardle
Rochdale
Littleborough 78616

Cornelius Chemical Co Ltd
Ibex House
Minories
London EC3
01-480 7525

Cornercroft (Plastics) Ltd
Ace Works
Parkside
Coventry Worcs
Coventry 23391

Corrosion Control Ltd
113a Corporation Street
Belfast BT1 3AE N. Ireland
Belfast 22004

Cosmocord Ltd
Acos Works
Eleanor Cross Road
Waltham Cross Herts
Waltham Cross 27331

Coubre & Scrutton (M & I) Ltd
430 Barking Road
London E13
01-476 4477

Courtaulds Engineering Ltd
PO Box 11
Foleshill Road
Coventry CV6 5AB Warwicks
Coventry 88771

Coventry Motor & Sundries Co Ltd
Spon End
Coventry CB1 3GY Warwicks
Coventry 20363

Cow (P B) (Plastics) Ltd
5 Falmouth Road
Slough Bucks
Slough 20262

Cow (P B) (Special Products) Ltd
Liverpool Road
Slough Bucks
Slough 20333

Cox (William J) (Sales) Ltd
London Road
Tring Herts
Tring 3286

Crabtree Engineering Group (Colne) Ltd
Green Works
Colne Lancs
Colne 5533

Crane Packing Ltd
Berwick Avenue, Trading Estate
Slough SL1 4QX Bucks
Slough 31122

Cravens Homalloy (Preston) Ltd
382 Blackpool Road
Preston PR2 2DP Lancs
Preston 29233

Staniforth Road
Darnall
Sheffield S9 4LL
Sheffield 49301

Crawford Plastics Ltd
Hazlewood Works
Arterial Road
Leigh-on-Sea Essex
Southend-on-Sea 524261

Cray Valley Products Ltd
St Mary Cray
Orpington BR5 3PP Kent
Orpington 32545

Creators Ltd
Albert Drive
Sheerwater
Woking Surrey
Woking 5981

Crew Electroplating Ltd & Cheshire Plastics
Town Bridge Works
Winsford Cheshire Winsford 2011
Winsford 2011

Critchley Bros
Brimscombe
Stroud GL5 2TH Glos
Brimscombe 2451

Crittal-Hope Ltd
Manor Works
Braintree Essex
Braintree 106

Halford Works
Smethwick
Birmingham 40
021-558 2191

Croda Chemicals Ltd
Cowich Hall
Snaith
Goole Yorks
Snaith 551

Cromwell (E M) & Co Ltd
Hadham Road
Bishops Stortford Herts
Bishops Stortford 2371

Crossley (Henry) (Packings) Ltd
Hill Mill
Astley Bridge
Bolton BL1 6PE Lancs
Bolton 41351

Curtis (Neville) & Co Ltd
Old Ford Goods Depot
Tredegar Road
London E3
01-980 5411

Curtis Plastics Ltd
Cradock Road
Skimpot Industrial Estate
Luton Beds
Luton 54215

Crystalate Mouldings Ltd
Crystalate House
Mill Lane
Tonbridge Kent
Tonbridge 2261

Cyanamid of Great Britain Ltd
Bush House
Aldwych
London WC2B 4PU
01-836 5411

Cyroma Plastics Ltd
Beaumont Road
Banbury Oxon
Banbury 50865

D B Plastics Ltd
37 Sun Street
Waltham Abbey Essex
Waltham Cross 25749

D M Industrial Plastics Ltd
Terrace Road
Walton-on-Thames Surrey
Walton-on-Thames 25710

D O & E Industries Ltd
42/44 Holmethorpe Avenue
Redhill Surrey
Redhill 64726

Dalau Specialised Plastics Ltd
Ford Road
Clacton-on-Sea Essex
Clacton-on-Sea 22391

Dale (John) Ltd
14/24 Brunswick Park Road
New Southgate
London N11
01-368 1272

Daleman (Richard) Ltd
325 Latimer Road
London W10
01-969 7455

Darpress Industrials Ltd
20 Nottingham Road
Somercotes Derbys
Leabrooks 2323

Davall (S) & Sons Ltd
1 Wadsworth Road
Perivale
Greenford Middx
01-998 1011

Davidson & Co (Coachbuilders)Ltd
Church Road
Liverpool 13
051-228 6377

Davidson Industries
Smitham Bridge
Hungerford Berks
Hungerford 2336

Davis (F J) (Motor Accessories) Ltd
18/20 Romeyn Road
Streatham
London SW16
01-769 2892

Dean & Furbisher (Darrington) Ltd
Pontefract Yorks
Pontefract 4796

Declon Foam Plastics Ltd
Cranbourne Road
Potters Bar Herts
Potters Bar 57151

Denco (Clacton) Ltd
355/9 Old Road
Clacton-on-Sea Essex
Clacton-on-Sea 22807

Denton & Co (Rubber Engineers) Ltd
South Shields Co Durham
South Shields 2363

Derwent Plastics Ltd
Bridge Works
Stamford Bridge YO4 1AL Yorks
Stamford Bridge 551

Devcon Ltd
Station Road
Theale
Reading Berks
Theale 304

Dewey Waters & Co Ltd
Coxs Green
Wrington
Bristol BS18 70S
Wrington 601

Dines Plastics Ltd
Whitehall Lane
Grays Essex
Grays Thurrock 4675

Dip Moulds Ltd
Trading Estate
Farnham Surrey
Farnham 21131

Display Craft Ltd
15/31 Morrish Road
London SW2
01-671 8262

Dixopak Ltd
Ivy Mill
Hensingham
Whitehaven Cumberland
Whitehaven 3761

D-Mac Ltd
Queen Elizabeth Avenue
Hillington Industrial Estate
Glasgow SW2
041-882 3364

Dobson (John) (Milnthorpe) Ltd
Bella Mill
Milnethorpe Westmorland

Donbox Co Ltd
West Tullos Industrial Estate
Aberdeen AB9 4AL
Aberdeen 27251

Dohm Industrial Ltd
Dohm House
Norwood Road
London SE24
01-674 7845

Dorman & Smith Ltd
Blackpool Road
Preston PR2 2DQ Lancs
Preston 28271

Dormex Industrial Plastics Ltd
Hatfield Street
Manchester M9 1QB
061-205 1545

Doulton Insulators Ltd
Tamworth Staffs
Tamworth 2113

Douthwaite (T N) Ltd
Blue Bell Yard Shiremoor Northumberland
Shiremoor 292

Dow Chemical Co (UK) Ltd
105 Wigmore Street
London W1
01-935 4441

Dowty Seals Ltd
Ashchurch
Tewksbury Glos
Tewkesbury 2441

Drelco Ltd
Valley Road Industrial Estate
St Albans Herts
St Albans 50881

Driclad Ltd
Staplehurst Road
Sittingbourne Kent
Sittingbourne 4422

Drix Plastics Ltd
Richmond Gardens
Portswood
Southampton Hants
Southampton 59666

Dryburgh (D) & Co Ltd
Fitzalan House
High Street
Ewell Surrey
01-394 0021

Dunbee Ltd
117/123 Great Portland Street
London W1
01-580 3264

Dunlop Co Ltd
(Polymer Engineering Division)
Evington Valley Road
Leicester
Leicester 36531

Dunlop Co Ltd
(Precision Rubbers Division)
Shepshed
Leicester
Shepshed 2151

Dugdale (S) Son & Co Ltd
Valley Mill
Sowerby Bridge
Yorks
Halifax 21381

Duple Group Sales Ltd
The Hyde
Hendon
London NW9
01-205 6412

Duple Motor Bodies (Northern) Ltd
Vicarage Lane
Blackpool Lancs
Blackpool 62251

Du Pont (United Kingdom) Ltd
Du Pont House
18 Bream's Buildings
Fetter Lane
London EC4
01-242 9044

Dura Co (Cheadle) Ltd
Harley Road
Sale Cheshire
061-973 2900

Durapipe & Fittings Ltd
Bentinck Road
West Drayton Middx
West Drayton 3631

Duraplex Plastics Ltd
201 Newport Road
London E10
01-556 9122

Duratube & Wire Ltd
Central Way
Faggs Road
Feltham Middx
01-890 3453

Duraweld Ltd
Sherwood Street
Scarborough Yorks
Scarborough 63538

Durham Raw Materials Ltd
1/4 Great Tower Street
London EC3
01-626 4333

Dust Covers Ltd
India House
75 Whitworth Street
Manchester M1 6HB
061-236 4809

Du Vergier (E) & Co Ltd
Duplast Works
Essex Road
Hoddesdon Herts
Hoddesdon 62217

East Anglia Plastics Ltd
(Cole (E R) Ltd)
Temple Works
Knight Road
Strood Kent
Medway 78441

E R F Engineering
(Plastics Division)
New Street
Bidulph Moor
Stoke-on-Trent Staffs
Bidulph 2433

East Coast Plastics Ltd
Laundry Lane
North Walsham Norfolk
North Walsham 3461

Eastwood Plastics Ltd
Arterial Road
Leigh-on-Sea Essex
Southend-on-Sea 525265

Ebonestos Industries Ltd
Excelsior Works
Rollins Street
London SE15
01-639 2080

Econopack Ltd
Industrial Estate
Heage Road
Ripley
Derby DE5 3FH
Ripley 3615

Edge (Wilfred) & Sons Ltd
Burscough Town
Ormskirk Lancs
Burscough 2239

Eeto Plastics Ltd
Market Street
Tottington
Bury Lancs
Tottington 2151

Efi-Astex Ltd
Craig-yr-Eos
Brecon Road
Merthyr Tydfil Wales
Merthyr Tydfil 3391

Egatube Ltd
St Asaph
Flintshire N Wales
St Asaph 2431

Ekco Plastics Ltd
Priory Crescent
Southend-on-Sea SS2 6PP Essex
Southend-on-Sea 49481

Elco Plastics Ltd
Desborough Park Road
High Wycombe Bucks
High Wycombe 21164

Electra Plastic Welders Ltd
6 Elder Street
Commercial Street
London E1
01-247 2238

Elford Plastics Ltd
Brookfield Works
Wood Street
Elland Yorks
Elland 3053

Elliott (E) Ltd
Bescot Crescent
Walsall Staffs
Walsall 28951

Elk & Co Ltd
Elkon Works
West Molesey Surrey
01-979 8441

Elm Plastics Ltd
22 Motney Road
Rainham
Gillingham Kent
Gillingham 31447

Elson & Robbins Ltd
Bennett Street
Long Eaton
Nottingham NG10 4HL
Longeaton 2225

Elta Plastics Ltd
Yarm Road
Stockton
Teeside Co Durham
Stockton 62299

Enalon Plastics Ltd
North Premier Works
Drayton Road
Tonbridge Kent
Tonbridge 3343

Enfield Plastics Ltd
Alexander Road
Enfield Middx
01-804 2833

Engineering & Glassfibre Developments Ltd
Colne Lancs
Colne 4145

Engineering Polymers
Evelyn House
30 Alderley Road
Wilmslow Cheshire
Wilmslow 29611

English Electric Reinforced Plastics Div.
Freckleton Works
Mill Lane
Warton
Preston PR4 1AR Lancs
Freckleton 521

English Electric Plastics Division
Dick Lane Works
Bradford BD3 7AU Yorks

Enkothene Ltd
30/31 Station Close
Potters Bar Herts

Ensekote Ltd
Thornecliffe
Sheffield S30 4YP
Ecclesfield 3171

Envopak Ltd
Envopak Works
Powerscroft Road
Sidcup Kent
01-300 7861

Espinasse Ltd
Riven Works
Bridgewater Road
Wembley Middx
01-902 9581

Essex Mica Co Ltd
Coronation House
Hainault
Ilford Essex
01-500 4166

Eurobung Ltd
Roe Head Mill
Far Common Road
Mirfield Yorks
Mirfield 3376

Evans (Frederick) Ltd
Plastic Works
Long Acre
Birmingham 7
021-327 3071

Evered & Co (Holdings) Ltd
Surrey Works
Lewisham Road
Smethwick
Warley Worcs
021-558 3191

Everready Plastic Co Ltd
Chesham Close
Romford Essex
Romford 64191

Ewarts Ltd
PO Box No 5
Dudley Worcs
Dudley 54411

Ewart Plastics Ltd
8 Long Street
Shoreditch
London E2
01-739 9253

Expanded Rubber & Plastics Ltd (BXL)
675 Mitcham Road
Croydon CR9 3AC Surrey
01-684 3622

Export Packing Service Group
Staplehurst Road
Sittingbourne Kent
Sittingbourne 4422

Extruda Ltd
Bury Street
Radcliffe
Manchester M26 9QB
061-736 3811

Extrudex Products (BXL)
Aycliffe Trading Estate
Darlington Co Durham
Aycliffe 5122

Eyre & Baxter (Stampcraft) Ltd
229 Derbyshire Lane
Sheffield S8 8SD
Sheffield 50135

F P T Industries Ltd
The Airport
Portsmouth PO3 5PE Hants

F W H Plastics Ltd
Park Farm Road
Folkeston Kent
Folkeston 56665

Falconcraft Ltd
83/95 Hainault Road
Romford RM5 3AH Essex

Fasweld Ltd
Grenofen House
Tavistock Devon
Tavistock 2625

Featly Products Ltd
Farrel Street
Salford M7 9TJ Lancs
061-832 6546

Fecher (M J) Ltd
Cuckoo House
Welcroft Road
Slough Bucks
Slough 21864

Ferguson (James) & Sons Ltd
Lea Park Works
St George's Road
Merton Abbey
London SW19
01-648 2283

Ferguson Shiers Ltd
Cheetham Street
Manchester
061-681 4041

Fern Plastic Products Ltd
Cross Street North
Wolverhampton Staffs
Wolverhampton 27459

Ferro (Great Britain) Ltd
Wombourne
Wolverhampton Staffs
Wombourne 4144

Fibre Form Ltd
Holloway Street
Lower Gornal
Dudley Worcs
Sedgley 3866

Fibrebond Ltd
Bridge Road
Woolston via Warrington Lancs
Warrington 34019

Fibre Glass Ltd
St Helens Lancs
St Helens 24022

Fibrenyle Ltd
Skylon House
Gosford Road
Beccles Suffolk
Beccles 2442

Fi-Glass Developments Ltd
Station Road
Edenbridge Kent
Edenbridge 3465

Finlay (William) (Belfast) Ltd
Ballygomartin Road
Belfast BT13 3NN
Belfast 45555

Finlock Group Ltd
Frant Road
Tunbridge Wells Kent
Tunbridge Wells 27161

Finlock Products Ltd
Lyde Road
Penmill Trading Estate
Yeovil Somerset
Yeovil 4372

Fisher (Harold) (Plastics) Ltd
Grove Mills
Gynn Lane
Honley
Nr Huddersfield Yorks
Huddersfield 61271

Fisons Scientific Apparatus Ltd
Bishops Meadow Road
Loughborough Leics
Loughborough 5781

Fitzgibbon & Murray Ltd
Smallfield Works
Station Road
Horley Surrey
Horley 4598

Flexible Packaging Division (BXL)
Huddersfield Road
Darton
Barnsley Yorks
Darton 3388

Flexible Reinforcements Ltd
20 Lord Street
Manchester M4 4FP
061-832 9966

Flexo Plywood Industries Ltd
Flexo Works
South Chingford
London E4
01-529 0177

Flight Equipment & Engineering Ltd
Asheridge Road
Chesham Bucks
Chesham 2563

Flomaster Ltd
Hopwood Lane
Halifax Yorks
Halifax 53748

Fluorocarbon Co Ltd
Caxton Hill
Hertford
Hertford 4697

Foamair Ltd
Manor Farm Road
Alperton Middx
01-998 4041

Foam Engineers Ltd
Dashwood Avenue
High Wycombe Bucks
High Wycombe 20711

Foam Plastics Packaging Ltd
Church Street
Dorking Surrey
Dorking 4696

Foley Packaging & Insulation Ltd
Hardy Road
Farlington
Portsmouth PO6 1JX Hants
Cosham 70128

Forbes (Kenneth) (Manchester) Ltd
Grove Lane
Didsbury
Manchester 20
061-445 3422

Forbes (Kenneth) (Plastics) Ltd
Denver
Downham Market Norfolk
Downham Market 3088

Fordham Pressings Ltd
Melbourne Works
Dudley Road
Wolverhampton WU2 4DS
Wolverhampton 23861

Fordath Ltd
Brandon Way
West Bromwich Staffs
021-553 1665

Fordham Pressings Ltd
Melbourne Works
Dudley Road
Wolverhampton WU2 4DS Staffs
Wolverhampton 23861

Forest Engineering (Notts) Ltd
Radford
Nottingham NG8 1PQ
Nottingham 73075

Formatrend Ltd
Fylde Road
Southport Lancs
Southport 86971

Formed Plastic Containers Ltd
Holmethorpe Avenue
Redhill Surrey
Redhill 64726

Formica Ltd
P O Box 2
De la Rue House
84–86 Regent Street
London W1
01-734 8020

Formold Ltd
Ruscombe Works
Nr Twyford Berks
Loddonvale 5443

Formwood Ltd
Tufthorn Avenue
Coleford Glos
Coleford 3305

Foster Bros Plastics Ltd
Lea Brook Tube Works
P O Box 20
Wednesbury Staffs
021-556 1343

Foster Plastics (Rainford) Ltd
Lords Fold
Rainford
St Helens Lancs
Rainford 2011

Foster (W H) & Sons Ltd
Cardale Street
Rowley Regis
Warley Worcs
Blackheath 1819

Fothergill & Harvey Ltd
Summit
Littleborough Lancs
Littleborough 78831

Fourfold Mouldings Ltd
Hollings Street Works
Cottingley
Bingley Yorks
Bingley 2852

Fraser & Borthwick Ltd
Barrmill Road
Galston Scotland
Galston 484

Fraser & Glass Ltd
995 High Road
North Finchley
London N12
01-445 0355

Freeman Chemicals Ltd
P O Box 8
Ellesmere Port
Wirral Cheshire
051-355 6171

Freeman (William) & Co Ltd
Suba-Seal Works
Staincross
Barnsley Yorks
Barnsley 4081

Frost (C B) & Co Ltd
Vesey Street
Birmingham 4
021-359 4324

Frys Diecastings Ltd
Prince Georges Road
London SW19
01-648 0371

Formpak Ltd
P O Box 1 BS
22 Hanover Square
London W1
01-629 8000

Futters (London) Ltd
16 Acton Lane
London NW10
01-965 4222

G C Plastics Ltd
Riversdale Road Works
Akew Road
Gateshead-upon-Tyne NE8 2JT
Gateshead 70184

G F W Plastics Ltd
Goldthorn Hill
Wolverhampton Staffs
Wolverhampton 39547

G K N Sankey Ltd (Plastics Division)
Wellington
Telford Salop
Wellington 4321

Manor Works
Spring Road
Ettingshall
Wolverhampton
Bilston 43771

G K N Screws & Fasteners Ltd
Kings Norton Division
Stirchley
Birmingham 30
021-458 6611

Gansolite Ltd
Haxby Road
York
York 25552

Garden Pools Ltd
Old Milberton Road
Leamington Spa Warwicks
Leamington 26827

Gazelle Plastics Ltd
Colley Lane Estate
Bridgewater Somerset
Bridgewater 8221

General Celluloid Co Ltd
Beaumont Road
Banbury Oxon
Banbury 50865

General Developments Co (Glasgow) Ltd
Watt Road
Hillington
Glasgow SW2
041-882 4601

General Foam Products Ltd
North Shields Northumberland
Northshields 74191

Genrista Ltd
Springfield Mills
Farsley
Nr Pudsey Yorks
Pudsey 70025

Getalit Ltd
183 Harts Lane
Barking Essex
01-594 5511

Giplex Ltd
12–22 Telford Way
East Acton
London W3 7XB
01-743 9416

Gilbern Sports Car (Components) Ltd
Pentwyn Works
Llantwit Fardre
Pontypridd Glam
Newton Llantwit 571

Gilchrist & Fisher Ltd
Kingsfisher Works
Western Way
Exeter Devon
Exeter 76407

Gina Plastics Ltd
Morris Road
London E14 6NT
01-987 2431

Girdlestone Electronics Ltd
Woodbridge IP12 1ER Suffolk
Woodbridge 2212

Glascade (G R P) Ltd
Quayside Road
Bitterne
Southampton Hants
Southampton 31323

Glasdon Ltd
117–123 Talbot Road
Blackpool FY1 3QY Lancs
Blackpool 22378

Glass Fibre Developments Ltd
Alpine Works
Newton Road
Crawley Sussex
Crawley 25692

Glassfibre Fabrications Ltd
Hadfield Mills
Hadfield
Hyde SK14 7EB
Glossop 3112

Glass Fibre Laminates Ltd
White Lund
Morecambe Lancs
Morecambe 4040

Glass Fibre Engineering Co Ltd
Trading Estate
Farnham Surrey
Farnham 21201

Gloster Engineering (Cheltenham) Ltd
338 Swindon Road
Cheltenham GL51 9JZ Glos
Cheltenham 20343

Glover Plastics Ltd
Edmont Street
Mossley
Ashton-under-Lyne Lancs
Mossley 2444

Glynwed Plastics Ltd
Metrex Division
Alexandra Works
West Bromwich
021-557 4221

Godfrey Insulation Ltd
J J Siddons Factory Estate
Hill Top
West Bromwich Staffs
021-556 0011

Goldring Manufacturing Co (Great Britain) Ltd
486–488 High Road
Leytonstone
London E11
01-539 8343

Goodall (E A) & Co Ltd
Albert Street
Redditch Worcs
Redditch 63341

Goodburn Plastics Ltd
Arundel Road Industrial Estate
Uxbridge Middlesex
Uxbridge 32256

Goodfellow Metals Ltd
Ruxley Towers
Claygate
Esher Surrey
Esher 65391

Goudies of Bothwell Ltd
29–35 Main Street
Bothwell
Glasgow
Bothwell 3396

Graham & Wylie Ltd
Dalmarnock Bridge
Rutherglen
Glasgow 041-447 4301

Graham (H G) & Son Ltd
Perseverence Mills
Cross Chancelor Street
Leeds LS6 2TF
Leeds 34806

Grant Plastics Co Ltd
Unit 12
The Mill Trading Estate
Acton Lane
Harlesdon
London NW10
01-965 9351

Greef (R W) & Co Ltd
Garrard House
31–45 Gresham Street
London EC2
01-606 8771

Green Bros (Geebro) Ltd
Summerheath Road
Hailsham Sussex
Hailsham 771

Greenbank Engineering Co Ltd
Orient Works
Gate Street
Blackburn BB1 3AJ
Blackburn 56401

Greenwood Rawlins & Co Ltd
S R S Works
Brook Road
Rayleigh SS6 7XW Essex
Rayleigh 4903

Griffiths (A E) (Smethwick) Ltd
Booth Street
Smethwick
Warley Worcs
021-558 1571

Griflex Products Ltd
Vauxhall Industrial Estate
Ruabon
Wrexham N Wales
Ruabon 2551

Grilon & Plastic Machinery Ltd
Leader House
117–120 Snargate Street
Dover Kent
Dover 2656

Grover (A F) & Co Ltd
Higham Road
Chesham Bucks
Chesham 4861

Guest Industrials Ltd
81 Gracechurch Street
London EC3
01-626 5631

H J B Plastics Ltd
P O Box 127
Abbey Park Road
Leicester LE4 5B2
Leicester 26301

Hadrian Plastics Ltd
10 Windsor Crescent
Newcastle-upon-Tyne NE6 3PL
Newcastle 811361

Haines & Sherman Ltd
The Runway
South Ruislip HA4 6SE Middlesex
01-845 3535

Haliplas Ltd
Tower Works
33 East Road
Burnt Oak
Edgeware Middx
01-952 5501

Hall & Co (South East) Ltd
39 Cherry Orchard Road
Croydon CR9 6AP Surrey
01-688 4444

Hallite Plastics Ltd
Sedgeley Works
Wigman Road
Aspley
Nottingham
Nottingham 292171

Halmatic Ltd
Brookside Road
Havant PO9 1JR Hants
Havant 6161

Harcostar Ltd
Windover Road
Huntingdon PE18 7EF
Huntingdon 2323

Hardura Group Ltd
P O Box 14
Great Harwood
Blackburn BB6 7NW Lancs
Blackburn 51411

Hardy (M W) (Mercantile) Ltd
Gillett House
55 Basinghall Street
London EC2V 5EB
01-606 4651

Hargill Ltd
Albion Works
Windsor Street
Cheltenham GL52 2DF Glos
Cheltenham 21688

Harrison (A T) & Co (Mouldings) Ltd
30–36 Queensway
Enfield Middx
01-804 5134

Harrison & Jones (Flexible Foam) Ltd
Bee Mill
Shaw Road
Royton
Oldham Lancs
061-633 2451

Harrison & Sons (High Wycombe) Ltd
Coates Lane
High Wycombe Bucks
High Wycombe 22321

Harrison Bros (Plastics) Ltd
375 Tyburn Road
Birmingham 24
021-327 2317

Harwin Engineers Ltd
Farlington
Portsmouth PO6 1RT Hants
Cosham 70451

Harvey (J J) (Manchester) Ltd
Oldham Street
Denton M34 3SL Lancs
061-336 3951

Hartley Baird Ltd
5 Fredericks Place
Old Jewry
London EC2R 8BY
01-606 4170

Harwood Prospect Mill Co Ltd
Albert Mill
Great Harwood
Blackburn BB6 7BG Lancs
Great Harwood 3423

Hayward Plastics Ltd
Chiltern Avenue
Amersham Bucks
Amersham 4101

Healey Injection Mouldings Ltd
Barnfield Industrial Estate
Tipton Staffs
021-557 3538

Healey Mouldings Ltd
Wolverhampton Road
Oldbury
Warley Worcs
021-552 2731

Heathrod Ardwin & Co Ltd
Oxonia Works
Abbingdon Berks
Abbingdon 1541

Heaven Dowsett & Co Ltd
197 Shady Lane
Great Barr
Birmingham 22A
021-357 9391

Hedges Reinforced Plastics Ltd
403 Edgeware Road
The Hyde
London NW7
01-205 4695

Hellerman Plastics
(Bowthorpe Heelerman Ltd)
Gatwick Road
Crawley Sussex
Crawley 28888

Hellyar (John) & Co Ltd
Colewood Road Industrial Estate
Swalecliffe
Whitstable Kent
Chestfield 2841

Helmets Ltd
Moat Factory
Wheathampstead Herts
Wheathampsted 2221

Henley Plastics Ltd
Fairview Works
Henley-on-Thames Oxon
Henley 3861

Hennessey Products (Hampshire) Ltd
Downley Road
Havant North Industrial Estate
Havant PO9 2NJ
Hants

Hercules Powder Co Ltd
1 Great Cumberland Place
London W1H 8AL
01-262 7766

Hermetite Products Ltd
Tavistock Road
West Drayton Middx
West Drayton 3731

Hettich & Walls Plastics Ltd
Victoria Road
Burgess Hill Sussex
Burgess Hill 5333

Henry & Thomas Ltd
Yeo Street
Bow Common
London E3
01-987 4545

High Density Plastics Ltd
Commercial Street
Todmorden Yorks
Todmorden 2457

High Peak Plastics Ltd
St Johns Works
Tideswell
Buxton Derbys
Tideswell 460

Hilma (London) Ltd
Heathmans Road
Parsons Green
London SW6
01-736 1316

Hills Precision Die Castings Ltd
Cateswell Road
Birmingham B11 3DX
021-706 0691

Hilton Plastics Ltd
Hanworth Road
Sunbury-on-Thames Middx
Sunbury 3325

Hindmarch Smith Plastics Ltd
The Mayphil
Battlebridge
Wickford Essex
Wickford 3088

Hispeed Plastics Ltd
Llandegai Industrial Estate
Bangor N Wales
Bangor 4281

Hoechst UK Ltd
Hoechst House
Kew Bridge
Brentford Middx
01-995 1355

Holliday (G J) & Co
Barclays Bank Chambers
Queen Street
Wellington Salop
Wellington 55726

Holloid Plastics Ltd
High Street
Odiham Hants
Odiham 2275

Holpak Ltd
Bessemer Road
Welwyn Garden City Herts
Welwyn Garden 28100

Holmbush Plastics Ltd
Holmbush Industrial Estate
St Austell Cornwall
St Austell 4891

Homogeneous Plastics Ltd
Gosport Works
145 Gosport Road
Walthamstow
London E17
01-520 4206

Honeywill-Atlas Ltd
84 Mill Lane
Carshalton Surrey
01-669 2261

Hozelock Ltd
Haddenham
Aylesbury Bucks
Haddenham 438

Huls (UK) Ltd
Eastbury House
30–40 Albert Embankment
London SE1
01-735 3303

Humpage & Moseley Ltd
86–88 Hospital Street
Birmingham 19
021-236 1144

Huntley Boorne & Stevens Ltd
Woodley
Reading RG5 4SL Berks
Woodley Park 5151

Hupfield Plastics Ltd
Wolsey Road
Hemel Hempstead Herts
Hemel Hempstead 56860

Hutchings (F A) (Kingston) Ltd
326 Ewell Road
Surbiton Surrey
01-397 6552

Hyde Plastics Ltd
Mynster House
8 Peterborough Road
Harrow Middx
01-864 0115

Hydralon Ltd
Britannia Road
Northam
Southampton SO1 1RH Hants
Southampton 24272

Hygienex Industries Ltd
College Road
Fishponds
Bristol BS16 HR
Bristol 651247

Hyam (I & J) Ltd
Brighton Mill
Spencer Street
Oldham Lancs
061-633 2261

Hysol Sterling Ltd
Sterling House
Heddon Street
London W1
01-734 9931

I C L Plastics Ltd
Grove Park Mills
188 Northwoodside Road
Glasgow
041-332 9078

I D Precision Plastics
Forest Row Sussex
Forest Row 2969

I to I Plastics Ltd
15 Thames Road
Barking Essex
01-594 0234

I M I Developments Ltd
Witton
Birmingham B6 7BA
021-356 4848

I T S Rubber Ltd
The Rubber Works
Petersfield Hants
Petersfield 2345

Illingworths (Plastics) Ltd
Coxmoor Road
Sutton-in-Ashfield Notts
Sutton-in-Ashfield 5291

Ilsen (H) Ltd
147 Coles Green Road
London NW2
01-450 6644

Imco Plastics Ltd
Imco Works
Glastonbury Somerset
Glastonbury 3534

Impetus Building Components Ltd
Kirkintilloch
Glasgow
041-776 5161

Imperial Chemical Industries Ltd (ICI)
Millbank
London SW1
01-834 4444

Imperial Chemical Industries Ltd (ICI)
(Dyestuffs Division)
Hexagon House
Blackley
Manchester 9
061-740 1460

Imperial Chemical Industries Ltd (ICI)
(Plastics Division)
Bessemer Road
Welwyn Garden City
Welwyn Garden 23400

Industrial Mouldings Ltd
Emscote Warwick
Warwick 44421

Industrial Plastic Extrusions Ltd
76 Rivington Street
London EC2
01-739 6288

Industrial Polymers Ltd
Berkeley Square House
London W1
01-629 8030

Industrial Reels Ltd
Lamberhead Industrial Estate
Wigan Lancs
Wigan 82611

Industrial Stoving Co Ltd
10 Towerfield Road
Shoeburyness
Southend-on-Sea Essex
Shoeburyness 2385

Ingram (Thomas A) & Co Ltd
Prescott Street
Birmingham 18
021-554 4576

Initial Plastics Industries Ltd
Barnet Trading Estate
Park Road
High Barnet Herts
01-440 2116

Injection Moulded Plastics Ltd
Marsh Lane
London Road
Ware Herts
Ware 5201

Insulating Components & Materials Ltd
Wellhead Lane
Perry Bar
Birmingham B42 2TB
021-356 4554

Insulating Materials Ltd
Park Works
Appleton Village
Widnes Lancs
Widnes 2897

Insulation Equipments Ltd
Salop Road
Oswestry Salop
Oswestry 2351

Insulators Ltd
40 Church Road
Paddock Wood
Tonbridge Kent
Paddock Wood 4488

Insulcap Ltd
197 Shady Lane
Great Barr
Birmingham 22A
021-357 9391

Intaplas Mouldings Ltd
Charleswood Place
East Grinstead Sussex
East Grinstead 24501

International Computers Ltd
Cadwell Lane
Hitchin Herts
Hitchin 52051

Intrama Ltd
Imperial Way
Watford WD2 4JR Herts
Watford 31034

Invenit Industrial Plastics Ltd
Maidstone Road
Paddock Wood
Tonbridge Kent
Paddock Wood 2185

Invicta Plastics Ltd
Oadby
Leicester LE2 5WD
Oadby 3356

Irvine (Martin) (Plastics) Ltd
22 Martlesham Heath
Ipswich IP5 7RD
Kesgrave 3804

Isere Nord
Cree House
Creechurch Lane
London EC3
01-283 1551

Island Plastics Ltd
Edward Street
Ryde Isle of Wight
Ryde 3777

J B C Plastics (Dewsbury) Ltd
Mill Street East
Dewsbury Yorks
Dewsbury 3060

Jackmore Ltd
Bourne End Bucks
Bourne End 24911

Jagger (Albert) Ltd
Centaur Works
Green Lane
Walsall WS2 9SJ
Walsall 27373

Jacobson van den Berg & Co (UK) Ltd
Jacoberg House
Emerald Street
London W1
01-405 9113

Jay Plastics Ltd
Tetbury Glos
Cirencester 2728

Jennings Bros (Harringay) Ltd
118 Warham Road
London N4
01-348 3535

Jeffries (P R) (Guernsey) Ltd
North Plantation Street
St Peter Port
Guernsey C I
Guernsey 21512

Jig Tools (Pentyrch) Ltd
Pontygwindy Trading Estate
Caerphilly CF8 3TF
Caerphilly 4161

Jobling (James A) & Co Ltd
Wear Glass Works
Sunderland SR4 6EJ
Sunderland 57251

Johnsen & Jorgensen (Plastics) Ltd
Grinstead Road
London SE8
01-692 1406

Judge International Ltd
PO Box 12
Brierley Hill Staffs
Lye 2126

Judge International Housewares Ltd
Brierley Hill Staffs
Lye 3911

Jupiter Plastics Ltd
Lea Bridge Road
Leyton
London E10
01-539 7826

K P Plastics (Bletchley) Ltd
Ward Road
Bilton Estate
Bletchley Bucks
Bletchley 4811

K W Chemicals Ltd
55/57 High Holborn
London WC1
01-242 7981

Kautex Ltd
Elstree Way
Elstree Herts
01-953 1777

Kalle UK Ltd
Hoechst House
Kew Bridge
Brentford Middx
01-995 1355

Karobes Ltd
Queensway Trading Estate
Leamington Spa Warwicks
Leamington Spa 23281

Kay–Metzeler Ltd
Bollington
Macclesfield Cheshire
Bollington 3041

Kaylis Chemical Ltd
Weston Street
Bolton Lancs
Bolton 26236

Kabi (Electrical & Plastics) Ltd
Cranbourne Road
Potters Bar Herts
Potters Bar 53444

Kedo Plastics Ltd
Marsh Road
Crediton Devon
Crediton 2932

Keeler Optical Products Ltd
Clewer Hill Road
Windsor Berks
Windsor 60464

Kelcoat Engineering Plastics Ltd
Barnfields Industrial Estate
Leek Staffs
Leek 3547

Keil (E) & Co Ltd
Russell Gardens
Wick Lane
Wickford Essex
Wickford 2295

Kenney Plastics Co Ltd
Main Street
Ratby Leicestershire
Kirby Muxloe 3095

Kent Mouldings
Foots Cray
Sidcup Kent
01-300 3333

Kent Plastics UK Ltd
Derrychara
Enniskillen N Ireland
Enniskillen 3131

Kenure (J F) Ltd
Central Way
North Feltham Trading Estate
Feltham Middx
01-890 2604

Kerr (J & W) (Plastics) Ltd
10 Virginia Street
Greenock Scotland
Greenock 25371

Ketch Plastics Ltd
Trent Valley Works
Lichfield Staffs
Lichfield 51321

Kew Laminates Ltd
35 Parker Street
Edgbaston
Birmingham 16
021-454 6407

King Packaging Ltd
Bessemer Road
Welwyn Garden City Herts
Welwyn Garden 28100

Kings Plastics Ltd
Houlton Street
Bristol BS2 9ON
Bristol 551446

Kingsley & Keith (Chemicals) Ltd
Suffolk House
George Street
Croydon CR9 3QL
01-686 0544

Kingston Plastics Ltd
Moseley Avenue
West Moseley Surrey
01-979 8331

Kirton Kayaks
Marsh Lane
Crediton Devon
Crediton 3295

L B (Plastics) Ltd
Firs Works
Nether Heage
Derby DE5 2JJ
Ambergate 2311

L G Plastics
Bath Road
Longwell Green
Bristol BS15 6DN
Bristol 673121

L G S Ltd
Townley Hill
Hanson Street
(Off Spring Vale)
Middleton Lancs
Middlteon 5523

L & P Plastics Ltd (Reed Paper Group)
Larkfield
Nr Maidstone Kent
Maidstone 77811

La Brecque Engineering Co Ltd
Pollard Moor Works
Padiham
Burnley Lancs
Padiham 71215

Laconite Ltd
Walton Bridge
Shepperton Middx
Walton-on-Thames 28944

Lacrinoid Products Ltd
Gidea Park Works
Stafford Avenue
Hornchurch RM11 2ET Essex
Hornchurch 52525

Lamb (N) (Scrap) Plastics
Irkdale Mill
Smedley Road
Colleyhurst
Manchester 10
061-205 1320

Lambert Bros (Walsall) Ltd
Walstead Road West
Walsall WS5 4BA Staffs
Walsall 21585

Lancasters Plastics Ltd
Tividale
Warley Worcs
021-557 3368

Lane (P H) Ltd
98-106 Manor Road
Wallington Surrey
01-669 4911

Langley (London) Ltd
Faraday Road
Crawley Sussex
Crawley 29091

Lanplas (Formerly Lancashire Plastics)
Irk Mill
Oldham Road
Middleton
Manchester M24 1AZ
061-643 6171

Laporte–Synres Ltd
Stallingborough
PO Box 35
Grimsby Lincs
Immingham 2464

Latter (A) & Co Ltd
43 South End
Croydon CR9 1AN Surrey
01-688 9335

Lawrence (I J) & Sons Ltd
Butler Street
Oldham Road
Manchester M4 6LE
061-205 3900

Lax & Shaw Ltd
69 South Accommodation Road
Leeds 10
Leeds 21568

Lenanton (John) & Son Ltd
8/36 West Ferry Road
Millwall
London E14 8JZ
01-987 1240

Lennig Chemicals Ltd
Lennig House
2 Masons Avenue
Croydon CR9 3RB Surrey

Lettercast Ltd
Barton Mill Estate
Maze Street
Bristol BS5 9TH
Bristol 556172

Levity Plastics Ltd
Reddicap Trading Estate
Sutton Coldfield Warwicks
021-329 4461

Lierman Ltd
Monarch Works
Picardy Manorway
Belvedere Kent
Erith 34410

Lincoln Plastics Ltd
Outer Circle Road
Lincoln
Lincoln 27217

Link 51 Ltd
Uxbridge Road
Hayes Middx
01-573 6585

Linecross Plastics Co Ltd
Barrowden Rutland
Morcott 816

Link Associates (R P B) Ltd
Vynters Manor
Crick
Nr Rugby Warwicks
Crick 547

Lintafoam Ltd
Kingsmead Road
Loudwater
High Wycombe Bucks
High Wycombe 21501

Litholite Mouldings Ltd
Sandown Road
Watford WD2 4DQ Herts
Watford 23377

Lloyds Cartons Ltd
Pollard Street East
Manchester M10 7EU
061-273 1585

Loftheath Plastics Ltd
Haslebury Plucknett
Crewkerne Somerset
Crewkerne 3604

London Aryid Plastics Ltd
Buckingham Avenue
Trading Estate
Slough Bucks
Slough 27661

London Association for the Blind
90 Peckham Road
London SE15
01-703 6153

London Bankside Products Ltd
The Runeay
South Ruislip HA4 6SE Middx
01-845 3535

London Varnish & Enamel Co Ltd
City Works
Carpenters Road
Stratford
London E15
01-534 4482

Long & Hambly Ltd
Empire Works
Slater Street
High Wycombe Bucks
High Wycombe 26141

Longwell Green Coachworks
Longwell Green
Bristol BS15 6DN
Bristol 673121

Lorival Ltd
Little Lever
Bolton BL3 1AR Lancs
Farnworth 72155

Lovell Bros & Stapley
14a Maple Road
Earlswood
Redhill Surrey
Redhill 65401

Low & Bonar (Textiles & Packaging) Ltd
63/73 King Street
Dundee DD1 9JA
Dundee 24111

Low (Gordon) (Plastics) Ltd
88 Place Road
Cowes Isle of Wight
Cowes 2151

Loyne (Belfast) Ltd
17 Ravenhill Road
Belfast 6 N Ireland
Belfast 58228

Loyne Ltd
Margaret Street
Ashton-under-Lyne Lancs
Ashton-under-Lyne 4551

Lucent Plastics Ltd
202 Cambridge Road
Kingston-upon-Thames
01-549 0669

Lukely Engineering & Moulding Co Ltd
180 Carisbrooke Road
Newport Isle of Wight
Newport (IOW) 2714

Luxus Ltd
Cheltenham House
61 High Street
Banbury Oxon
Banbury 50541

Lynester Products Ltd
Nelson Lane
Warwick
Warwick 44221

M C M Tools Ltd
715 Kings Road
Kingstanding
Birmingham B44 9HS

M P Plastics (Maurice Powell Ltd)
Saffron Walden Essex
Saffron Walden 3631

Mackins & Wood Ltd
Jury Lane
Sidlesham
Nr Chichester Sussex
Sidlesham 484

McCrae & Drew Ltd
Colinslee Works
Neilston Road
Paisley Scotland
041-884 2286

M A M Rubber Manufacturing Co Ltd
Pound Lane
London NW10
01-459 1187

MacEchern & Co Ltd
5 High Street
Chislehurst BR7 5AH Kent
01-467 1103

Mackie (James) & Sons Ltd
Albert Foundry
Belfast BT12 7ED N Ireland
Belfast 27771

McMurdo Instrument Co Ltd
Rodney Road
Portsmouth Hants
Portsmouth 35361

Maison Fittings Ltd
3/5 Washington Street
Liverpool 1
051-709 4819

Manhattan Equipment Ltd
Moorswater
Liskeard PL14 4LG Cornwall
Liskeard 2336

Manuplastics Ltd
South Down Works
Kingston Road
London SW20
01-542 3421

Marbon UK Ltd
Box 13
Clayton
Manchester M11 4RJ
061-223 2002

Mardel (G B) Ltd
Croft Works
Croft Road
Thame Oxon
Thame 3431

Marglass Ltd
Sherborne Dorset
Sherborne 3722

Marley Extrusions Ltd
Lenham
Maidstone Kent
Harrietsham 391

Marley Floor Tile Co Ltd
London Road
Riverhead
Sevenoaks Kent
Sevenoaks 55255

Marley Foam Ltd
Dickley Lane
Lenham
Maidstone Kent
Harrietsham 491

Marshall (C & C) Ltd
Ponswood Industrial Estate
Hastings Sussex
Hastings 7691

Marshall (C F) & Sons Ltd
Leigh-on-Sea
Southend-on-Sea SS9 4EH Essex
Southend 524261

Marshall of Cambridge (Engineering) Ltd
Airport Works
Newmarket Road
Cambridge CB5 8RX
Cambridge 56291

Marston Excelsior Ltd
Wobaston Road
Wolverhampton WV10 6OJ Staffs
Fordhouses 3361

Martin–Golod Products Ltd
Maybury Gardens
Willisden High Street
London NW10
01-459 3888

Martins Birmingham Ltd
New North Road
Exmouth EX8 1EL Devon
Exmouth 5601

Martin's Products Ltd
92 Turnmill Street
London EC1
01-253 6070

Marton Plastics
Casdon Works
Mitcham Road
Blackpool FY4 4QW Lancs
Blackpool 62288

Matthew (R) & Sons
Trowbridge Road
Westbury Wilts
Westbury 2644

Matthews & Yates Ltd
Cyclone Works
Swinton
Manchester
061-794 2273

May & Baker Ltd
Dagenham RM10 7XS Essex
01-592 3060

Mayfair Plastics Ltd
Church Street
Dorking Surrey
Dorking 4696

Mawson Taylor Ltd
Pioneer Mills
Milltown Street
Radcliffe M26 9WX Lancs
061-723 2831

Measom Freer & Co Ltd
Injectoid Works
Pullman Road
Wigston
Leicester LE8 2DE
Leicester 881588

Meldrum (A W) Thermoplastics Ltd
Skimpot Estate
Luton LU4 0JH Beds
Luton 54101

Mellowes Orb Engineering Ltd
Westhulme Street
Featherstall Road
Oldham Lancs
061-624 5295

Lido Works
Grotton
Nr Oldham Lancs
061-624 5607

Mentmore Manufacturing Co Ltd
Platignum House
Six Hills Way
Stevenage SG1 2AY Herts
Stevenage 2488

Mercury Signs Ltd
Burton Road
Norwich NOR 84N
Norwich 49325

Merlin Reinforced Plastics Ltd
Morriston
Swansea SA6 8QD S Wales
Swansea 71731

Merriott Mouldings Ltd
Merriott Somerset
Crewkerne 2457

Merrivale Manufacturing Co Ltd
Bensham Grove
Thornton Heath
CR4 8YE

Metal Box Co Ltd
37 Baker Street
London W1A 1AN
01-486 5577

Metal & Plastic Products Co (Balsall Common) Ltd (The)
9 George Road
Erdington
Birmingham 23
021-373 3092

Metal Pressings Ltd
Oare
Faversham Kent
Faversham 3225

Metropolitan Flexible Products Ltd
Queens Street
Great Harwood
Blackburn Lancs
Great Harwood 2171

Metropolitan Plastics Ltd
Faraday Way
St Mary Cray
Orpington Kent
Orpington 31631

Mica & Micanite Supplies Ltd
Mica House
Barnsbury Square
London N1
01-607 3032

Micanite & Insulators Ltd
Empire Works
Blackhorse Lane
London E17
01-527 5500

Mica Products Ltd
Golden Green
Tonbridge Kent
Hadlow 701

Mickleover Transport Ltd
Twyford Works
Whitby Avenue
London NW10 7SH
01-965 7788

Middlesex Oil & Chemical Works Ltd
West Drayton Middlesex
West Drayton 5181

Middleton Plastics Ltd
Battledown
Cheltenham GL52 6OY Glos
Cheltenham 53481

Midland Fibreglass Products Ltd
The Old Mill
Sutton-in-Ashfield NG17 4JT Notts
Sutton-in-Ashfield 5272

Midland Industries Ltd
Bailieborough Co Cavan
Ireland
Bailieborough 75

Midland Silicones Ltd
Reading Bridge House
Reading RG1 8PW Berks
Reading 57251

Microplas Ltd
James Estate
132 Western Road
Mitcham CR4 3XD Surrey

Miles Redfern Ltd
Watling Street
Dunstable LU6 3QP Beds
Dunstable 64244

Mill Plastics Ltd
The Old Mill
Standom
Ware, Herts
Ware 821551

Minerva Moulding Ltd
Raans Road
Amersham Bucks
Amersham 3981

Minnesota Mining & Manufacturing Co Ltd
3M House
Wigmore Street
London W1A 1ET
01-486 5522

Minto & Turner Ltd
45 Chorlton Street
Manchester M1 3FX
061-236 8273

Mining & Chemical Products Ltd
Alperton
Wembley HA0 4PE Middx
01-902 1191

Mitchell (Andrew) & Co
356 Amulree Street
Glasgow E2
041-778 5461

Mitchell (W A) & Smith Ltd
Church Path
Church Road
Mitcham CR4 3YE Surrey
01-648 4684

Mitra (Plastics) Ltd
Arthur Street
Oswestry
Salop
Oswestry 3225

Mobbs Miller Ltd
Arthur Street
Northampton
Northampton 31804

Modern Fittings Ltd
17 Platt Street
Bolton Lancs
Bolton 25952

Modern Plastics Ltd
2 South Side
The Green
London N15
01-808 8765

Modern Veneering Co Ltd
Coronation Road
High Wycombe Bucks
High Wycombe 30473

Monk (Anthony) (Engineering) Ltd
Eastfield Side
Sutton-in-Ashfield NG17 4JR Notts
Sutton-in-Ashfield 2522

Monks (M) & Co Ltd
Forward Works
Woolston Warrington Lancs
Warrington 35944

Monopol Plastics Ltd
Dennis Road
Tanhouse Industrial Estate
Widnes Lancs
051-424 4121

Monsanto Chemicals Ltd
Monsanto House
10/18 Victoria Street
London SW1
01-222 5678

Montgomery (Daniel) & Son Ltd
Kirkintillock
Glasgow
041-776 5276

Moon Aircraft Ltd
Clift Works
Chippenham Wilts
Box 488

Moore (Harold) & Son Ltd
Bailey Street
Sheffield S1 3BR
Sheffield 27311

Moorside Machining Co Ltd
Ebor Mills
Dubb Lane
Bingley BD16 2BS Yorks
Bingley 2211

Morane Plastic Co Ltd
Gresham Road
Staines Middx
Staines 51985

Morgan & Grundy Ltd
Cowley
Uxbridge Middx
Uxbridge 38551

Morgan, Berkeley & Co Ltd
Hersham Green
Walton-on-Thames Surrey
Walton-on-Thames 28421

Morning (Croydon) Ltd
Hanworth (Croydon) Ltd
Hanworth Lane
Chertsey Surrey
Chertsey 4301

Moral Ltd (Morane Plastic Co Ltd)
Gresham Road
Staines Middx
Staines 51985

Moseley (David) & Sons Ltd
Mancunian Way
Ardwick
Manchester M12 6HL
061-273 3341

Moss (Robert) Ltd
Langford Lane
Kidlington OX5 1HX Oxon
Kidlington 3073

Mosses & Mitchell Ltd
Weydon Lane
Farnham Surrey
Farnham 21236

Motools Ltd
18 St Johns Road
Ryde Isle of Wight
Ryde (IOW) 4036

Moulded Fastners Ltd
Vestry Estates
Otford Road
Sevenoaks Kent
Sevenoaks 56176

Moulded Plastics (Birmingham) Ltd
Reddicap Trading Estate
Sutton Coldfield Warwicks
021-355 1086

Moulding Powders Ltd
Lamberhead Industrial Estate
Pemberton
Wigan Lancs
Wigan 82921

Moulding Service Ltd
56/58 Spencer Street
Birmingham B18 6DS
021-236 1166

Moulton Mouldings Ltd
High Street
Moulton
Spalding PE12 6PN Lincs
Spalding 566

Multicraft Ltd
Glynn Works
New Lane
Havant Hants
Havant 2486

Mundel Cork & Plastics Ltd
Croydon CR9 4AR Surrey
01-688 4141

Muntz Plastics
Alexandra Works
West Bromwich Staffs
021-557 4221

Mycalex & Rosite Ltd
Ashcroft Road
Cirencester GL7 1QY Glos
Cirencester 2551

Muehlstein–Northwestern Ltd
7 Cork Street
London W1X 1PB
01-734 4787

Mundet Cork & Plastics Ltd
Vicarage Works
Vicarage Road
Croydon Surrey
01-688 4141

N B Mouldings Ltd
Tavistock Street
Dunstable LU6 1NG Beds
Dunstable 62462

Nairn Coated Products
(Division of Nairn Williamson Ltd)
Lune Mills
Lancaster
Lancaster 65222

Nairn International Ltd
Path Office
Nether Street
Kirkcaldy Scotland
Kirkcaldy 61111

National Adhesives & Resins Ltd
Galvin Road
Slough Bucks
Slough 29191

National Plastics Ltd
Avenue Works
Walthamstow Avenue
London E4 8SY
01-527 2323

Negri Bossi Evans Ltd
143 Goldsmith Avenue
Portsmouth PO4 8RE Hants
Portsmouth 32233

Neosid Ltd
Howardsgate
Welwyn Garden City Herts
Welwyn Garden 25011

Netlon Ltd
Kelly Street
Mill Hill
Blackburn BB2 4PJ Lancs
Blackburn 55541

Nettle Accessories Ltd
Warren Street
Stockport SK1 1XF
061-480 8181

Newalls Insulation & Chemical Co Ltd
Washington Co Durham
Washington 3333

Newmore Plastics Ltd
Henfaes Lane
Welshpool Montgomeryshire
Welshpool 2671

Newtown Industries Ltd
Newtown
Lymington SO4 8RA Hants
Lymington (Hants) 2181

Nickwood Plastics Ltd
85 Buckingham Palace Road
London SW1
01-834 5345

Nico Manufacturing Co Ltd
Oxford Road
Clacton-on-Sea Essex
Clacton-on-Sea 22333

Nightingale Signs (Blackburn) Ltd
Feniscowles
Blackburn Lancs
Blackburn 21207

Non-Sag Seating Co Ltd
Dashwood Avenue
High Wycombe Bucks
High Wycombe 20711

Norbert Plastics Ltd
Townley Mill
Hanson Street
Middleton
Manchester
061-643 5523

Norris Cheshire Ltd
Meadow Mills
Water Street
Stockport Cheshire
061-480 9508

Norseman (Cables & Extrusions) Ltd
Brantham Hill
Manningtree Essex
Manningtree 2103

Norsk Hydro (UK) Ltd
Queens House
2 Holly Road
Twickenham Middlesex
01-892 9025

North (Robert) & Sons Ltd
18 Rudolph Road
Bushey
Watford WD1 3DY Herts
01-950 2223

North East Glassfibre Works Ltd
Tullos
Aberdeen AB1 4BS Scotland

North Hill Plastics Ltd
49 Grayling Road
London N16
01-800 3773

North West Plastics Ltd
Parr Bridge Works
Mosley Common Road
Boothstown
Worsley
Manchester M28 4AJ
061-790 4433

Northern (Neon) Lights Blackpool L
Chorley Road
Blackpool Lancs
Blackpool 31462

Norton Plastics Ltd
Heanor Road
Ilkeston Derbys
Ilkeston 3121

Noton (S) Ltd
Blackhorse Lane
Walthamstow
London E17 6DU
01-527 2262

Notreme Engineering Developments
248 Uttoxeter Road
Longton
Stoke-on-Trent ST3 5QL Staffs
Stoke-on-Trent 33353

Novadel Ltd
St Anns Crescent
London SW18
01-874 7761

Novoplas Ltd
Greenland Mills
Bradford-on-Avon Wilts
Bradford-on-Avon 2971

Nutshell Cap Co Ltd
Star Works
Church Street
Bury BL9 6AX Lancs
061-764 1972

Nutt & Co Ltd
Acorn Works
Broad Lane
Pudsey Yorks
Pudsey 4202

Nylonic Engineering Co Ltd
Woodcock Hill
Rickmansworth WD3 1PN Herts
Rickmansworth 76261

Oates Ltd
Station Sawmills
Gateford Road
Worksop Notts
Worksop 4351

Olympic Plastics Ltd
Fourth Way
Wembley Middx
01-902 7933

Omega Plastics Ltd
Northwick Road
Canvey Island Essex
Canvey 2163

Optical Products Ltd
370/373 Station Road
Forest Gate
London E7
01-534 1766

Orbex Ltd
Phoenix Mill
(Duke Street)
Failsworth
Manchester M35 9DS
061-681 3720

Ormond Brassfoundry Ltd
Westminster Road
Birmingham 20
021-554 1026

Orr (William) (Foam) Ltd
Britannia Wharf
Baldwin Terrace
Wharf Road London N1
01-226 4224

Osma Plastics Ltd
Rigby Lane
Dawley Road
Hayes Middx
01-573 7799

Owen (Henry) & Sons Ltd
72/77 Caroline Street
Birmingham 3
021-236 2181

Osprey Plastics Ltd
PO Box 5
Gladstone Lane
Scarborough Yorks
Scarborough 63376

Oxley Developments Co Ltd
Priory Park
Ulverston Lancs
Ulverston 2621

P & B Plastics Ltd
Shaw Health Mill
Stockport SK3 8BN Cheshire
061-480 4007

P D I Ltd
Hampton Street
Birmingham B19 3LT
021-236 3866

P F (Technical Mouldings) Ltd
Redsand Works
Sanderstead Road
South Croydon Surrey
01-686 0856

P H Plastics Ltd
Conway Road
Mochdre
Colwyn Bay Denbighs
Colwyn Bay 44332

P R M (London) Ltd
19 Pine Walk
Surbiton Surrey
01-399 4431

P & S (Cirencester) Ltd
Love Lane Estate
Cirencester Glos
Cirencester 2205

Pacra Plastics Ltd
Abbey Road
Shepley
Huddersfield Yorks
Kirkburton 2953

Page (Charles) & Co Ltd
52 Grosvenor Gardens
Victoria
London SW1
01-730 8151

Paklite Ltd
Denbigh Road
Bletchley Bucks
Bletchley 5621

Palgrave Brown Colchester
Haven Road
Colchester Essex
Colchester 78374

Pancolite Plastics Ltd
312 King Street
Hammersmith
London W6
01-748 3592

Paragon Plastics Ltd
Cross Bank
Balby
Doncaster Yorks
Doncaster 51131

Paramount Plastics Products Ltd
5/9 Annerley Station Road
London SE20
01-778 2201

Parglas Ltd
Barton Manor
Bristol BS2 0RP
Bristol 552325

Park Lane Plastics (Aldridge) Ltd
Morford Road
Northgate
Aldridge Staffs
Aldridge 54150

Parkinson Cowan Ltd
Terminal House
Grosvenor Gardens
London SW1
01-730 0111

Parnall & Sons Ltd (Plastics Division)
Lodge Causeway
Fishponds
Bristol BS16 3JU
Bristol 654271

Parsons Bros Ltd
Sutton Road
Hull Yorks
Hull 851102

Paton, Calvert & Co Ltd
Binns Road
Liverpool L13 1BU
051-228 2721

Pearce Road Signs Ltd
Insignia House
New Cross Road
London SE14
01-692 6611

Peel Engineering Co Ltd
Viking Works
Peel Isle of Man
Peel 342

Peerless Plastics Ltd
Litchfield Road Industrial Estate
Tamworth Staffs
Tamworth 68383

Pendred (Norman) & Co Ltd
9/13 Gladiator Street
London SE23
01-690 4841

Pendry (Plastics) Ltd
River Works
Brent Crescent
London NW10
01-965 7932

Performance Plastics Ltd
Thorpe End
Melton Mowbray Leics
Melton Mowbray 4538

Perkins (E) & Co Ltd
19 Selbourne Street
Walsall Staffs
Walsall 27963

Perks (Mark) Ltd
11/12 Anne Road
Smethwick
Warley Worcs
021-558 2991

Permafence Ltd
28 Blacamoor Lane
Maidenhead Berks
Maidenhead 29133

Permali Ltd
Bristol Road
Gloucester GL1 5SU
Gloucester 28282

Permanoid Ltd
Vincent Works
New Islington
Manchester M4 7JX
061-205 1371

Permark Ltd
Cranleigh Surrey
Cranleigh 3021

Perry (E S) Ltd
Osmiroid Works
Fareham Road
Gosport PO1 0AL Hants

Pershore Mouldings Ltd
Trading Estate
Pershore Worcs
Pershore 460

Peter Plastics Ltd
Chain Bridge Road Estate
Blaydon Co Durham
Blaydon 2005

Petitdemange (R) Ltd
Harbour Road
Oulton Broad
Lowestoft Suffolk
Lowestoft 62021

Pharmaceutical Plastics Ltd
Outer Circle Road
Lincoln
Lincoln 27217

Pinchin Johnson & Associates Ltd
(Industrial Division)
93/97 New Cavendish Street
London W1
01-580 0831

Phillips Petroleum UK Ltd
Portland House
Stag Place
London SW1
01-828 9766

Phipps Plastic Products Ltd
Bagnall Street (Eagle Lane)
Great Bridge
Tipton Staffs
021-557 1113

Phoenix Rotational Mouldings Ltd
Pipegate
Market Drayton TF9 4HY
Salop
Pipe Gate 441

Phoenix Rubber Ltd
Pipe Gate
Market Drayton TF9 4HY
Salop
Pipe Gate 441

Phoenix Timber Co Ltd
Phoenix House
New Road
Rainham RM13 8RJ Essex

Phosco Ltd
Hoe Lane
Ware Herts
Ware 3466

Pickup (John) Ltd
Richmonde Works
Olive Lane
Darwen BB3 0EU Lancs
Darwen 71037

Pioneer Plastic Containers Ltd
Great South West Road
Feltham Middx
01-890 2288

Pillar Plastics Ltd
Dallow Road
Luton Beds
Luton 27612

Pioneer Oilsealing & Moulding Co Ltd
Barrowford
Nelson Lancs
Nelson 62241

Plasmic Ltd
38 Crawley Road
London N22 6AG
01-889 2543

Plasro Plastics Ltd
38 Wates Way
Mitcham CR4 4HR Surrey
01-640 1251

Plastanol Ltd
Crabtree Manorway
Belvedere Kent
Erith 31631

Plastestrip Ltd
Trenance Mill
St Austell Cornwall
St Austell 4771

Plastic Display Units (Strand) Ltd
26 Charing Cross Road
London WC2
01-836 0396

Plasticable Ltd
Hawley Lane
Farnborough Hants
Farnborough 41385

Plastic Coating Equipment Ltd
Industrial Estate
By-Pass
Guildford Surrey
Guildford 64611

Plastic Constructions Ltd
Seeleys Road
Greet
Birmingham B11 2LP
021-773 1331

Plastic Coatings Ltd
Woodbridge Estate
Guildford Surrey
Guildford 64611

Plastic Dipping Co Ltd (The)
Swan Lane
Sandhurst
Camberley Surrey
Yateley 2467

Plastic Engineers Ltd
Treforest Industrial Estate
Pontypridd Glam
Treforest 2371

Plastic Extruders Ltd
Units 11 & 12
Heron Trading Estate
Wickford Essex
Wickford 2534

Plastic Dip Mouldings Ltd
Terrace Road
Walton-on-Thames Surrey
Walton-on-Thames 25710

Plastic Rotational Mouldings Ltd
Industrial Estate
Irvine
Ayrshire Scotland
Irvine 2475

Plastic Supplies (Bristol) Ltd
Epstein Building
Mivart Street
Bristol BS5 6JF
Bristol 551282

Plastic Tube & Conduit Co Ltd
Sterling Works
Aldermarston
Reading RG7 5QD Berks
Woolhampton 2346

Plasticisers Ltd
Old Mills
Drighlington
Bradford BD11 1BY Yorks
Drighlington 2202

Plastics Design & Engineering Ltd
Common Lane
Culcheth
Warrington Lancs
Culcheth 4221

Plastics Fabrication Co Ltd
Pounsley Road
Dunton Green
Sevenoaks Kent
Sevenoaks 51341

Plastics Fabrication & Printing Ltd
Leconsfield Road
London N5 2SA
01-226 5847

Plastics (Manchester) Ltd
26 Buxton Street
London Road
Manchester M1 2QB
061-273 4567

Plastics Marketing Co Ltd
Pounsley Road
Dunton Green
Sevenoaks Kent
Sevenoaks 51341

Plastics (Wednesbury) Ltd
Bright Street
Wednesbury Staffs
021-526 2962

Plastics Reinforced (Liverpool) Ltd
Strand Road
Bootle L20 4BO
051-992 5281

Plastifilm Products Ltd
South Road
Kingswood
Bristol BS15 2JN

Plastiglide Products Ltd
Masons Road
Stratford-upon-Avon Warwicks
Stratford 5181

Plastikade (Manchester) Ltd
New Mill
Park Road
Dukinfield Cheshire
061-330 7311

Plastshapes Ltd
Hillend Industrial Estate
Dunfermline Fife
Inverkeithing 3651

Plastix & Metal Engraving Co Ltd
18-20 Prudhoe Street
North Shields Northumberland
North Shields 72885

Plastona (John Waddington) Ltd
40 Wakefield Road
Leeds 10
Leeds 72244

Plastotype
56 Stamford Street
London SE 1
01-928 5207

Plex (Engineering) Ltd
Victoria Road
Burgess Hill Sussex
Burgess Hill 5629

Plessey Co Ltd (Components Group) (The)
Kembrey Street
Swindon Wilts
Swindon 6211

Plextrude Ltd
145 Frimley Road
Camberley Surrey
Camberley 63161

Plucknett (C J) & Co Ltd
16 Warren Lane
Woolwich
London SE18
01-854 1253

Plutec Ltd
Seymore Road
Nuneaton Warwicks
Nuneaton 4492

Plysu Products Ltd
Station Road
Woburn Sands
Bletchley Bucks
Woburn Sands 2311

Poet Plastics Ltd
Factory Unit 3
First Avenue
Bletchley Bucks
Bletchley 76136

Polybags Ltd
197 Ealing Road
Wembley Middx
01-902 1100

Polyglass Ltd
11 South Road
Morecambe Lancs
Morecambe 3003

Polygrow Plastics
Borehamgate House
King Street
Sudbury Suffolk
Sudbury 2263

Polymon Developments Ltd
Bewley House
2 Swallow Place
London W1
01-629 5615

Polypenco
Gate House
Welwyn Garden City Herts
Welwyn Garden 21221

Polyservices (Bedford) Ltd
Taylors Road
Stotford
Hitchin Herts
Stotford 238

Polytubes (Nylaflex) Ltd
17 Forset Road
Dorridge
Solihull Warwicks
Knowle 5633

Polythene Drums Ltd
Wigan Road
Skelmersdale Lancs
Tawd Vale 3641

Polytop Plastics Ltd
Bluebridge Industrial Estates Ltd
Halstead Essex
Halstead 3225

Pontefract Box Co Ltd
King Street
Pontefract Yorks
Pontefract 2026

Poplar Playthings Ltd
Brackla Industrial Estate
Coity
Bridgend Glam
Bridgend 3761

Poppe Rubber & Tyre Co Ltd
Sherland Road
Twickenham Middx
01-892 2271

Poron Insulation Ltd
Poron Works
Torpoint PL11 3AX Cornwall
Millbrook 551

Portex Ltd
Hythe Kent
Hythe 66863

Portobello Fabrications Ltd
Coleford Road
Sheffield S9 5PE
Sheffield 42781

Powell (E & F) Ltd
12 William Street
Carshalton Surrey
01-647 3099

Power Plastics Ltd
Daux Road
Billingshurst Sussex
Billingshurst 2916

Precision Mouldings (Plastics) Ltd
74 Buckingham Avenue
Trading Estate
Slough Bucks
Slough 21339

Precision Plastics Ltd
248 Bordesley Green
Birmingham 9
021-772 2191

Precision Products (Leeds) Ltd
79 Kirkstall Road
Leeds LS3 1LS
Leeds 36661

Premier Injection Mouldings Ltd
New Road
Newhaven Sussex
Newhaven 3224

Preci-Spark Ltd
Chapel Street
Syston
Leicester LE7 8HN
Syston 4781

Precision Units (Dorset) Ltd
Seacrest Works
Church Road
Parkstone Dorset
Parkstone 1664

Pressure Sealed Plastics Ltd
PO Box No 5
Wheat Bridge Road
Chesterfield Derbys
Chesterfield 70121

Preston (J) Ltd
202–208 West Street
Sheffield S1 3TJ
Sheffield 77545

Priestly Studios Ltd
Commercial Road
Gloucester GL1 2EQ
Gloucester 22281

Prior (John) Engineering Ltd
Peel Mill
Market Street
Shawforth
Rochdale Lancs
Whitworth 2391

Prior (John) Plastics Ltd
Conway Road
Mochdre
Colwyn Bay N Wales
Colwyn Bay 44332

Pritchard (Plastics) Ltd
Kings Hill
Bude EX23 8QN Cornwall
Bude 3211

Pritex (Plastics) Ltd
Station Mills
Wellington Somerset
Wellington 2216

Prodorite Ltd
Eagle Works
Leabrook
Wednesbury Staffs
021-556 1821

Protected Conductors Ltd
468 Staniforth Road
Sheffield S9 3FW
Sheffield 49379

Punfield & Barstow (Mouldings) Ltd
Honeypot Lane
London NW9
01-204 4141

Pye of Cambridge Ltd
St Andrews Road
Cambridge CB4 1DL
Cambridge 58985

Pyke (M C) & Associates
Fforchneol House
Cwnaman
Aberdare Wales
Aberdare 3859

Pylon Developments
Staplehurst Road
Sittingbourne Kent
Sittingbourne 4019

QD Plastics (Glasgow) Ltd
Broadmeadows Estate
Dumbarton Scotland
Dumbarton 2076

QED Design & Development Ltd
Woodside
Commonside
Keston Kent
Farnborough (Kent) 53753

Quinton (T H) Ltd
Hamlet Works
Brook Road
Rayleigh Essex
Rayleigh 3247

RF Development Co Ltd
Priory Lane
St Neots Hunts
St Neots 2506

Radiamp Co Ltd
Lea Road Trading Estate
Waltham Abbey Essex
Waltham Cross 25536

Radnall (F A) & Co Ltd
Vauxhall Works
Dartmouth Street
Birmingham 7
021-359 1341

Railko Ltd
Loudwater
High Wycombe Bucks
Bourne End 22551

Rainbow (S H) Ltd
Glendale Gardens
Leigh-on-Sea Essex
Southend-on-Sea 79068

Raleigh Industries Ltd
Lenton Boulevard
Nottingham NG7 2DD
Nottingham 77761

Range Valley Engineering Ltd
Ceramyl Works
Diggle
Oldham Lancs
Saddleworth 3345

Ranton & Co Ltd
Commerce Road
Brentford Middx
01-560 8151

Ravenscroft Plastics Co
Green Lane
Newton
Tewkesbury Glos
Tewkesbury 2272

Ray Engineering Co Ltd
Southmead
Bristol BS10 5EB
Bristol 626074

Ray Mouldings Ltd
3 Plant House
Craven Avenue
Ealing
London W5 2SY
01-567 5314

Rayburn Plastics Ltd
Whitehouse Street
Walsall Staffs
Walsall 27359

Raychem Ltd
Cheney Manor Trading Estate
Swindon Wilts
Swindon 27146

Raylite Supplies Ltd
Cotmanhay Road
Ilkeston Derbys
Ilkeston 5653

Rector (M R) Ltd
20 Lord Street
Manchester M4 4FQ
061-832 9966

Reddiplex Ltd
Enfield Industrial Estate
Redditch Worcs
Redditch 67277

Rediweld Ltd
Alpine Works
Newton Road
Crawley Sussex
Crawley 25692

Redland Pipes Ltd
Redland House
Castle Gate
Reigate Surrey
Reigate 42488

Redland Quilting Ltd
19a Bristol Road
Bridgwater Somerset
Bridgwater 2249

Regent Plastics Ltd
Crayford Road
Crayford
Dartford Kent
Crayford 26671

Re-fab Ltd
Bolton Street Works
Oldham Lancs
061-624 4133

Regina Industries Ltd
Victoria Road
Stoke-on-Trent ST4 2HX
Stoke-on-Trent 23217

Regis Machinery (Sussex) Ltd
Felpham Road
Felpham
Bognor Regis Sussex
Bognor Regis 22002

Rehau-Plastiks Ltd
690 Stirling Road
Slough SL1 4SU Bucks
Slough 26839

Reinforced & Microwave Plastics Group (BAC) Ltd
Six Hills Way
Stevenage Herts
Stevenage 2422

Reinforced Plastics Applications (Swansea) Ltd
Gorseinon Road
Gorseinon
Swansea S Wales
Gorseinon 3551

Rendar Instruments Ltd
Victoria Road
Burgess Hill Sussex
Burgess Hill 2642

Resinform Ltd
Child Lane
Roberttown
Liversedge Yorks
Heckmondwike 2007

Resinous Chemicals
Portland Road
Newcastle-upon-Tyne NE2 1BL
Newcastle 25151

Resistant Equipment Ltd
76 Rivington Street
London EC2
01-739 6288

Resistoid Ltd
Back Manor Street
off Fountain Street
Bury Lancs
061-764 3851

Resoid Ltd
Reading RG5 4SN Berks
Sonning 2351

Rexpak
Bridge Street
Golborne Lancs
Ashton-in-Makerfield 78833

Richardson (A & R)
1 Kenninghall Road
Edmonton
London N18
01-807 5868

Ridout (D H) & Son Ltd
Stedman Road
Southbourne
Bournemouth Hants
Bournemouth 43362

Roanoid Ltd
131–137 Renfrew Road
Paisley Scotland
041-889 5432

Roanoid Plastics Ltd
21 Montrose Avenue
Glasgow SW2
041-882 4401

Roberts (Alfred) & Sons Ltd
Deykin Avenue
Witton
Birmingham B6 7HN
021-327 1181

Robinson Thermoforming
(E S & A Robinson Ltd)
Yate
Bristol BS17 5AA
Chipping Sodbury 2256

Robinson Waxed Paper Co Ltd
Filwood Road
Fishponds
Bristol BS16 3RY
Bristol 656232

Rocel Ltd
Little Heath Works
Old Church Road
Coventry CV6 7DW Warwicks
Coventry 88031

Rock Electrical Accessories Ltd
6 Commerce Road
Brentford Middx
01-560 8151

Rolinx Ltd
Ledson Road
Bagley
Wythenshawe
Manchester M23 9WP
061-998 5353

Rollite Products (Bridlington) Ltd
Clarence Drive
Filey Yorks
Filey 2224

Rolls-Royce (Composite Materials) Ltd
Avonmouth
Bristol BS11 9DU
Avonmouth 4821

Ronder Plastic Assemblies Co
Faldo Road
Barton
Bedford Beds
Shillington 675

Rondopack Plastics Ltd
48 Coldharbour Lane
Harpenden Herts
Harpenden 60221

Rootes Mouldings Ltd
367–8 Buckingham Avenue
Trading Estate
Slough Bucks
Slough 24461

Rootes Plastics Ltd
367–8 Buckingham Avenue
Trading Estate
Slough Bucks
Slough 24461

Rosedale Associated Manufacturers Ltd
Tower Building
20 Eastbourne Terrace
London W2
01-262 1851

Ross Chemical & Storage (Plastics) Ltd
Portland Works
Barrmill Road
Glaston Ayrshire
Galston 484

Ross Thermoplastics Ltd
Windmill Industrial Estate
Fowey Cornwall
Fowey 2304

Ross Warmafoam Ltd
Power House
Formby
Liverpool L37 6AJ Lancs
Formby 72181

Rotafoil Ltd
Bridge Road Works
Lymington Hants
Lymington 3288

Rotalac Co Ltd (The)
(Branch of Garford-Lilley Industries (Ltd))
Canal Road
Timperley
Altrincham Cheshire
061-928 3336

Rotary Castings Ltd
Palk Road
Wellingborough Northants
Wellingborough 2145

Rowley (Plastics) Ltd
1257 Pershore Road
Stirchley
Birmingham B30 2YT
021-458 2719

Rubbarite Ltd
Caledonia Works
Henry Street
Liverpool L13 1a7 Lancs
051-228 2361

Rubberplas (Birmingham) Ltd
Facet Road
King's Norton
Birmingham 30
021-458 5245

Rubber Plastics Ltd
Willington Square
Wallsend-on-Tyne Northumberland
Wallsend 625441

Rubert & Co Ltd
Acru Works
Demmings Road
Cheadle Cheshire
061-428 5855

Rudic Products Ltd
Moorside Road
Winnall
Winchester Hants
Winchester 3254

Rustless Curtain Rod Co Ltd
New Lane Mills
Cleckheaton BD19 6LQ Yorks
Cleckheaton 3492

Rutherford Plastics Ltd
133 Blyth Road
Hayes Middx
01-573 6145

Rydgeway Plastics Ltd
No 2 Industrial Estate
Howard Chase
Basildon Essex
Basildon 3371

Rutland Plastics Ltd
Coldoverton Road
Oakham LE15 6NU Rutland
Oakham 2178

Ryford Ltd
Reddicap Trading Estate
Sutton Coldfield
Birmingham
021-354 7661

SBD Construction Products Ltd
Maple Cross
Rickmansworth Herts
Rickmansworth 77311

SIC Plastics Ltd
1 Grangeway
London NW6
01-624 0858

SOM Plastics Ltd
Winchester Avenue
Denny Scotland
Denny 766

SOM Plastic Mouldings Ltd
Grosvenor Works
Tulketh Street
Southport Lancs
Southport 56871

STC (Standard Telephones & Cables) Ltd
Moulding Division
Footscray
Sidcup Kent
01-300 3333

Sail Craft Ltd
Waterside
Brightlingsea
Colchester Essex
Brightlingsea 2117

Sales Aids (Sedlescombe)
Sedlescombe
Battle Sussex
Sedlescombe 397

Salon (Nelson) Ltd
Hollin Bank
Brierfield
Nelson Lancs
Nelson 65217

Salter Packaging Ltd
Priestley Way
Crawley Sussex
Crawley 25126

Salter Plastics Ltd
Vestry Estates
Otford Road
Sevenoaks Kent
Sevenoaks 56176

Sanenwood Products Ltd
Lowther Road
Sheffield S6 2DQ Yorks
Sheffield 344611

Saro Products Ltd
Whippingham
East Cowes
Isle of Wight Hants
Cowes 2421

Screenprints (Vacuum Formers) Ltd
Hawkwell
Hockley Essex
Hockley 2795

Schenectady-Midland Ltd
Four Ashes
Wolverhampton Staffs
Standeford 555

Scott Bader Co Ltd
Wollaston
Wellingborough NN9 7RL Northants
Wollaston 495

Schulman (A) Inc Ltd
Airport Works
Eastern Road
Portsmouth Hants
Portsmouth 62212

Seaforth Plastics
60 Oakhurst Road
Southend-on-Sea SS2 5DU Essex
Southend-on-Sea 67043

Securos Ltd
Northumbrian Way
Killingworth
Newcastle-upon-Tyne NE12 0EH
Newcastle 665413

Sefton Mills Ltd
Sefton
Liverpool 23
051-626 4172

Selcol Products Ltd
Hall Street
Long Melford Sussex
Long Melford 705

Sendale (Plastics) Ltd
Ramsey Road
St Ives Hunts
St Ives 2095

Serk (R & D)
Gloucester Trading Estate
Hucclecote
Gloucester
Gloucester 68666

Severn Plastics Ltd
Stonehouse Glos
Stonehouse 2217

Shadbolt (F R) & Sons Ltd
North Circular Road
London E4 8PZ
01-527 6441

Sharna-Ware (Manufacturing) Ltd
Lumb Mill
Droylsden
Manchester
061-370 3467

Sharp (Anthony J) (Intramatic) Ltd
1–3 Belmont Road
Whitstable Kent
Whitstable 4949

Shawinigan Ltd
118 Southwark Street
London SE1
01-928 2765

Shaw Hathernware
Hathern Station Works
Loughborough Leics
Hathern 273

Shaw Munster Ltd
Jayesco Works
31 Commercial Street
Birmingham 1
021-643 7427

Sheffield Smelting Co Ltd
PO Box 28
Royds Mills
Windsor Street
Sheffield S4 7WD Yorks
Sheffield 20966

Shell Chemicals (UK) Ltd
(Industrial Chemicals)
Shell Centre
Downstream Buildings
London SE1 7PG
01-934 1234

Sherborne G R P Mouldings
Priestlands Lane
Sherborne Dorset
Sherborne 3246

Short Bros & Harland Ltd
Queens Island
Belfast BT3 9DZ
Belfast 58444

Shrewsbury Plastics & Engineering Ltd
Lancaster Road
Harlescott
Shrewsbury SYN 3NQ Salop
Shrewsbury 51318

Siegrist Orel Ltd
Star Lane
Westwood Industrial Estate
Margate Kent
Thanet 63571

Sign Equipment Ltd
Victoria Road
Burgess Hill Sussex
Burgess Hill 5611

Silent Channel Products Ltd
Ferrars Road
Huntingdon
Huntingdon 2191

Silleck Engineering Ltd
Eaglescliffe Industrial Estate
Stockton-on-Tees Teeside
Eaglescliffe 780984

Silvaflame Co Ltd
72 Warstone Lane
Birmingham 18
021-236 8551

Silver (E) (B S) Ltd
11 Anne Road
Smethwick
Warley Worcs
021-588 2991

Sim (L A) & Co Ltd
Regal House
Twickenham Middx
01-892 8042

Sima Plastics Ltd
Houldsworth
Reddish
Stockport Cheshire
061-432 5354

Simco Plastics Ltd
85 Ruston Street
Birmingham 16
021-454 5918

Sindall Concrete Products Ltd
347 Cherry Hinton Road
Cambridge
Cambridge 4809

Singleton Flint & Co Ltd
Newland Works
Lincoln
Lincoln 24542

Smith (Stanley) & Co
Worple Road
Isleworth Middx
01-560 3931

Smith (Wildrid) Ltd
Gemini House
High Street
Edgeware Middx
01-952 6655

Snow (J E) Plastics Ltd
Snolon Works
Derby Road
Clay Cross
Chesterfield Derbys
Clay Cross 3514

Solport Bros Ltd
Portia House
Goring Street
Goring-by-Sea Sussex
Worthing 44861

Souplex Ltd
West Gate
Morecambe Lancs
Morecambe 1717

Southern Industries Agency
2 Strongbow Road
Eltham
London SE9
01-859 0644

Spa Plastics Ltd
Alma Road
Chesham Bucks
Chesham 4951

South Western Plastics Ltd
148 Newfoundland Road
St Pauls
Bristol
Bristol 551050

Specialised Mouldings Ltd
Redwongs Way
Huntingdon Trading Estate
Huntingdon
Huntingdon 3537

Soplaril (Great Britain) Ltd
Fulton Road
Exhibition Grounds
Wembley Middx
01-902 5971

Spencer Knight & Co Ltd
Britannia Mill
Station Road
Mirfield Yorks
Mirfield 3831

Spicer Cowan Ltd
19 New Bridge Street
London EC4
01-353 4211

Splintex Ltd
Nightingale Road
Hanwell
London W7 IDQ
01-567 6711

Speglestein (S) & Son Ltd
80–84 Wallis Road
London E9
01-985 7177

Stadium Ltd
30–36 Queensway
Enfield Middx
01-804 4343

Stainless Steel Fabricators Ltd
Gale Street
Syke
Rochdale Lancs
Rochdale 44546

Standard Engineering Plastics Ltd
Stenplas Works
Shiffnal Street
Bolton Lancs
Bolton 21927

Stanley (Alfred) & Sons Ltd
Wednesbury Road
Walsall WS1 3RX
Walsall 25623

Stanton (A E) Plastics (Southern) Ltd
Brockhampton Lane
Havant Hants
Havant 6227

Stantone Plastics Ltd
Bates Road
Harold Wood Essex
Ingrebourne 40061

Stanway Screens Ltd
Oil Croft Orchard
Bredon
Tewkesbury Glos
Bredon 378

Stechford Mouldings Ltd
Northcote Road
Stechford
Birmingham 33
021-783 3061

Stelling (G H) & Sons Ltd
De Havilland No 1 Factory
Airport
Eastern Road
Portsmouth Hants
Portsmouth 60661

Stephenson Blake & Co Ltd
Upper Allen Street
Sheffield S3 7AY Yorks
Sheffield 77842

Stephens (Plastics) Ltd
Hawthorn
Corsham Wilts
Hawthorn 0221

Sterilin Ltd
12–14 Hill Rise
Richmond Surrey
01-940 9982

Sterling Moulding Materials Ltd
Sterling House
Heddon Street
London W1
01-734 7080

Sterling Varnish Co Ltd (The)
Fraser Road
Trafford Park
Manchester 17
061-872 0282

Stevens (Michael S) Ltd
60 George Street
Richmond Surrey
01-940 9941

Stewart (P B) Ltd
Thameside
Windsor Berks
Windsor 69232

Stewart Plastics Ltd
Canford Works
Purley Way
Croydon CR9 4BS Surrey
01-686 2231

Stewarts and Lloyds Plastics
St Peters Road
Huntingdon
Huntingdon 2121

Stones Plating Co Ltd
26–32 Voltaire Road
London SW4
01-720 1007

Storey Bros & Co Ltd
Parkville House
Bridge Street
Pinner Middx
01-866 8861

Storey Bros & Co Ltd
White Cross
Lancaster
Lancaster 3232

Straco Ltd
London Lane
Upton on Severn Worcs
Upton on Severn 2284

Strebor Plastics
Lido Works
Grotton
Oldham Lancs
061-624 5607

Streetly Manufacturing Co Ltd
Streetly Works
Sutton Coldfield Warwicks
021-353 2411

Stress Free Plastics Ltd
Redmarsh Drive
Thornton
Cleveleys
Blackpool FY5 4HP Lancs
Cleveleys 6835

Stuart, Kinney & Co Ltd
11 Argyle Street
London W1A 4ES
01-734 9837

Structoplast Ltd
Ford Aerodrome
Ford
Arundel Sussex
Littlehampton 6955

Studio Plastic & Metal Components Ltd
183–187 Park Lane
Aston
Birmingham B6 5DG
021-359 4531

Submarine Products Ltd
Bridge End
Hexham Northumberland
Hexham 3166

Sundt Plastics Ltd
Cameron Road
Chesham Bucks
Chesham 2551

Sundt Plastics Ltd
New Walk
Hanham
Bristol
Bristol 672241

Sunnytoys (Distributors) Ltd
Cowley Works
Cowley Road
Blackpool Lancs
Blackpool 46313

Sunstan Industrial Mouldings Ltd
Sunplas Works
Henshaw Lane
Yeadon
nr Leeds Yorks
Rawdon 4731

Suntex Safety Glass Industries Ltd
Thorney Lane
Iver Bucks
Uxbridge 34970

Supa-Ware (Southend) Ltd
Manilla Road
Southend-on-Sea SS1 2TS Essex
Southend 68059

Superide Ltd
Windsor Mill
Hollingwood
Oldham Lancs
061-681 2206

Super Oil Seals & Gaskets Ltd
Studley Road
Redditch Worcs
Redditch 64292

Sure Form Plastics Ltd
Industrial Estate
Cheney Manor
Swindon Wilts
Swindon 6751

Surface Coating Synthetics Ltd
Whitby Avenue
Park Royal
London NW10
01-965 5129

Surfleet (Plastics) Products
Plumtree Works
Old Melton Road
Plumtree
Nottingham NG12 5NH
Plumtree 3631

Swales (William A) & Co Ltd
PO Box 10
Dock Road
North Shields Northumberland
North Shields 72151

Swansea Plastics & Engineering Ltd
Pipe House Wharf
Morfa Road
Swansea Glamorgan
Swansea 51098

Swift (S M) (Exeter) Ltd
Pound Lane
Exmouth EX8 4NN Devon
Exmouth 4771

Sylpon Manufacturing Co Ltd
73 Maygrove Road
London NW6 2EL
01-624 9364

Symbol Signs
Chantry Lane
Grimsby Lincs
Grimsby 56154

Synthetic Resins Ltd
Frodsham House
Edwards Lane
Speke
Liverpool L24 9HX
051-486 1395

TAC Construction Materials Ltd
PO Box 22
Trafford Park
Manchester M17 1RU
061-872 2181

TPT Ltd
Romiley
Stockport SK6 4DY
061-430 6061

TT Containers Ltd
Rye Lane
Otford
Sevenoaks Kent
Otford (Kent) 2169

TT Mouldings Ltd
Bruce Grove
Wickford Essex
Wickford 2148

Tadsown Ltd
Westwood House
Warwick Street
Leamington Spa
Leamington Spa 21271

Tallon Plastics Ltd
Feeder Road
Bristol BS2 0SG
Bristol 78055

Tanerone Ltd
30–33 Rutherford Close
Eastwood
Southend-on-Sea Essex
Southend 522532

Tar Residuals Ltd
Plantation House
Mincing Lane
London EC3
01-626 3494

Taylor (C F) (Plastics) Ltd
Moorside Road
Winnal Trading Estate
Winchester Hants
Winchester 62261

Taylor Industries Ltd
Rowlands Gill Co Durham
Rowlands Gill 2207

Tecalemit (Engineering) Ltd
Plymouth PL6 8LA Devon
Plymouth 62844

Technical Treatments Ltd
Rye Lane
Otford
Sevenoaks Kent
Otford 2169

Technichrome Products (Plastics) Ltd
15 Kingsbury Terrace
Islington
London N1
01-254 8651

Techno-Plast Ltd
2–10 Windsor Road
Slough Bucks
Slough 34367

Technovac
Carlisle Street
Sheffield S4 7LP Yorks
Sheffield 25045

Tekta Packaging Ltd
Waverley Works
Queen Street
Market Rasen Lincs
Market Rasen 2360

Telco Ltd
Alma Road
Ponders End
Enfield Middx
01-804 1282

Telcon Plastics Ltd
Green Street Green
Orpington BR6 6BH Kent
Farnborough 55685

Temple Buildings Products Ltd
Temple Mill
Liphook Hants
Passfield 281

Tenaplas Ltd
Upper Basildon
Pangbourne RG8 8ST Berks
Upper Basildon 333

Tennant (Charles) & Co Ltd
214 Bath Street
Glasgow C2
041-332 6442

Tenoosa Ltd
Dulwich Road
Radford
Nottingham NG7 3DN
Nottingham 76036

Terminal Insulators Ltd
Wilbury Way
Hitchin Herts
Hitchin 50851

Tesa Tapes Ltd
Dolman House
Ascot Road
Bedfont
Feltham Middx
Ashford (Middx) 59131

Textile Bonding Ltd
off Midland Road
Higham Ferrers Northants
Rushden 4571

Textile Mouldings Ltd
Globe Works
Accrington Lancs
Accrington 35123

Thermalon Ltd
Berkeley Square House
London W1
01-629 8030

Therma-Plast Ltd
Rushey Lane
Tyseley
Birmingham B11 2BT
021-706 6363

Thermoplastic Coatings Ltd
Albion Industrial Estate
Oldbury Road
West Bromwich Staffs
021-533 3014

Thermo Plastics Ltd
Luton Road
Dunstable LU5 4LN Beds
Dunstable 64255

Thetford Moulded Products Ltd
Mill Lane
Thetford Norfolk
Thetford 4266

Thomas & Vines Ltd
Maple Cross
Denham Way
Rickmansworth Herts
Rickmansworth 75111

Thomson (T) Sons & Co (Barrhead) Ltd
Fereneze Works
Barrhead
Glasgow
041-881 1038

Thurgarolle (Successors) Ltd
Rothwell Road
Kettering Northants
Kettering 4422

Thames Estuary Plastics Ltd
289 Kiln Road
Thundersley
Benfleet SS7 1QS Essex
Southend-on-Sea 559076

Tyne (William) (Plastics) Ltd
Eastfield Drive
Penicuik Scotland
Penicuik 2451

Tinker Grayson Ltd
Britannia Mills
Stoney Battery
Huddersfield Yorks
Huddersfield 55221

Tod (W & J) Ltd
Ferrybridge
Weymouth Dorset
Weymouth 3434

Tomkinson (F W) Ltd
York Terrace
Hockley Hill
Birmingham 18
01-554 3341

Toone Plastics Ltd
Kenilworth Drive
Oadby Leics
Oadby 3226

Torpey (Sylvester) & Sons Ltd
Foster Street
Liverpool 12O 8EX
051-922 1877

Townstal Products Ltd
Industrial Estate
Dartmouth Devon
Dartmouth 2041

Trade Laminators Ltd
Burowhill
Wiveliscombe Somerset
Lydeard St Lawrence 309

Transatlantic Plastics Ltd
45 Victoria Road
Surbiton Surrey
01-399 5271

Tratt Plastics Ltd
Stanley House
St Chads Place
London WC1
01-837·0094

Tresco Plastics Ltd
High Street
Earls Barton
Northampton
Earls Barton 376

Trig Engineering Ltd
33b Bridgwater Road
North Petherton
Bridgwater Somerset
North Petherton 781

Triplex Foundries Group Ltd
Upper Church Lane
Tipton Staffs
021-557 1293

Triton Plastics Ltd
No 22 Woods Lane Factory Centre
Cradley Heath
Warley Worcs
Cradley Heath 69560

Tropical Packers (Cheshire) Ltd
Albany Mill
Canal Street
Congleton Cheshire
Congleton 2401

Troviplast Ltd
College House
29–31 Wrights Lane
London W8 5BR
01-937 0117

Truform Plastics Ltd
Hospital Hill
Waterside
Chesham Bucks
Chesham 4657

Tryka Ltd
Farrell Street
Salford M7 9TN Lancs
061-832 6546

Tuckers (Sheffield) Ltd
Shoreham House
Shoreham Street
Sheffield S1 4SR Yorks
Sheffield 29691

Tufnol Ltd
PO Box 376
Perry Barr
Wellhead Lane
Birmingham B42 2TB
021-356 6218

Tullis (John) & Son Ltd
Tullibody
Alloa Clackmannanshire
Alloa 3314

Turbro Ltd
Underbank Mill
Daniel Street
Whitworth
Rochdale Lancs
Whitworth 2345

Turner Brothers Asbestos Co Ltd
Spotland
Rochdale Lancs
Rochdale 47422

Turner & Hughes Ltd
64A High Street
Wimbledon Common
London SW19
01-946 9967

Tye (John) & Sons (Packages) Ltd
The Sachet Centre
717 North Circular Road
London NW2
01-450 7232

Tygadure (Division of Fothergill & Harvey Ltd)
Littleborough Lancs
Littleborough 78831

Tyne Plastics Ltd
Team Valley Trading Estate
Gateshead Co Durham
Gateshead 876093

UECL Ward Brooke
Fassetts Road
Loudwater Bucks
Lane End 26233

UG Closures & Plastics Ltd
Staines House
158–162 High Street
Staines Middx
Staines 57486

UK Optical Bausch & Lomb Ltd
Mill Hill
Bittacy Hill
London NW7
01-346 2660

UK Plastics Ltd
Swan Street
Petersfield Hants
Petersfield 2291

USI Engineering Ltd
Burtonwood
Warrington Lancs
Newton-le-Willows 5131

US Industrial Chemicals Co
9 Oatlands Chase
Weybridge Surrey
Walton-on-Thames 28495

Ulster Plastics Ltd
Welwyn Garden City Herts
Welwyn Garden 26338

Ultra Electronics (Components) Ltd
Fassetts Road
High Wycombe Bucks
High Wycombe 26233

Union Carbide UK Ltd (Chemicals Divisio
Peter House
Oxford Street
Manchester 1
061-236 2226

Uniroyal Ltd (Tyre & General Product Divisio
PO Box 47
Castle Mills
Edinburgh 3
031-229 7351

Unit Buying Services Ltd
The Runway
Ruislip HA4 6SE Middx
01-845 3535

United Moulders Ltd
Fernhurst
nr Haslemere Surrey
Fernhurst 456

United-Carr Ltd
Wallingford Road Industrial Estate
Uxbridge Middx
Uxbridge 38681

Unitex
Halfpenny Lane
Knaresborough HG5 0PP Yorks
Knaresborough 2677

VMG Plastics Ltd
8 Southfields
Welwyn Garden City Herts
Welwyn Garden 29481

Vacmobile (Ireland) Ltd
Priory Works
Belsize Road
London NW6 4BU
01-328 0136

Vacuum Formers Ltd
London Road
Macclesfield Cheshire
Green Hills 2889

Vacuum Research Ltd
Burrell Way
Thetford Norfolk
Thetford 2466

Vactite Wire Co Ltd
Linacre Lane
Bootle L20 6AE Lancs
051-922 1661

Vale (J S)
129 Duddeston Mill Road
Birmingham B7 4SR
021-359 3407

Valley Plastic Coaters Ltd
Chain Bridge Road Estate
Blaydon Co Durham
Blaydon 2005

Vanguard Plastics Ltd
Ely Cambs
Ely 3421

Varicol Signs Ltd
Bayton Road
Exhall
Coventry Warwicks
Bedworth 5411

Vectis Laminates Ltd
Temple Mill
Passfield
Liphook Hants
Passfield 281

Varnish Industries Ltd
Oakenbottom Road
Bolton BL2 6DP Lancs
Bolton 23507

Venesta International-Vencel Ltd
West Street
Erith Kent
Erith 36922

Vero Precision Engineering Ltd
Southmill Road
Regents Park
Southampton Hants
Southampton 71061

Vicsons Ltd
148 Pinner Road
Harrow Middx
01-427 0706

Victoria Plating Co Ltd
Plastics Division
65 Merton High Street
London SW19
01-542 2861

Victoria International Plastics Ltd
South Block
Cricklewood Trading Estate
Claremont Road
London NW2
01-452 1021

Victor Plastics (Manchester) Ltd
Cheltenham House
Cheltenham Street
Pendleton
Salford M6 6NT Lancs
061-736 4538

Viking Industrial Plastics Ltd
73 Grosvenor Street
London W1X 9DD
01-629 8368

Vinablend Ltd
Ash Street
Leicester LE5 0DG
Leicester 22527

Vinatex Ltd
New Lane
Havant P09 2NQ Hants
Havant 6350

Vinyl Compositions Ltd
Grimshaw Lane
Bollington
Macclesfield SK10 5JF Cheshire
Bollington 2485

Vinyl Products Ltd
Mill Lane
Carshalton Surrey
01-669 4422

Vinyl Reprocessors Ltd
Grimshaw Lane
Bollington
Macclesfield SK10 5JF Cheshire
Bollington 2066

Visijar Laboratories Ltd
Pegasus Road
Croydon CR9 4PR Surrey
01-686 6341

Viskase Ltd
185 London Road
Croydon CR9 2TT Surrey
01-686 2921

Vitafoam Ltd
Middleton
Manchester M24 2DB
061-643 4301

Vitafoam (Bradford) Ltd
Castle Mills
Apperley Road
Bradford Yorks
Bradford 612631

Vitafoam (Wraystone) Ltd
Lea Bridge Road
London E10
01-539 9041

Vitalam (Mandleberg) Ltd
Seaford Road
Pendleton
Salford 6 Lancs
061-736 5343

Vitamol (Clyde) Ltd
Porterfield Road
Renfrew Scotland
041-886 2384

Vitamol Precision Ltd
Audenshaw
Manchester M34 5FE
061-336 5931

Vitesta Ltd
Walker Street
Rochdale Lancs
Rochdale 46931

Vogue Plastics (Liverpool) Ltd
Horby Boulevard
Bootle 20 Lancs
051-922 8181

Vulcascot Ltd
43 Wales Farm Road
North Acton
London W3
01-992 8862

WB Industrial Plastics Ltd
Navigation Road
Burslem
Stoke-on-Trent ST6 3BO Staffs
Stoke-on-Trent 88661

WCB Containers Ltd
Stamford Works
Bayley Street
Stalybridge SK15 1QQ 061-330 6511

Wade Couplings Ltd
Argyle Street
Birmingham B7 5TN
021-327 4077

Walker Litherland Plastic Ltd
Whitehall Mill
Darwen BB3 2LR Lancs
Darwen 72155

Walker of Manchester
Little Newton Street
Manchester M4 6FR
061-236 8213

Wall & Leigh Thermoplastics
Friar Park Road
Wednesbury Staffs
021-556 2161

Wallington & Weston & Co Ltd
Vallis Mills
Frome Somerset
Frome 3271

Walsh (S & J) (Plastics) Ltd
Hereford Road
Blackburn BB6 7NW Lancs
Blackburn 51411

Wandleside Warren Wire Co Ltd
Dunmurry
Belfast N Ireland
Dunmurry 2535

Wandex Distributors (London) Ltd
Polymer House
32–38 Rivington Street
London EC2
01-739 9651

Ward Adams Co Ltd (The)
Powder Mills
Leigh
Tonbridge Kent
Hildenborough 3585

Ward & Goldstone Ltd
107 Frederick Road
Salford M6 6AP Lancs
061-736 5822

Ward (Thomas W) Ltd
Albion Works
Savile Street
Sheffield S4 7UL
Sheffield 26311

Wardle (Bernard) (Everflex) Ltd
Caernarvon N Wales
Caernarvon 3431

Warne (William) & Co Ltd
Gascoigne Road
Barking Essex
01-594 3800

Warren (F R) & Co Ltd
79 Ashley Down Road
Bristol 7
Bristol 45051

Waterside Plastics Ltd
Todmorden Lancs
Todmorden 2471

Watson Industrial Plastics
Hobart Road
Princes End
Tipton Staffs
021-557 3741

Wear Ventilator & General Sheet Metal Co Ltd
Bonnersfield
Sunderland SR6 0AB Co Durham
Sunderland 4169

Webster Products (Chiswick) Ltd
192B Chiswick High Road
London W4
01-994 4213

Weil (Joseph) & Son Ltd
Friars House
39–41 New Broad Street
London EC2
01-588 5052

Weldex Plastics Ltd
34a High Street
High Wycombe Bucks
High Wycombe 27346

Wells Hinton Plastics
Willow Wood Works
London Road
Amersham Bucks
Amersham 7910

Welsh Mouldings (1964) Ltd
105–107 Cannon Street
London EC4
01-626 0454

Welsh Products Ltd
Goat Mill Road
Dowlais
Merthyr Tydfil Glam
Merthyr Tydfil 2401

Welsh Trust (Rhigos) Ltd
Hirwaun Industrial Estate
Aberdare Glam
Aberdare 421

Weltonhurst Ltd
Blackburn Road
Darwen BB3 1QJ
Darwen BB3 1QJ Lancs
Darwen 72124

Welwyn Plastics (1955) Ltd
Woodside Road
Welwyn North Herts
Welwyn 4484

Westbrook Marine Co Ltd
St Michaels Trading Estate
Bridport Dorset
Bridport 3664

West Country Converters Ltd
Station Road
Warminster Wilts
Warminster 2026

Western Laminates Ltd
New Road
Brixham TQ5 8NF Devon
Brixham 4634

Western Pressings Ltd
Brackla Industrial Estate
Bridgend Glam
Bridgend 4067

Westfield Engineering Co (Marine) Ltd
Cabot Lane
Creekmoor
Poole BH17 7DA Dorset
Broadstone 4300

Westford Plastics & Engineering Ltd
Westford
Wellington Somerset
Wellington 2377

Westinghouse Brake & Signal Co Ltd
82 York Way
London N1 9AJ
01-837 6432

Westpole Products Ltd
Claverings Industrial Estate
Montague Road
Edmonston
London N9 0AP
01-807 8978

Westway Models Ltd
236 Woodhouse Road
London N12
01-368 2912

Wetty Oates & Co Ltd
Gateford Road
Worksop Notts
Workshop 4351

Whaley Welding Co Ltd
Phoenix Sidings
Bishopton Road
Stockton-on-Tees Teeside TS19 0AD
Stockton 62531

Wheatley (A) Ltd
Reynolds Mill
Newbridge Lane
Stockport SK1 2NR Cheshire
061-480 6319

Wheway Frames Ltd
Grays Lane
Moreton-in-Marsh Glos
Moreton-in-Marsh 838

Whiteley Electrical Radio Co Ltd
Victoria Street
Mansfield Notts
Mansfield 24762

Whitson (James) & Co Ltd
High Street
Yiewsley
West Drayton Middx
West Drayton 3771

Widnes Foundry & Engineering Co Ltd
Lugsdale Road
Widnes Lancs
Widnes 2889

Wilkie & Paul Ltd
Grove Works
Slateford Road
Edinburgh 3
031-443 2384

Willamot Industrial Mouldings Ltd
Butchers Road
London E16
01-476 3151

Wilmot Breeden Ltd
Amington Road
Birmingham B25 8EW
021-706 3344

Willmott Son & Phillips Ltd
52A Blackstock Road
Finsbury Park
London N4
01-226 5257

Willow Plastics Engineering Co Ltd
4 Victoria Road
Hendon
London NW3
01-203 0353

Wilson (F E) (Plastics Ltd)
Newhall Street
Sutton Coldfield Warwicks
021-354 1318

Wilson (J) & Co Ltd
1A St Mary's Road
London SE15
01-639 5919

Wilson Sherrif Ltd
Airport Service Road
Portsmouth P03 5PD Hants
Portsmouth 65655

Wincanton Transport & Engineering Co Ltd
Station Road
Wincanton Somerset
Wincanton 2021

Windshields of Worcester Ltd
Raglan Street
Barbourne Worcester
Worcester 27111

Witco Chemical Co Ltd
Bush House
Aldwych
London WC2
01-836 6473

Witton Moulded Plastics Ltd
Electric Avenue
Witton
Birmingham B6 7JP
021-327 1941

Wokingham Plastics Ltd
Fishponds Road
Wokingham RG11 2QH Berks
Wokingham 2271

Woodland Plastics
Portmore
Lymington Hants
Lymington 2181

Woolen & Co Ltd
19 Love Street
Sheffield S3 8NZ
Sheffield 25871

Wooley (Frederick) Ltd
Farm Street
Hockley
Birmingham B19 2TY

Wragby Plastics Ltd
Wragby Lincs
Wragby 383

Wye Plastics Ltd
Madley
Nr Hereford
Madley 484

Wyllie-Young Ltd
College Milton
East Kilbride Lanarks
East Kilbride 21463

Yorkshire Dyeware & Chemical Co Ltd (The)
27 Kirkstall Road
Leeds LS3 1LL
Leeds 38881

Xenit Products Ltd
95 Farnham Road
Slough Bucks
Slough 21289

Xlon Products Ltd
323A Kennington Road
London SE11
01-735 8551

Zephyr Plastic Products Ltd
72 Blackburn Road
Birstall
Leeds
Batley 4740

Zonex Ltd
123 Talbot Road
Blackpool Lancs
Blackpool 45287

Glossary of Terms

ABS: see acrylonitrile butadiene styrene.

Accelerator: a catalyst, or a substance added to a catalyst, to accelerate a chemical action, thus reducing the time or temperature required for processing.

Acetal: thermoplastics material produced by polymerisation of formaldehyde and possessing high softening point and numerous good physical properties resulting in its use for bearing, gears, bushes, etc.

Acetal copolymers: thermoplastic materials produced by polymerisation of formaldehyde with other monomers, as opposed to polyacetal (qv).

Acrylic resins: thermoplastic material produced by polymerisation of an acrylic resin. Can be extremely clear, *eg* 'Perspex' (polymethylmethacrylate).

Acrylonitrile: acrylic monomer containing nitrogen used with styrene, styrene and butadiene, or butadiene to produce thermoplastics and rubbers.

Arcylonitrile butadiene styrene: thermoplastic material produced by copolymerisation of acrylonitrile, butadiene and styrene. Exceptionally high impact strength and good resistance to heat distortion and chemicals, suitable for electroplating.

Activator: alternative term for an 'accelerator' (qv).

Air-slip forming: bubble assist vacuum forming (qv).

Alkyd resins: saturated and unsaturated polyester resins used to produce polyurethanes and for laminating and casting. Thermosetting alkyd moulding material has various moulding advantages over aminos and phenolics. See diallyl phthalate.

Allyl resin: resin used to produce thermosetting moulding materials and laminates.

Aminoplastics: thermosets produced by condensation of formaldehyde with aminos, such as melamine and urea.

Antioxidant: compound which stabilises the composition to which it is added by preventing oxidation.

Antistatic agent: material which reduces the tendency of a composition to accumulate static electrical charges.

Bakelite: one of the earliest plastics. Thermosetting resins of the phenol-formaldehyde type with high electrical insulation properties. Used with 'fillers' as moulding materials and in laminates.

Blow moulding: process used to make one-piece hollow plastics products, *eg* bottles. There are various techniques but they are normally based on the use of extrusion or modified injection moulding equipment.

Boss: a protrusion on a component which adds strength and also may facilitate alignment during assembly.

Bottom force: see force.

Bubble assist: refinement of the vacuum forming (qv) method of forming plastics sheet in which the sheet is prestretched by air pumped into the vacuum box prior to applying the vacuum.

Butadiene: readily polymerised gas, used with acrylonitrile or styrene to produce rubbers; or with acrylonitrile and styrene to produce ABS resins.

CAB: see cellulose acetate butyrate.

Calendering: production of plastics sheet of high quality and to close tolerances by use of a roller mill comprising three or more rollers. Most products calendered at present are either PVC or ABS or a mixture of both.

Carbon fibre: reinforcing material which when mixed with a suitable matrix produces a remarkably stiff and strong composite, with low friction and wear characteristics and weight advantages.

Casein plastics: tough non-flammable thermoplastics or thermosets produced from rennet casein, extruded to form rod or tube or extruded and compression moulded to form sheet. Used for products such as buttons in which dimensional stability is not important.

Cast plastics: liquid resins with hardening agents poured into moulds and hardened with or without heat. Includes certain phenol formaldehyde, acrylic, epoxide, polyester, nylon and polyurethane resins.

Catalyst: substance which promotes and changes the rate of a chemical reaction without undergoing permanent change in its own composition.

Celluloid: (trade name) plastics consisting of celluse nitrate and camphor. Very flammable. Produced as sheet, rod and tube. One of the first plastics.

Cellulose acetate: cellulose ester used with plasticiser as an injection moulding and extruding material. Not so highly flammable as cellulose nitrate. Used as sheet for packaging, also to produce pens, combs, knife-handles, etc.

Cellulose acetate butyrate (CAB): plastics similar to cellulose acetate but with better moisture, solvent and oil resistance and dimensional stability.

Cellulose nitrate: (nitrocellulose) plastics often known by its trade name Celluloid (qv).

Cellulose propionate: plastics produced in a similar manner to cellulose acetate. Used in the USA for portable radio and domestic equipment housings.

Cellulosic plastics: plastics based on modified cellulose (organic fibre). Divided into mainly cellulose esters and cellulose ethers.

Chlorinated polyether: thermoplastic material mainly used where corrosion resistance is important, *eg* control valves, tank and pipe linings.

Chlorinated polyethylene: thermoplastics whose properties depend on the degree of chlorination, up to an optimum 27% for maximum flexibility. Used for footwear mouldings, gaskets, weather protective sheeting, toys, textile interlinings.

Compression moulding: process mainly used to manufacture thermosetting plastics products. It employs a split mould which is heated, opened to receive the moulding material and then closed under pressure.

Continuous lamination: producing continuous lengths of reinforced plastics sheet by passing the reinforcement, in the form of woven fabric or random mat, through a tank containing resin, then compacting through rollers and curing in an oven.

Copolymer: material produced by polymerising two or more different monomers (qv).

Crosslinking (plastics): chemical bonding of thermosetting molecules when they are cured.

DAIP: diallyl isophthalate (qv).

DAP: diallyl phthalate (qv).

Diallyl isophthalate (DAIP): plastics alkyd moulding compound used in electrical components for rockets and supersonic aircraft. Similar to diallyl phthalate (qv) but with better heat resistance and more expensive.

Diallyl phthalate (DAP): plastics alkyd moulding material whose main areas of use are in the field of miniaturised electrical and electronic components. See also diallyl isophthalate.

Dip coating: method of coating products with PVC, as an alternative to painting or stove-enamelling; involves preheating the product, dipping into PVC paste (plastisol), draining off and curing in an oven.

Dip moulding: plastics forming process in which a heated male former is dipped into a bath of plastics and the coating left on after withdrawal of the former is cured and stripped from the former. Industrial gloves, handle grips, etc, are made in this way.

Direct screw transfer: a moulding system used for thermosetting plastics, it is based on the injection moulding process used for thermoplastics.

DMC: dough moulding compound (qv).

Dough moulding compound (DMC): composition of polyester resin and glass fibre (or other fibre) reinforced plastics for compression moulding.

Drape forming: type of vacuum forming (qv) process in which heated plastics sheet is pulled down over a male mould, or the mould is pushed into the sheet, and the vacuum applied. Drape forming is used mainly to produce a certain degree of performing.

DST: direct screw transfer (qv).

Emulsion: a relatively stable suspension of one liquid in another, in which it will not dissolve.

Epoxy resins (epoxides): plastics available as liquid and solid resins and moulding compounds, into which large amounts of filler can be incorporated. Used for laminating, coating, adhesives, casting, encapsulation and compression moulding. Applications include cast and encapsulated electrical components, laminated plastics vessels, pipework and printed circuit boards.

EPS: expanded polystyrene (qv).

Ethylcellulose: cellulose plastics with similar properties to other cellulosics but with lower density and exhibiting outstanding toughness at low temperatures.

Ethylene vinyl acetate copolymers (EVA): polyolefin plastics whose properties resemble polyethylene but are more easily processed. EVA films are used for food packaging, and as outer layers for greenhouses. Tubing is used for medical and beverage dispensing. Properties include outstanding flexibility, clarity, resilience and low temperature characteristics.

EVA: see ethylene vinyl acetate copolymers.

Expandable polystyrene: a form of polystyrene containing a volatile blowing agent such as pentane and, on heating (usually by steam), it can be expanded to fill a suitable mould.

Expanded polystyrene: cellular plastics used in building, as boards for insulation (and even as an infill in various concrete structures) and in packaging, chiefly in the form of mouldings shaped to fit round components to act as a container and cushioning agent.

Extenders: substances used for filling out adhesive and plastics compositions. In PVC compositions, extenders are added to the plasticiser mixture to impart particular properties or reduce cost.

FEP: see fluoroethylene propylene.

Filament winding: method of producing spherical or cylindrical reinforced plastics mouldings by winding continuous filament reinforcement onto a mandrel previously coated with resin.

Filler: a reinforcing material, fillers such as glass fibre, woodflour, asbestos fibre, etc, are incorporated in plastics to improve properties.

Fillet: internal rounded corners in a moulded product, employed to increase strength in that area where a sharp change in section would cause weakness.

Floating weight: term applied to the top door of an enclosed mixer, it floats on a cushion of air in the pressure cylinder.

Flood lubrication: a copious flow of oil through a journal bearing, necessary when the journal is hot to provide adequate lubrication and prevent overheating of the oil.

Fluidised bed coating: method of coating products with plastics. Involves passing air through fine plastics powder particles so they are agitated and behave as a fluid, then dipping the heated product into them and finally sintering or curing the coating.

Fluorocarbons: a group of thermoplastic resins which are very strong, tough and stable with excellent temperature and electrical properties. Used for electrical components and as bearing materials. Group includes FEP, PTFE and PCTFE (qv).

Fluoroethylene propylene (FEP): thermoplastic of the fluorocarbon group with similar properties to PTFE (qv) but can be injection moulded and extruded.

Force: term used in plastics processing to denote the complementary parts of a mould which convey the moulding pressure to the material in the mould cavity; hence 'top force' and 'bottom force' are the top and bottom halves of a mould.

Formaldehyde (formalin): gas, usually used in a water solution, to produce plastics with phenol, urea or melamine. When polymerised by itself it produces polyacetal.

Formalin: water solution of formaldehyde (qv).

Friction nip: an ingoing nip in which the surface speed of one roll is greater than the other.

Friction welding: spin welding (qv).

Gate: restricted orifice in a mould between the channel transporting the material to be moulded and the actual mould cavity itself.

Glass reinforced plastics: see reinforced plastics.

GP polystyrene: general purpose polystyrene as opposed to high impact polystyrene (qv).

GRP: glass reinforced plastics, see reinforced plastics.

Hand lay-up: method of producing reinforced plastics mouldings manually. Suited to large mouldings or low quantities. Involves coating male or female mould with release agent, then a gel coat to give good surface finish, then adding layers of reinforcement and resin to produce laminate.

HDPE: high density polyethylene, see polyethylene.

Heatsealing: method of welding thermoplastic materials, principally used for sealing polyethylene film in the manufacture of bags.

High density polyethylene: polyethylene (qv) with a density of 0.936-0.965.

High impact polystyrene: plastics with a higher impact strength than general purpose polystyrene due to the presence of a modifier, such as rubber.

H I polystyrene: high impact polystyrene (qv).

Homopolymer: polymer produced from a single type of monomer (qv) as opposed to a copolymer (qv).

Hot gas welding: method for welding thick thermoplastic sheeting, used where high frequency welding and heatsealing are impractical. Involves heating the welding area by a jet of hot air or inert gas and often employing a filler rod of similar material to the sheet.

Hot melt strength: ability of a thermoplastic, when heated to just below melting point, to be stretched and formed.

Injection moulding: technique for moulding thermoplastics similar to the pressure diecasting process. It consists basically of two operations, the softening of the thermoplastic by heat and its injection or transfer under pressure into a closed mould.

Intermix: the overlapping action of the maximum diameters of the rotors in a mixer (in a similar manner to a pair of gears).

Ionomers: transparent thermoplastic materials which exhibit toughness, low friction characteristics and resistance to alkalis, solvents, and oils, but not acids. Used for medical phials, tool handles, road marking 'cats eyes'.

Laminate: a composite, usually sheet or tube, produced by bonding layers of reinforcement, *eg* paper, fabric, glass fibre, with a resin.

Latex: fine dispersion or emulsion of a plastics or rubber in water (plural —*latices*).

Latices: plural of latex (qv).

LDPE: low density polyethylene, see polyethylene.

Leathercloth: cloth with a woven or knitted base which has been coated with a plastics, and which has an embossed surface giving a leather-like appearance.

Low density polyethylene: polyethylene (qv) with a density of 0.91–0.93.

Matched die moulding: method of moulding reinforced plastics products in which a preform of the reinforcing material is placed in a mould, the required amount of resin is added and the mould is closed. Similar to the compression moulding of dough moulding compounds (qv).

Matched metal moulding: see matched die moulding.

Matrix: see reinforced plastics.

Melamine formaldehyde: aminoplastics used principally for moulded tableware (cups, plates, saucers, bowls) but cellulose-filled grades used for saucepan handles, ashtrays and electrical accessories requiring good heat resistance.

MF: melamine formaldehyde (qv).

Mix hard: a plastics formulation which includes little or no plasticiser.

Monomer: a compound or molecule which can be used to produce polymers or copolymers (qv).

Nip: the ingoing action of two rollers running in opposite directions.

Nitrocellulose: see Celluloid.

Nylon: generic name (not trade name) for high molecular weight polyamide plastics. Nylons are distinguished by numbers (which refer to the carbon atoms present in the monomers used to produce them) and although each type has properties common to the group, it also has certain properties specific to itself. The types now in common use include: nylon 6.6; nylon, 6; nylon 6.10; nylon 11; and nylon 12. Used for electrical insulation, brushes, gear and bearings, machine parts, filament and yarn, textiles, etc.

Parison: the length of plastics tube which emerges from an extruder, and which is blown into shape in the blow moulding process.

PCTFE: polychlorotrifluoroethylene (qv).

Perspex: (trade name) polymethyl methacrylate (qv).

PETP: Polyethylene terephthalate (qv).

Phenol formaldehyde: thermosetting plastics which can be formulated with many and varied fillers which impart distinct characteristics to the finished product.

Phenolics: phenol formaldehyde (qv).

Phenoxy: thermoplastic epoxy type of material characterised by good clarity, rigidity and toughness. It can be injection moulded, extruded and blow moulded.

Pinch-off line: the line on a blow moulded article where the mould has closed around the parison.

Plasticiser: a substance incorporated in plastics compositions to promote ease of processing and improve the flexibility of the finished product.

Plug assist: vacuum forming (qv) process in which the sheet is prestretched by a male forming tool (plug) into a female mould before applying vacuum. Used for deep drawn products.

PMMA: polymethyl methacrylate (qv).

Polyacetal: thermoplastic material produced by polymerisation of formaldehyde alone (see acetal).

Polyamides: group of thermoplastic polymers, certain types of which are referred to as nylons (qv).

Polycarbonate: thermoplastic material which is transparent and exhibits good mechanical properties over a wide range of temperatures up to 145°C.

Polychlorotrifluoroethylene (PCTFE): a fluorocarbon plastics with similar properties to PTFE but is easier to fabricate.

Polyester resins: thermosetting plastics used for laminating, encapsulation and casting, and as dough moulding compounds (qv) for compression moulding. Mainly used as liquid binders for glass fibre (and other fibre) reinforced plastics.

Polyethylene (polythene): the most widely known and still perhaps the most important thermoplastic. Generally divided into two types, low density (LD) and high density (HD), although the actual difference in density is quite small, but HD polyethylene is noticeably harder and stiffer and the two-types exhibit different permeability characteristics and chemical resistance. Used to manufacture containers of all kinds.

Polyethylene terephthalate: specialist plastics with a combination of hardness, low coefficient of friction and reasonable electrical properties.

Polymer: substance composed of large molecules built up from two or more (sometimes thousands) of identical simpler molecules.

Polymerisation: the process of combining two or more molecules of the same compound (the monomer) to form a new substance with properties different from the monomer. **Copolymers** are combinations of different molecules. **Natural polymers** include natural rubber, cellulose, casein. **Synthetic polymers** include plastics, synthetic rubbers and synthetic resins.

Polymethyl methacrylate (PMMA): thermoplastic material with exceptional optical clarity and resistance to outdoor exposure, used in the form of cast sheets, blocks and rods, and as injection moulding and extrusion material. This acrylic material is sometimes known by the trade name 'Perspex'.

Poly 4-methylpentene-1 (TPX): often known by its trade name TPX, this thermoplastic is a transparent polyolefin.

Polyolefins: a group of thermoplastics the most important of which are polyethylene and polypropylene (qv), but also includes EVA and TPX.

Polyphenylene oxide (PPO): thermoplastic polymer with good electrical and mechanical properties and excellent dimensional stability over a wide range of temperatures. Uses include: medical equipment, electronic components, hot water controls, etc.

Polyphenylene sulphide: relatively new plastics available in both thermosetting and thermoplastic forms. Excellent resistance to acids, alkalis and solvents and has good temperature resistance. Used as mouldings and for powder coating.

Polypropylene: thermoplastic of the polyolefins group available in light colours. Numerous applications include packaging film, domestic utensils, pipe fittings, bottles, etc.

Polystyrene: a tough, white thermoplastic with high insulation power. Used for impregnation of electrical coils, surface coatings, etc. Types available include general purpose or unmodified polystyrene, high impact or toughened polystyrene, expanded polystyrene, or the copolymers SAN and ABS (qv).

Polysulphone (polysulfone): thermoplastic copolymer with high tensile strength, good electrical properties and excellent resistance to acids, alkalis and oils, and these properties are retained for long periods at high temperatures. Produced in the USA. Mainly electrical and electronic uses.

Polytetrafluoroethylene (PTFE): fluorocarbon plastics with outstanding chemical inertness and heat resistance. Processing is not possible by usual methods and mouldings are produced by sintering, it can be extruded but with difficulty. Uses include high duty bearings, seals, valves, gaskets, linings, pipes in chemical plant, non-stick cooking surfaces, etc.

Polythene: common term for polyethylene (qv).

Polyurethane foams: cellular plastics basically of three types: **flexible urethane foams** used for upholstery, matresses, textile linings, acoustic insulation, etc; **rigid urethane foams** used for heat insulation (*eg* refrigeration), cold stores, building-panel cores, etc; and **semi-flexible foams** used for arm rests, packaging, etc.

Polyurethane resins: polymers available as thermoplastics or thermosets, soft rubber-like to hard brittle. Thermoplastic urethanes can be processed by conventional thermoplastic techniques or other, more expensive, types can be cast. Uses include: solid truck tyres, shock absorbers, bushings, bearings, gears, belting, shoe heels, etc, depending on type.

Polyvinyl acetate (PVA): plastics material whose main uses are as adhesives (book-binding, etc), emulsion paints, paper sizes.

Polyvinyl chloride (PVC): probably the most versatile thermoplastic material, being available in many forms over a wide range of hardness and flexibility. Varying the use of plasticisers in PVC can make it more suitable for numerous applications. Uses include: film-like coatings, wall and floor tiles, domestic appliance components, dolls, gramophone records, gloves, hoses, etc.

Polyvinyl chloride acetate: thermoplastic material with similar properties to PVC but improved solubility and lower softening temperature.

PP: polypropylene (qv).

PPO: polyphenylene oxide (qv).

PPS: polyphenylene sulphide (qv).

Pressure bag compacting: compacting reinforced plastics moulding by applying air pressure inside a heated vessel, as opposed to vacuum bag compacting (qv).

Promotor: alternative term for an 'accelerator' (qv).

PS: polystyrene (qv).

PVA: polyvinyl acetate (qv).

PVC: polyvinyl chloride (qv).

Reinforced plastics: a wide range of products produced from reinforcing fibre (notably glass, but others such as asbestos and carbon fibres, etc, are used) and synthetic resin (called a 'matrix'). Both thermosetting and thermoplastic polymers are used as a matrix. Unsaturated polyester resins are still usually looked upon as the major matrix material, but many other plastics can be and are used extremely successfully (the best known being nylon).

Rigid PVC: polyvinyl chloride (qv) composition with little or no plasticiser. Used for calendering, injection moulding, extrusion, etc.

Rotational moulding: process for forming hollow plastics bodies in which a measured quantity of polymer is placed inside a split female mould, the mould is heated and rotated in at least two planes dispersing the polymer over all parts of the mould, and then cooled whilst still rotating.

Short shot: term used in injection moulding when insufficient material is injected into the mould, which results in an incomplete moulding being produced.

Silicones: polymers which can take the form of fluids for lubrication, rubbers or plastics. Silicone resins are employed in electrical insulation applications, including component encapsulation. Silicone rubber is unique among synthetic elastomers being the only one with useful operating temperature, and resistant to weather, ozone, chemicals, with a 'non-stick' surface. Silicones are used as release agents, because of incompatability with most materials, and as lubricants and additives.

Sink mark: sunken area in a moulding, usually occurring where there is a rib or boss on the reverse side. Some materials with a high coefficient of expansion such as polypropylene, are more prone to this phenomenon than others, such as polystyrene. Sink marks can be disguised by grained or textured surfaces.

Sinter coating: see fluidised bed coating.

Spray coating: method of coating large products with thermoplastic. Involves heating area to be coated with flame-gun, spraying on plastics powder and finally sealing with flame-gun.

Spray lay-up: method of mechanising 'hand lay-up' (qv) of reinforced plastics by spraying reinforcement and resin onto mould.

Solid urethanes: see polyurethane resins.

Spinneret: die containing several small holes through which plastics is forced to form fibres.

Spin welding: welding circular thermoplastics products by pressing two parts into close contact and rotating one of them. Friction developed at the interface gives a fast, reliable and pressure tight joint.

Stabiliser: an additive which confers a high degree of stability to a solution, mixture, etc. In plastics, a compound incorporated in order to eliminate deterioration due to heat, light, oxidation.

Synthetic resin: usually used to describe most plastics polymers.

Thermoforming: a process in which a thermoplastic sheet, foil film or other regular form is heated, deformed by a shaping force and allowed to set in the new shape on cooling. Mainly vacuum forming (qv).

Thermoplastic: a plastics material which can be softened by heating and formed to the desired shape which is retained on cooling. Process can be repeated without damage to material.

Thermosetting: a plastics material which undergoes an irreversible chemical change when heated (termed 'curing') and forms an insoluble infusible mass which will not soften on heating but will decompose. Curing usually takes place under heat and pressure but some polymers can be cured without either.

Top force: see force.

Toughened polystyrene: see high impact polystyrene.

TPS: toughened polystyrene, see high impact polystyrene.

TPX: trade name for poly 4-methylpentene-1 (qv).

Transfer moulding: technique for moulding thermosetting plastics used to overcome problems inherent in the compression moulding process, such as the need for delicate inserts. The material, under heat and pressure in the transfer pot, softens and flows into the mould. Thus pins and inserts are not subjected to direct pressure from the press ram.

UF: urea formaldehyde (qv).

Unmodified polystyrene: general purpose polystyrene (qv).

Urea formaldehyde: aminoplastics made by condensing formaldehyde with urea. Uses include domestic electrical fittings (switch covers, plugs, fuse holders etc) bottle caps, knobs and handles.

Urethane elastomers: see polyurethane resins.

Vacuum bag compacting: when moulding reinforced plastics by hand lay-up (qv) the moulding can be compacted by enclosing it and the mould in a bag from which the air is then evacuated thus compacting by atmospheric pressure, as opposed to hand rolling or pressure bag (qv) compacting.

Vacuum forming: method of shaping thermoplastic sheet, in which the sheet is softened by heat and then deformed by reducing the air pressure on one side of it, with atmospheric pressure on the other side providing the shaping force against a male or female mould. The basic process is capable of considerable refinement and variation, such as drape forming, plug assist vacuum forming, bubble assist vacuum forming (qv).

Wet lay-up: hand lay-up (qv).

XPS: expanded polystyrene (qv).
